M000219852

Northwestern Handbook
of Surgical Procedures
2nd Edition

Nathaniel J. Soper, M.D.
*Department of Surgery, Feinberg School of Medicine,
Northwestern University, Chicago, Illinois, U.S.A.*

Dixon B. Kaufman, M.D., Ph.D.
*Department of Surgery, Feinberg School of Medicine,
Northwestern University, Chicago, Illinois, U.S.A.*

Illustrations by Simon Kimm, M.D.

AUSTIN, TEXAS
USA

VADEMECUM
Northwestern Handbook of Surgical Procedures, 2nd Edition
LANDES BIOSCIENCE
Austin, Texas USA

Please address all inquiries to the Publisher:
Landes Bioscience, 1806 Rio Grande, Austin, Texas 78701, USA
Phone: 512/ 637 6050; FAX: 512/ 637 6079

ISBN: 978-1-57059-707-7

Library of Congress Cataloging-in-Publication Data

Soper, Nathaniel J.
 Northwestern handbook of surgical procedures / Nathaniel J. Soper, Dixon B. Kaufman ; Illustrations by Simon Kimm. -- 2nd ed.
    p. ; cm. -- (Vademecum)
 Handbook of surgical procedures
 ISBN 978-1-57059-707-7
 1. Surgery, Operative--Handbooks, manuals, etc. I. Kaufman, Dixon B. II. Northwestern University (Evanston, Ill.) III. Title. IV. Title: Handbook of surgical procedures. V. Series: Vademecum.
 [DNLM: 1. Surgical Procedures, Operative--methods--Handbooks. WO 39]
 RD37.B45 2011
 617'.91--dc22
                        2010053771

Dedications

To my wife, Cindy, and my three sons, who have been supportive of my academic career throughout our lives. Also, to the many surgical trainees who have enriched my life by allowing me to share my joy of operating and the teaching of surgical techniques.

—N.J.S.

To all my colleagues that appear on these pages, and to those that do not, who have advanced the field of operative surgery, making it a more perfect therapy for those that count on our skills to enhance their well being; and above all, to Katina.

—D.B.K.

Contents

Editors

Malcolm DeCamp, M.D.
Cardiothoracic Surgery
Chapters 25, 27, 85-92, 94

Mark K. Eskandari, M.D.
Vascular Surgery
Chapters 106, 107, 109, 115, 116, 118-120

Dixon B. Kaufman, M.D., Ph.D.
Transplantation
Chapters 99, 104

David M. Mahvi, M.D.
Gastrointestinal and Surgical Oncology
Chapters 12, 13, 15, 17, 68

Thomas Mustoe, M.D.
Plastic Surgery
Chapters 82, 83

Marleta Reynolds, M.D.
Pediatric Surgery

Nathaniel J. Soper, M.D.
Gastrointestinal
Chapters 7, 26

Cord Sturgeon, M.D.
Endocrine
Chapters 60-65

Department of Surgery, Feinberg School of Medicine, Northwestern University, Chicago, Illinois, U.S.A.

Illustrations by Simon Kimm, M.D.

Contributors

Michael Abecassis, M.D.
Chapters 100-102

Katherine A. Barsness, M.D.
Chapter 125

Richard H. Bell Jr., M.D.
Foreword

David Bentrem, M.D.
Chapters 11, 14, 16, 18

Kevin Bethke, M.D.
Chapters 69, 70

Matthew G. Blum, M.D.
Chapter 27

Anne-Marie Boller, M.D.
Chapters 42, 43, 47, 48, 50, 54, 56, 58

Eric Cheon, M.D.
Chapter 11

Anthony C. Chin, M.D.
Chapter 124

John J. Coyle, M.D.
Chapter 32

Alberto de Hoyos, M.D.
Chapters 25, 85-90, 93, 94

Gregory Dumanian, M.D.
Chapters 81, 84

Dina Elaraj, M.D.
Chapters 60-65

Jonathan Fryer, M.D.
Chapters 10, 103

Robert D. Galiano, M.D.
Chapter 80

Amy L. Halverson, M.D.
Before You Start,
Chapters 49, 51, 52, 55, 66

Nora Hansen, M.D.
Chapters 71, 73-75

Wilson Hartz, M.D.
Chapters 2, 46

Amanda Hayman, M.D.
Chapter 14

Eric Hungness, M.D.
Chapters 1, 5, 22-24, 28, 35

Jacqueline Jeruss, M.D., Ph.D.
Chapters 71, 73-75

Seema A. Khan, M.D.
Chapters 57, 72

Melina R. Kibbe, M.D.
Chapters 110, 111

John Kim, M.D., FACS
Chapter 79

Seth Krantz, M.D.
Chapter 18

Richard Lee, M.D., M.B.A.
Chapter 95

Joseph R. Leventhal, M.D., Ph.D.
Chapters 97, 98

Mary Beth Madonna, M.D.
Chapter 126

S. Chris Malaisrie, M.D.
Chapter 95

Patrick M. McCarthy, M.D.
Chapter 95

Edwin McGee, M.D.
Chapter 95

Mark D. Morasch, M.D.
Chapter 105

Alexander P. Nagle, M.D.
Chapters 9, 34, 40, 45

William H. Pearce, M.D.
Chapters 113, 123

Jay B. Prystowsky, M.D.
Chapters 4, 6, 8, 33

Carla Pugh, M.D.
Chapters 31, 36, 37, 39

Heron E. Rodriguez, M.D.
Chapters 108, 112, 114, 117, 121, 122

David Rothstein, M.D.
Chapter 127

Michael B. Shapiro, M.D.
Chapters 21, 59

Anton I. Skaro, M.D., Ph.D.
Chapter 96

Steven J. Stryker, M.D.
Chapters 38, 44, 53, 67

Mark Toyama, M.D.
Chapters 3, 19, 20, 41

Jeffrey D. Wayne, M.D.
Chapters 29, 30, 76-78

Foreword

I am delighted that Dr. Soper and Dr. Kaufman have edited a second edition of the *Northwestern Handbook of Surgical Procedures*. This book fulfills a need in surgical education to improve the teaching of operative skills. Over the past few years, many in the academic surgical community have been re-evaluating the effectiveness of residency training. In 2006, the Surgical Council on Resident Education (SCORE) was formed by six academic and administrative organizations with oversight and/or interest in surgical graduate education. One of the major findings of this group has been that the operative experience of U.S. general surgery residents, even in some common, essential operations is inadequate. Among other things, this means that we must do a better job of preparing residents to have a maximal learning experience in the operating room when they do have the opportunity to perform a procedure.

The Northwestern Handbook of Surgical Procedures is designed to be reviewed prior to performing or participating in an operation. The authors of the book have identified the key steps of performing each procedure, to provide a framework to the learner for understanding the tasks and the sequence of those tasks necessary for successful performance. Breaking a complex performance like an operation into a series of steps is a technique well validated in the educational literature and provides the basic scaffold upon which the surgeon-in-training can add nuances and variations that are encountered in the course of experience, ultimately building a strong mental model or image of the operation.

I would urge residents or students who are using this book to discuss the steps of the procedure with faculty members before the start of an operation to understand if your views of the key tasks are aligned. It also would be helpful to discuss with the faculty member which steps you believe you understand or have mastered as opposed to those steps which seem unclear to you or about which you have concerns as to your ability. This kind of active participation in the course of an operation will make your educational experience a richer one.

Surgery is a changing field and it is difficult to know how technology will change our approach as the years go by. No doubt, you will see major changes in surgery over the course of your career. In the meantime, you should strive for the best technical grounding you can build for yourself. This requires a combination of analytic understanding and technical practice. This book aims to help you to approach an operation with an intellectual framework. I wish you success in the learning and practice of surgery.

Richard H. Bell, Jr., M.D.
Assistant Executive Director, The American Board of Surgery, Inc.
Adjunct Professor of Surgery, Northwestern University

STOP!

Before You Start: Optimizing Operative Care of the Surgical Patient

Amy L. Halverson

Optimizing the outcome of a surgical patient starts prior to the patient entering the operating room. The following questions should be addressed prior to surgery.

1. Has this patient had appropriate evaluation for comorbidities, including cardiac risk and sleep apnea?
2. What is the appropriate antibiotic prophylaxis?
3. What is appropriate DVT prophylaxis?
4. Are the available blood products appropriate for the estimated blood loss?
5. Has the surgical site been marked? Depending on your institution's policy, midline, single organ procedures, as well as endoscopies without intended laterality may not require site marking.

Safety and efficiency in the OR rely on good communication among the operating room team members. Having all team members introduce themselves facilitates subsequent communication. Prior to commencing the operation the following issues should be discussed.

1. Correct patient, site and procedure are confirmed.
2. Surgeon reviews the critical portions of the procedure as appropriate, duration of procedure and estimated blood loss.
3. Anesthesia staff review issues critical to the patient.
4. Nursing staff reviews equipment availability and other concerns.
5. Confirm that appropriate antibiotics have been given.
6. Confirm that necessary imaging is available.

At the conclusion of the procedure the following should be verified:

1. The name of procedure as recorded.
2. Instrument counts are correct.
3. The specimen is correctly labeled.
4. Postoperative recovery and care plan for the patient.

SECTION 1: GASTROINTESTINAL

Section Editors: Nathaniel J. Soper
and David M. Mahvi

1

Exploratory Laparotomy: Open

Eric Hungness

Indications

Open exploratory laparotomy is indicated where a surgically correctable problem may exist in the abdomen. The most common indications for open exploratory laparotomy include conditions of acute intra-abdominal infection and acute traumatic injuries. Open exploration is particularly useful when questions arise concerning the integrity or the condition of the bowel. Whereas CT can provide very accurate anatomic information regarding retroperitoneal and solid organ structures, it is much less reliable for evaluation of the bowel. Diagnostic laparoscopy may be considered because it is less invasive; however, it also has lower diagnostic accuracy for evaluating the intestine. An advantage of open laparotomy is the ability to address the primary problem, whatever it might be.

Preop

Prior to exploratory laparotomy the patient should have appropriate venous access and should (if possible) be well-resuscitated. It is advantageous to place a Foley catheter prior to abdominal exploration. When performing exploratory laparotomy for blunt trauma have adequate operative suction (two suctions), lighting, and carefully position the patient such that the chest and/or mediastinum can be accessed intraoperatively. Antibiotic prophylaxis should be instituted prior to the incision. Choice of agent should be based on the pathogens likely to be encountered. Second-generation cephalosporins or other agents that cover aerobic and anaerobic enteric pathogens are frequently used. General endotracheal anesthesia is required, along with good muscle relaxation.

Procedure

Step 1. The patient is placed in the supine position. A midline abdominal incision is made from the xiphoid to the pubis. When rapid abdominal access is required in traumatic situations, the incision can be made most rapidly with 2-3 scalpel passes. The first pass cuts through the subcutaneous tissue down to the level of the fascia. A second pass of the scalpel can be used to incise the fascia in the midline. The peritoneum can then be entered using a scissors. It is best to complete the fascial incision prior to incising the peritoneal cavity as any tamponade-effect will be released once the peritoneal cavity is entered.

Step 2. On entering the abdominal cavity, pay attention to where bleeding or contamination appears to be arising. It is best not to be distracted by bowel injury/ contamination in the setting of massive hemoperitoneum. The peritoneal cavity is packed with laparotomy pads in the four quadrants of the abdomen, but packing should be done first in the quadrant that is most likely to be the source of the bleeding.

Northwestern Handbook of Surgical Procedures, 2nd Edition, edited by Nathaniel J. Soper and Dixon B. Kaufman. ©2011 Landes Bioscience.

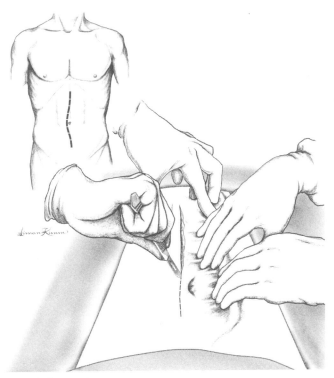

Figure 1.1. Exploratory laparoscopy.

Bowel injuries with ongoing enteric leakage can be controlled with temporary mass ligatures or application of noncrushing clamps.

Step 3. If there is massive bleeding, temporary control of hemorrhage can be obtained with compression of the aorta at the diaphragmatic hiatus. This can be performed using digital pressure, pressure from a Richardson retractor (back of one blade), or with an aortic occluder.

Step 4. Once hemorrhage is stabilized, the surgeon should explore the four corners of the abdomen while removing the temporary packs (if applicable). It is important to utilize a systematic approach to ensure that all intra-abdominal structures are visualized. Particular care should be taken to ensure that relatively inaccessible areas (diaphragm, lesser sac, pelvis) are carefully evaluated. Warm saline irrigation of the relevant quadrant should be performed while the exposure is optimized.

Step 5. Exploration is begun in the left upper quadrant. It is important to visualize the diaphragm, spleen, stomach, and gastroesophageal junction. Appropriate control measures or repacking should be instituted as applicable if specific injuries are identified.

Step 6. Attention is next directed to the right upper quadrant, taking care to visualize the diaphragm, the diaphragmatic surface of the liver, the integrity and condition of the gallbladder, the lateral aspect of the liver, and the undersurface of the liver.

Step 7. The right lower quadrant of the abdomen is visualized next, paying particular attention for the presence of any bowel or bladder perforation or retroperitoneal hematoma in the area of the iliac vessels.

Step 8. In examining the left lower quadrant, attention is directed to assessing the integrity and condition of the sigmoid colon and looking for evidence of retroperitoneal injury.

Step 9. The integrity of the small bowel is next determined by "running" the small bowel from the ligament of Treitz to the ileocecal valve. The surgeon should make a mental note as to whether there is any evidence of a central retroperitoneal hematoma. Both sides of the small bowel should be examined. The mesentery is inspected simultaneously.

Step 10. The colon is inspected beginning at the cecum with evaluation of the appendix and periappendiceal structures. Inspection continues by examining the cecum and continuing up the right colon carefully examining the hepatic flexure. Examination continues by assessing the transverse colon with the omentum reflected cephalad. Complete evaluation of the splenic flexure may require division of the splenocolic ligament. Evaluation continues by inspecting the left colon and sigmoid colon. Complete evaluation of the right, left, or sigmoid colon may require mobilization of these structures by division of the lateral peritoneal attachments (white line of Toldt) and medial reflection.

Step 11. The anterior surface of the stomach and duodenum should be examined next. In the process, the surgeon should pay particular attention to whether there is any evidence of blood or inflammation in the lesser sac by closely examining the lesser omentum.

Step 12. The pancreas and lesser sac are evaluated next by entering the lesser sac. This is accomplished by making an incision on the undersurface of the omentum just cephalad to the transverse colon. This is most easily accomplished to the left side of the midline. Both the anterior surface of the pancreas as well as the posterior aspect of the stomach can be inspected through this incision.

Step 13. If duodenal injury is suspected the duodenum can be mobilized by performing a Kocher maneuver (lateral incision of retroperitoneum and medial reflection of the duodenum). This also allows visualization of the right renal vein and inferior vena cava. The third portion of the duodenum can be visualized by performing a Cattel maneuver (division of the lateral attachment of the cecum and medio-cephalad cecal reflection).

Step 14. If bowel or solid organ injuries are identified they should be addressed prior to closure.

Step 15. At the completion of the procedure the abdomen should be irrigated with warm saline solution (antibiotics are not required in this fluid).

Step 16. The fascia is closed with running 0-monofilament sutures beginning at the superior and inferior aspects of the incision and meeting in the middle. This skin is either closed or left open as dictated by the intraoperative findings.

Postop

Careful postoperative management and evaluation of the fluid balance should be performed. In instances of trauma or infection that require significant resuscitation, abdominal compartment syndrome can occur with severe hemodynamic and metabolic consequences. The surgical incision should be examined on a daily basis and opened if there is evidence of infection present.

Complications

The most common complication of laparotomy is wound infection. Inadequate exploration can result in missed injuries. Wound dehiscence can also occur. Hypothermic coagulopathy can complicate prolonged exploratory laparotomy in many patients. Abdominal compartment syndrome can occur.

Follow-Up

The patient should be followed until wounds are healed. Long-term follow-up depends on the nature of the underlying disease/injury.

Acknowledgment

The editors and author wish to acknowledge Michael A. West for contributing to the previous version of this chapter.

Chapter 2

Inguinal Hernia Repair with Mesh: Open

Wilson Hartz

2

Indications

Open repair is indicated for primary or recurrent inguinal hernia in patients who are suitable operative risks.

Preop

General, spinal, epidural, or monitored local anesthesia with sedation may be chosen as anesthetic techniques, and the choice should be discussed with the patient. Intravenous prophylactic antibiotic is given 30 minutes prior to the skin incision. Deep vein thrombosis prophylaxis should be used in patients with risk factors for thromboembolism.

Procedure

Step 1. The patient is placed in the supine position. An oblique incision over the inguinal canal is made, using the pubic tubercle as a guide for the medial end of the incision.

Step 2. After opening the external oblique fascia, the ilioinguinal nerve is identified and preserved.

Step 3. Blunt or sharp dissection and a finger are used to surround and isolate the spermatic cord at the pubic tubercle. A Penrose drain is placed around the cord. During this dissection, the genitofemoral nerve should be identified and protected.

Step 4. The hernia sac is identified and separated completely from cord structures back to the level of the internal inguinal ring. In the case of a direct inguinal hernia, the internal ring should be examined to exclude the possibility of an additional indirect sac.

Step 5. For an indirect hernia, the sac is ordinarily treated by high ligation and excision of the sac or inversion into the internal inguinal ring. If the hernia is a sliding hernia, the sac can be inverted back into the internal ring. For direct hernias, the sac is inverted into the fascial defect. If desired, a mesh plug can be used to maintain reduction of the sac by placing it over the sac and securing it to the circumference of the defect

Step 6. Place an onlay patch of mesh over the inguinal canal. The spermatic cord is brought through a "key hole." The mesh should be secured with sutures or staples medially at the pubic tubercle, laterally into muscle beyond the external ring, superiorly to the conjoint tendon, and inferiorly to the shelving edge of the inguinal ligament.

Step 7. The external oblique fascia is closed, taking care not to make the external ring too tight, which can cause venous outflow obstruction from the testicle. The skin is closed with subcuticular suture.

Northwestern Handbook of Surgical Procedures, 2nd Edition, edited by Nathaniel J. Soper and Dixon B. Kaufman. ©2011 Landes Bioscience.

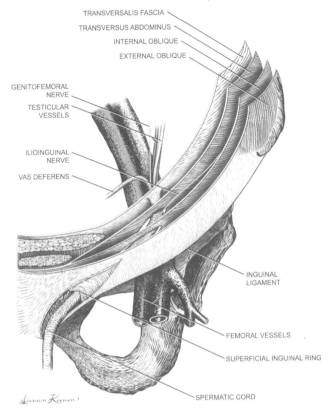

Figure 2.1. Inguinal hernia repair. Anatomy.

Postop

Patients can be discharged within a few hours after surgery if stable. Normal activities can generally resume in 2-4 weeks, although strenuous activity and heavy lifting are generally avoided for about 4-6 weeks.

Complications

Injury to the ilioinguinal nerve or genitofemoral nerve can result in chronic groin pain. Hematoma or seroma may occur. Infections may occur in the wound. If the mesh becomes exposed or infected, it may need to be removed. Mesh can migrate or erode. Recurrence occurs in 1-2% of patients operated for the first time.

Follow-Up

Patients may be seen periodically until they return to full activity.

Acknowledgment

The editors and author wish to acknowledge Ermilo Barrera, Jr. for contributing to the previous version of this chapter.

Chapter 3

Inguinal Hernia Laparoscopic Repair: Extraperitoneal Approach

Mark Toyama

3

Indications

Laparoscopic inguinal hernia repair is particularly indicated for recurrent or bilateral hernias. Its role in the management of first-time unilateral hernias is debatable.

Preop

Preoperative prophylactic antibiotic should be given intravenously 30 minutes prior to skin incision.

Procedure

Step 1. The patient is placed in the supine position with arms tucked at the sides. A Foley catheter is inserted into the bladder. The surgeon stands on the side opposite the hernia. Monitor(s) are placed at the foot of the table. The skin incision is placed just inferior to the umbilicus and dissection is carried down to the rectus sheath.

Step 2. A small incision is made in the anterior rectus sheath.

Step 3. The rectus abdominis muscle is bluntly dissected to expose the posterior rectus sheath.

Step 4. Blunt dissection is done to develop the space between the back side of the rectus muscle and the peritoneum. A finger or small retractor works well for this.

Step 5. A balloon dissector is then placed into the preperitoneal space and carefully advanced inferiorly to the level of the symphysis pubis. The balloon is inflated under direct vision of the laparoscope, creating a working area in the preperitoneal space.

Step 6. The balloon is removed and a 10 mm or 12 mm blunt trocar is placed into the preperitoneal space and secured. The preperitoneal space is insufflated with CO_2.

Step 7. A 30° laparoscope is placed into the preperitoneal space.

Step 8. Additional ports are placed. A combination of 2 mm and 5 mm ports can be use in a variety of configurations. The ports should be placed under direct vision, taking care to avoid puncturing the thin peritoneum.

Step 9. The preperitoneal space is bluntly dissected, reducing indirect, direct, or femoral hernias back into the peritoneal cavity.

Step 10. Very large indirect hernia sacs can be divided and the proximal end secured with a ligature, leaving the distal portion of the sac open and in situ. Care is taken to preserve all spermatic cord structures in men.

Northwestern Handbook of Surgical Procedures, 2nd Edition, edited by Nathaniel J. Soper and Dixon B. Kaufman. ©2011 Landes Bioscience.

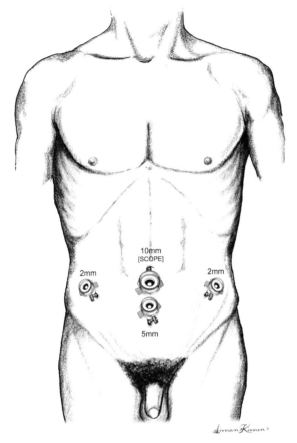

Figure 3.1. Laparoscopic inguinal hernia repair. Trocar placement.

Step 11. A large piece of mesh is then placed into the preperitoneal space and oriented to cover the direct, indirect, and femoral spaces.

Step 12. The mesh is secured with a tacking or stapling device to prevent mesh migration. The number of tacks required is variable, but this is done in such a manner as to avoid injuring or incorporating the ilioinguinal, iliohypogastric, lateral femorocutaneous, or genitofemoral nerves as well as any vascular or cord structures.

Step 13. After the mesh is placed and secured, the preperitoneum is desufflated under direct vision to ensure that the mesh remains flat and in the appropriate position.

Step 14. The skin incisions are closed with subcuticular sutures.

Postop

Postoperative manangement is similar to that of open hernia repair. The Foley catheter is removed before the patient leaves the operating room. Patients are discharged home when they can tolerate oral intake and void.

3

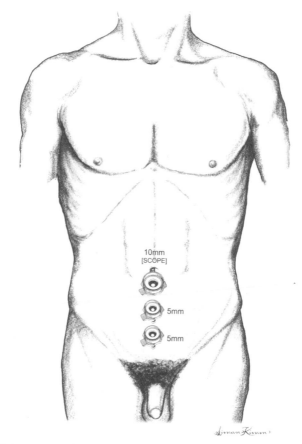

Figure 3.2. Laparoscopic inguinal hernia repair. Trocar placement.

Complications

A number of injuries are possible during laparoscopic preperitoneal hernia repair. These include nerve injury, vascular injury, bladder injury, colon or small bowel injury, testicular devascularization, and vas deferens injury. Urinary retention and/or infection may occur.

Seromas or hematomas may form in the dissected preperitoneal space. Pubic/pelvic osteitis may occur.

Wound infections are relatively rare. Mesh complications include infection, migration, and erosion. Finally, there is about a 2-5% chance of hernia recurrence.

3

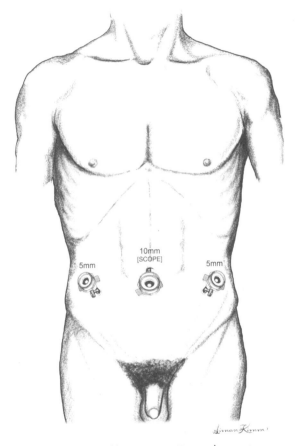

Figure 3.3. Laparoscopic inguinal hernia repair. Trocar placement.

Follow-Up

Patients are followed in the office approximately 2 weeks after their operation. Patients are instructed to avoid heavy lifting and straining for approximately 4-6 weeks after the operation.

Chapter 4

Ventral Hernia Repair: Open

Jay B. Prystowsky

Indications

In general, ventral hernias should be repaired in patients who are good operative risks to avoid the possibility of strangulation. Repair is definitely indicated in the presence of symptoms (pain, nausea, vomiting, etc.) or if the hernia cannot be reduced.

Preop

Patients should be given a preoperative systemic antibiotic for wound infection prophylaxis 30 minutes before operation. In cases where there is likely to be colon in the hernia sac or adhesions to the colon, a mechanical and pharmacologic bowel prep is indicated. Patients should be treated with sequential compression devices or subcutaneous heparin according to their preoperative risk factors for thromboembolism. Most procedures should be done under general anesthesia.

Procedure

Step 1. A skin incision should be made that will expose the full length of the hernia defect. In cases where a hernia has occurred in a portion of a previous incision, it is important to have adequate exposure to examine the remainder of the previous incision for possible defects.

Step 2. The subcutaneous tissue is dissected down to the hernia sac, at which point subcutaneous flaps are raised all around the hernia until normal fascia can be identified on all sides of the hernia. The peritoneal sac is then opened. Often, there is redundant peritoneum that can be excised back to expose the fascial edges of the defect. It is critical to definitively identify fascia at all margins of the defect.

Step 3. An assessment is made of the amount of tension that would be created with a primary repair of the defect. In general, defects more than 2-3 cm in diameter will not be amenable to primary closure.

Step 4. If primary closure is entertained, relaxation incisions can be made in the anterior rectus sheath or other anterior layers of abdominal fascia to decrease closure tension.

Step 5. Primary closure can be performed with direct approximation or a pants-over-vest method.

Step 6. If mesh is to be placed, the mesh should be cut in the shape of the defect but about 2-3 cm larger in all dimensions. The mesh is sewn to the underside of the fascial edges of the defect using interrupted vertical mattress sutures of a monofilament suture. It is critical that mesh is secured to overlying fascia and that there is sufficient overlap of mesh over the edges of the fascial defect.

Northwestern Handbook of Surgical Procedures, 2nd Edition, edited by Nathaniel J. Soper and Dixon B. Kaufman. ©2011 Landes Bioscience.

Step 7. If the subcutaneous fat can be closed over the repair without tension, this should be done with absorbable sutures. Often, this is not possible, If there will be a subcutaneous cavity over the repair, it is best to place a closed suction drain over the repair and bring it out through a separate stab wound in the skin.

Step 8. The skin is closed with staples or suture.

Postop

If an extensive lysis of adhesions has been performed, it may be appropriate to leave a nasogastric tube in place until bowel activity has returned. Ordinarily, however, early feeding can be initiated. The patient should be able to ambulate on the day following surgery in most cases.

Complications

The most important early complication of ventral hernia repair is wound infection, which can present a major problem if mesh is exposed or involved. The most significant late complication is hernia recurrence.

Follow-Up

The patient should be followed until healing is complete and normal activity resumed. The patient should be instructed about the risk and signs of recurrence and asked to return as needed should symptoms develop.

Acknowledgment

The editors and author wish to acknowledge John J. Coyle for contributing to the previous version of this chapter.

Chapter 5

Ventral Hernia Repair: Laparoscopic

Eric Hungness

Indications

In general, ventral hernias should be repaired in patients who are good operative risks to avoid the possibility of strangulation. Repair is definitely indicated in the presence of symptoms (pain, nausea, vomiting, etc.) or if the hernia cannot be reduced.

Preop

The alternative of open ventral hernia repair should be discussed with the patient. If the patient chooses laparoscopic surgery, a careful review of previous operations and examination of the abdomen is carried out to help plan potential access sites. On the day of surgery, patients should be given a preoperative systemic antibiotic for wound infection prophylaxis 30 minutes before operation. In cases where there is likely to be colon in the hernia sac or adhesions to the colon, a mechanical and pharmacologic bowel prep prior to surgery is indicated. Patients should be treated with sequential compression devices or subcutaneous heparin according to their preoperative risk factors for thromboembolism. A Foley catheter and nasogastric tube are placed immediately after the induction of anesthesia.

Procedure

Step 1. The entire abdomen is prepped and draped in sterile fashion. A sterile plastic barrier is utilized to avoid contact of the prosthetic material with exposed skin.

Step 2. Access is first obtained away from prior surgical sites, on the side opposite previously dissected areas. For example, if a patient has had a low anterior resection and has an incisional hernia, access should first be obtained on the right side of the abdomen to avoid placement of the initial operating port through adhesions.

Either a Veress needle or open technique can be used for initial access to the peritoneal cavity. Veress needle access can be difficult away from the midline. If Veress needle access is initially unsuccessful, the surgeon should have a low threshold for converting to an open access technique (e.g., Hasson cannula).

Step 3. An angled laparoscope is used to permit the surgeon to see around the edges of adhesions. The abdomen is explored and adhesions are assessed. Sites are selected for subsequent port placement. In general, two ports are placed on the same side as the first trocar, and at least one other port is placed on the contralateral side to facilitate the later securing of the mesh.

Northwestern Handbook of Surgical Procedures, 2nd Edition, edited by Nathaniel J. Soper and Dixon B. Kaufman. ©2011 Landes Bioscience.

Step 4. Adhesions are divided using either sharp dissection with electrosurgical cautery (staying away from bowel) or ultrasonic shears. Traction is the key to facilitate division of adhesions, and using two hands to dissect helps in manipulation of bowel and tissues. Occasionally, initial adhesiolysis must be done through one port to "clear" space for placement of subsequent ports. Special care must be taken to avoid injury to bowel.

Step 5. Once all hernia contents have been reduced and the edges of the defect are well-exposed, the defect is transilluminated from the abdomen and the defect margins are marked with a pen on the plastic barrier drape.

Step 6. An appropriately sized piece of polytetrafluoroethylene (PTFE) mesh is selected. It must overlap the edges of the defect by at least 2-3 cm circumferentially when the abdomen is insufflated. When the mesh is laid on the plastic barrier drape ("nonadhesion-forming" side of the mesh against the plastic barrier drape), the previously made ink marks identifying the defect edges are transferred to the prosthetic material. This aids in trimming of the mesh to 2-3 cm beyond the edges of the defect.

Step 7. At least four, but preferably six or eight, nonabsorbable stay sutures are placed circumferentially around the edges of the mesh, spaced equidistantly. The sites of suture placement are marked on the abdominal wall for future passage of the sutures.

Step 8. The mesh with the sutures is passed through the largest port (generally a 12 mm port except for the smallest mesh which can be placed through a 10 mm port) by rolling the mesh as tightly as possible.

Step 9. The mesh is oriented properly and unfurled in this orientation.

Step 10. The suture passer (disposable or reusable) is passed through the previously marked skin sites and each of the suture ends is grasped. For each suture site, the suture passer must be passed twice through the same skin puncture, but different fascial sites (1 cm apart) so that the suture ends can be tied external to the fascia.

Step 11. Once this is completed, the mesh is secured circumferentially using a laparoscopic tacking device. Tacks are placed 1-1.5 cm apart. Special care should be taken to avoid plication of the mesh.

Step 12. The abdomen is inspected for hemostasis and any bowel that was dissected is examined for leakage or injury. If there are no problems, all ports are removed under direct visualization to assure that there is no port site bleeding.

Step 13. The fascia is closed at port sites larger than 5 mm. Skin is closed at all port sites with absorbable subcuticular suture and/or sterile tapes. Drains are not used.

Postop

In general, pain is fairly significant in the first 24-48 hours and ileus is not uncommon. Patients are generally hospitalized for one or two nights. Seroma formation in the previous hernia soft tissue defect is common. While this can be alarming to the patient, nothing should be done unless the seroma is symptomatic or signs of infection appear. In general, seromas and hematomas will resolve in 3-4 weeks. If drainage is required, this can be done percutaneously, but should be avoided if possible. Vigorous physical activity should be limited for 2 weeks while tissue ingrowth occurs, but there is no limitation necessary thereafter.

Complications

Occult bowel injury is a serious potential complication. Patients who do not seem to be recovering appropriately within 24-48 hours or who demonstrate signs of peritonitis (fever, elevated white blood cell count) should have an abdominal CT scan and possible urgent return to surgery.

Follow-Up

The patient should be followed until healing is complete and normal activity resumed. The patient should be instructed about the risk and signs of recurrence and asked to return as needed should symptoms develop.

Acknowledgment

The editors and author wish to acknowledge Kenric M. Murayama for contributing to the previous version of this chapter.

Chapter 6

Cholecystectomy with Cholangiography: Open

Jay B. Prystowsky

Indications

The indications for open cholecystectomy are the same as for laparoscopic cholecystectomy AND inability to perform laparoscopic cholecystectomy (which, in general, is the procedure of choice). Indications for cholecystectomy include symptomatic cholelithiasis (acute or chronic cholecystitis), gallstone pancreatitis, acalculous cholecystitis, or choledocholithiasis.

Preop

Antibiotics are administered in cases of acute disease, choledocholithiasis, or age >65 years.

Procedure

Step 1. A right subcostal incision is performed.

Step 2. The costal margin is retracted cephalad; the hepatic flexure of the colon and the duodenum are retracted inferiorly.

Step 3. Grasping the fundus of the gallbladder with a clamp, it is lifted anteriorly and away from the liver.

Step 4. The peritoneum overlying the gallbladder is incised with cautery within a few millimeters of the liver.

Step 5. Progressively retracting it away from the liver, the gallbladder is dissected from Glisson's capsule in the gallbladder fossa, moving downward towards the porta hepatis. It is important to dissect close to the wall of the gallbladder.

Step 6. The cystic artery and cystic duct are identified.

Step 7. The cystic duct is dissected down to its junction with the common duct.

Step 8. The common duct immediately proximal and distal to the entrance of the cystic duct is identified to verify anatomy.

Steps 9-15. describe intraoperative cholangiography, which may be performed in selected cases. Indications for cholangiography generally include: elevated liver enzymes, stone in common bile duct either documented preoperatively or discovered by palpation intraoperatively, dilated common bile duct, recent gallstone pancreatitis, or difficulty dissecting or identifying biliary anatomy.

Step 9. To prepare for cholangiography, a ligature is placed proximally at the junction of the cystic duct and gallbladder.

Step 10. A small opening is made in the cystic duct and a cholangiocatheter (4-5 F) is passed into the duct for about 1-2 cm.

Step 11. The catheter is secured with a ligature or clip. Two 30 ml syringes are attached to the catheter with a three-way stopcock and extension tubing. One is filled with saline, the other with contrast diluted 50%. Saline is injected to confirm there are no leaks at

Northwestern Handbook of Surgical Procedures, 2nd Edition, edited by Nathaniel J. Soper and Dixon B. Kaufman. ©2011 Landes Bioscience.

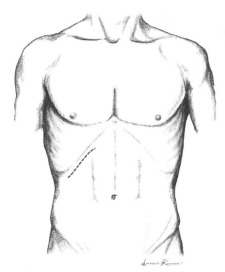

Figure 6.1. Cholecystectomy, open. Incision.

the site of catheter entrance into the cystic duct. It should be possible to aspirate bile if the catheter is properly positioned. Before injecting dye, air bubbles should be eliminated from the catheter and tubing.

Step 12. The patient is then placed in the Trendelenburg position and tilted to the right (to bring the common duct "off" the spinal column).

Step 13. Contrast is injected under fluoroscopic guidance.

Step 14. Easy flow of contrast distally into the duodenum and proximally into the right and left biliary radicals along with absence of filling defects constitutes a normal exam.

Step 15. The catheter is withdrawn and the cystic duct is ligated distal to the catheter entrance site. The cystic duct may then be transected.

Step 16. The cystic artery is ligated with nonabsorbable suture and transected between ligatures. The gallbladder is removed.

Step 17. The abdominal wall is closed in layers.

Postop

Diet may usually be instituted within 24 hours. Parenteral narcotics for pain are switched to oral prior to discharge.

Complications

Major complications include injury to the common bile duct and bile leak from the cystic duct stump; other surgical complications include wound infection and postoperative bleeding.

Follow-Up

Patients should be seen at 1-2 weeks and again at approximately 6 weeks. Most patients experience excellent relief of pain; 5% of patients will continue to have discomfort as they experienced preoperatively (postcholecystectomy syndrome).

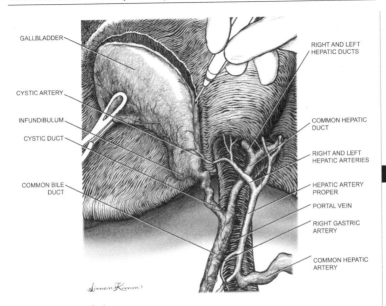

Figure 6.2. Open cholecystectomy.

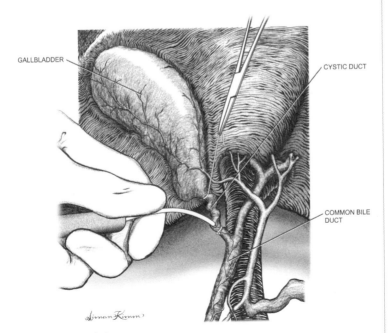

Figure 6.3. Open cholangiogram.

Chapter 7

Cholecystectomy with Cholangiogram: Laparoscopic

Nathaniel J. Soper

Indications

The indications for laparoscopic cholecystectomy include symptomatic gallstone disease (chronic cholecystitis or acute cholecystitis) or acute acalculous cholecystitis. Cholangiography may be done in selective cases. In general, the indications for cholangiography include choledocholithiasis, dilated common bile duct, recent gallstone pancreatitis without preoperative ERCP, or confusion about the anatomical orientation intraoperatively.

Preop

A first- or second-generation cephalosporin or an antibiotic of equivalent coverage is given 30 minutes prior to surgery. An orogastric tube is placed to decompress the stomach.

Procedure

Step 1. The entire abdomen is prepped and draped in standard sterile fashion. Access is gained at the umbilicus by either the Veress needle technique (closed technique) or open technique. We prefer the Hasson technique, using a 10 mm port.

Step 2. A 5- or 10-mm angled (usually, 30°) laparoscope is inserted and an exploratory laparoscopy is performed.

Step 3. The patient is placed in reverse Trendelenburg position and the other trocars are placed under direct visualization. A 10 mm trocar is placed in the subxiphoid epigastric region; a 5 mm trocar is placed in the right subcostal, midclavicular line; and a 5 mm trocar is placed in the right subcostal, anterior axillary line location.

Step 4. If the patient has had acute cholecystitis, there may be adhesions to the gallbladder. These adhesions can usually be swept away bluntly. The duodenum and/or colon may be adherent to the surface of the gallbladder. Therefore, while electrosurgical cautery can generally be used to facilitate the dissection of adhesions, cautery should be avoided if the duodenum or hepatic flexure of the colon is in proximity.

Step 5. The fundus of the gallbladder is grasped with an instrument placed through the right anterior axillary line port, and the tip of the gallbladder is retracted cephalad. The infundibulum of the gallbladder is retracted caudad and to the patient's right with a second grasper that is placed through the midclavicular port.

Step 6. A Maryland dissector placed through the epigastric port is used to clear the peritoneum over the infundibulum and hepatocystic triangle. The L-hook cautery also works well to dissect the neck of the gallbladder away from its bed. Dissection is undertaken both from the medial aspect, as well as from the lateral aspect while utilizing

Northwestern Handbook of Surgical Procedures, 2nd Edition, edited by Nathaniel J. Soper and Dixon B. Kaufman. ©2011 Landes Bioscience.

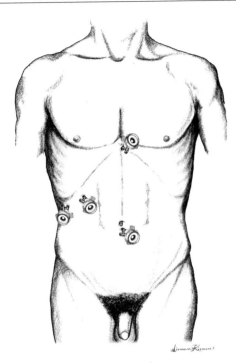

Figure 7.1. Laparoscopic cholecystectomy. Port placement.

countertraction on the infundibulum using a grasper manipulated with the surgeon's left hand. Using this technique, the lower part of the gallbladder is dissected away from the liver and a 'window', through which can be seen the liver, is created. There should be two, and only two, structures crossing this window; the cystic duct and cystic artery. This is the 'critical view of safety', which should be displayed prior to cutting or clipping any structures.

Step 7. Prior to dividing the cystic duct, a decision must be made regarding the need for a cholangiogram. If a cholangiogram is to be performed, a clip is placed across the cystic duct near the infundibulum. A transverse opening is created in the cystic duct. A "flash" of bile confirms that the opening is in the cystic duct. A cholangiocatheter is placed into the cystic duct and threaded distally toward the common bile duct. The catheter options include a balloon or straight catheter, and either a cholangiocatheter clamp or clips can be used to secure the catheter.

Two 30 ml syringes are attached to the catheter with a three-way stopcock and extension tubing. One is filled with saline, the other with contrast diluted 50%. Saline is injected to confirm there are no leaks at the site of catheter entrance into the cystic duct. It should be possible to aspirate bile if the catheter is properly positioned. Before injecting dye, air bubbles should be aspirated from the catheter and any extension tubing. Fluoroscopy or multiple static films can be used to verify the presence or absence of common bile duct stones and to display the biliary anatomy. If there are no stones, the operation can proceed. If there are common duct stones, there are four options:

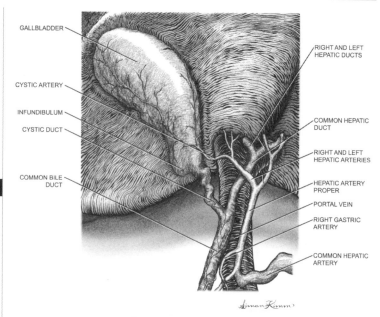

Figure 7.2. Laparoscopic cholecystectomy. Anatomy.

1. Complete cholecystectomy and perform postoperative ERCP;
2. Complete cholecystectomy and perform intraoperative ERCP;
3. Laparoscopic common bile duct exploration; or
4. Convert to open common bile duct exploration.

This decision depends on availability of a skilled endoscopist and the surgeon's experience with laparoscopic common bile duct exploration (see Chapter 9).

If the cholangiogram has been completed and a common bile duct exploration is not necessary or has been completed, the cystic duct is divided between clips (ordinarily, two clips are placed on the end toward the common bile duct).

Step 8. The cystic artery is cleared of surrounding attachments. Special care is taken to ensure that the artery is not the right hepatic artery by following it and observing its termination in the gallbladder. When this has been verified, the cystic artery is divided in continuity between clips.

Step 9. The infundibulum of the gallbladder is retracted anteriorly and cephalad, progressively dissecting the gallbladder from the liver bed using cautery and blunt or sharp dissection. Just before dividing the last attachments, the gallbladder bed is examined for hemostasis and the clips are examined for appropriate location and security. The right upper quadrant is irrigated and aspirated dry.

Step 10. Once amputated the gallbladder is placed into a specimen retrieval bag and removed through the periumbilical port. If the gallbladder is exceedingly large, full of gallstones, or contains large stones, it may not be easy to remove the gallbladder. Options include crushing the stones inside the gallbladder with a clamp, removing some of the stones/stone fragments to help decompress the gallbladder, and/or by enlarging the port incision.

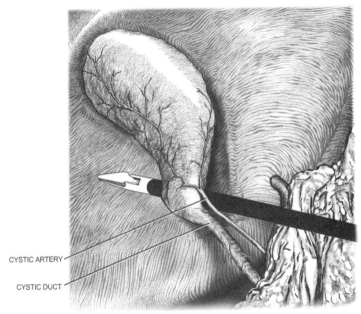

CYSTIC ARTERY

CYSTIC DUCT

Figure 7.3. Laparoscopic cholecystectomy.

Step 11. The 10 mm incisions are closed at the fascial level. All skin incisions are closed with absorbable subcuticular sutures.

Additional Steps. Conversion to open should occur if the anatomy is unclear, there is excessive bleeding, or if a complication such as common duct injury occurs. If the gallbladder is perforated during dissection, additional care should be taken to remove all of the spilled stones and clean up any spilled bile.

Postop

Patients are started on clear liquids on the evening of surgery and may have their diet advance ad libitum. Patients are either sent home the day of surgery or in 23 hours.

Complications

Major complications include bleeding, common duct injury, leakage of bile from the cystic duct stump, duodenal injury, or other bowel injury.

Follow-Up

The patient should be seen in 1-2 weeks to examine wounds and be seen later by either the surgeon or referring physician to confirm resolution of preoperative symptoms.

Acknowledgment

The editors and author wish to acknowledge Kenric M. Murayama for contributing to the previous version of this chapter.

Chapter 8

Common Bile Duct Exploration: Open

Jay B. Prystowsky

Indications

In general, open common duct exploration is indicated when stones are discovered by cholangiography during open cholecystectomy. It may be indicated when stones are discovered during laparoscopic cholecystectomy, and the surgeon is not familiar with the technique of laparoscopic duct exploration. Palpable stones in the common bile duct at the time of open cholecystectomy are another indication. An alternative therapy for stones in the common bile duct is postoperative endoscopic extraction via ERCP. Common duct exploration should be strongly considered when stones are large or multiple or there are anatomic considerations that would make the stones not amenable to endoscopic extraction.

Preop

Antibiotic prophylaxis is indicated. The early steps of the operation are described under open cholecystectomy with cholangiography.

Procedure

Step 1. Once the common duct has been identified, its anterior wall should be exposed for about 2.5-3 cm; care should be taken to avoid dissection along its lateral walls since that is where its blood supply exists.

Step 2. A #15 blade is used to create a small rent in the anterior wall of the duct, and Potts scissors are used to enlarge the rent in a longitudinal fashion for about 2 cm; stay sutures are placed on either side of the common bile duct incision to keep the aperture open.

Step 3. Randall stone forceps are passed distally and then proximally to clear the duct of stones by directly grasping them.

Step 4. A choledochoscope is useful to identify residual stones and assist in their extraction.

Step 5. An appropriately sized T-tube is placed into the common duct, and the common duct closed over the tube with a series of interrupted 4-0 absorbable sutures.

Step 6. A cholangiogram is performed to ascertain that the duct is clear of stones.

Step 7. A drain is placed near the common bile duct opening and brought out through a separate stab incision.

The remainder of the case proceeds as for open cholecystectomy.

Northwestern Handbook of Surgical Procedures, 2nd Edition, edited by Nathaniel J. Soper and Dixon B. Kaufman. ©2011 Landes Bioscience.

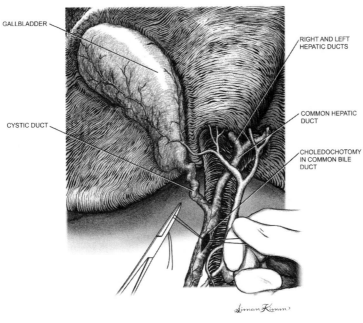

GALLBLADDER

RIGHT AND LEFT
HEPATIC DUCTS

COMMON HEPATIC
DUCT

CYSTIC DUCT

CHOLEDOCHOTOMY
IN COMMON BILE
DUCT

8

Figure 8.1. Common bile duct exploration.

Postop
The peritoneal drain can be removed in 24-48 hours if there is no bile leakage. The T-tube is initially placed to gravity drainage. Before discharge, a cholangiogram through the tube should be performed. If negative, the tube can be capped and the patient discharged.

Complications
Complications related to the T-tube are predominantly dislodgement or kinking. Retained stones may be present on follow-up cholangiogram and may require removal endoscopically or through the T-tube tract.

Follow-Up
About 2 weeks after surgery, the T-tube may be removed.

Chapter 9

Common Bile Duct Exploration: Laparoscopic

Alexander P. Nagle

Indications

Laparoscopic common bile duct exploration is indicated for the presence of common bile duct stone(s) during laparoscopic cholecystectomy. Usually stones are detected after an intraoperative cholangiogram (see Chapter 7).

Preop

Preoperative preparation is the same as for laparoscopic cholecystectomy with cholangiogram. The patient should be on a fluoroscopy-capable operating room table. A choledochoscope should be available. The initial steps of the operation are described in the chapter on laparoscopic cholecystectomy with cholangiogram. Two options for laparoscopic common bile duct exploration are possible: transcystic duct exploration of the common bile duct or choledochotomy (similar to open common bile duct exploration).

Procedure

Laparoscopic Transcystic Duct Exploration of the Common Bile Duct

Step 1. After the intraoperative cholangiogram reveals presence of common bile duct stones, transcystic duct exploration can be undertaken via the same hole in the cystic duct created for the cholangiogram. However, a larger hole with dilation of the cystic duct may be necessary to remove stones.

Step 2. A balloon catheter setup is utilized to dilate the cystic duct. The cystic duct should be gradually dilated over a period of 3-5 minutes. The cystic duct should never be dilated to a diameter larger than the common bile duct. The cystic duct needs to be dilated so that it is at least as large as the largest stone to be removed.

Step 3. Choledochoscopy can be performed through the cystic duct incision to visualize and localize the common duct stone(s). A choledochoscope is used that has a working channel of at least 1.2 mm. Body-temperature saline is used to irrigate the common bile duct to aid in visualization. If a stone is encountered, it should be removed before looking for more stones since failure to do so can result in stones first visualized floating up into the proximal bile ducts in the liver.

Step 4. To remove stones, a straight #4 wire basket (2.4 F) is preferable and should be threaded through the working channel of the choledochoscope. The wire basket is passed beyond the stone and opened. The stone is entrapped when the basket is withdrawn. Once entrapped, the stone should be gently grasped and the basket pulled snugly up against the end of the choledochoscope. Both the basket and choledochoscope are withdrawn completely as a unit. This process is repeated until all stones are completely removed.

Northwestern Handbook of Surgical Procedures, 2nd Edition, edited by Nathaniel J. Soper and Dixon B. Kaufman. ©2011 Landes Bioscience.

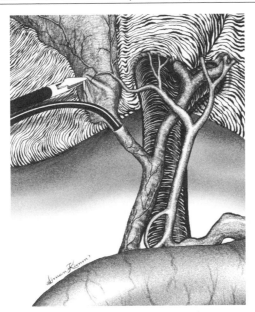

Figure 9.1. Laparoscopic transcystic common bile duct exploration.

Step 5. A completion cholangiogram is performed by reinserting a catheter through the cystic duct and securing it in place so that there is no leakage of contrast.

Step 6. A cystic duct tube for drainage can be inserted if there is concern for retained stones.

Laparoscopic Choledochotomy

This procedure is performed much as an open common duct exploration is performed. It requires the surgeon to have the capability to perform intracorporeal suturing and knot-tying. It is best to perform the common duct exploration before removal of the gallbladder since the clamps on the gallbladder can be used to retract the liver and to place traction on the common bile duct.

Step 1. Side-by-side stay sutures of 5-0 monofilament are placed about 2 mm apart in the wall of the common bile duct, just below the cystic duct-common bile duct junction. A longitudinal choledochotomy approximately 1 cm in length is created using microdissection laparoscopic shears. Any bile leakage is aspirated.

Step 2. The common bile duct is irrigated with body-temperature sterile saline to try to "float" any gallstones out via the choledochotomy.

Step 3. The choledochoscope is placed through the choledochotomy, and the method for stone retrieval/removal is similar to that described above.

Step 4. Once all stones have been removed, a T-tube of appropriate size is fashioned and passed into the abdominal cavity. The T-tube is placed into the choledochotomy and the common duct closed around the tube with a series of interrupted 4-0 absorbable sutures. In either procedure, a suction drain is placed to monitor for leakage of bile from the cystic duct closure or the choledochotomy.

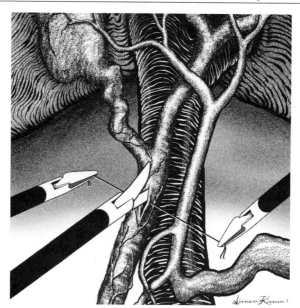

Figure 9.2. Laparoscopic choledochotomy.

Postop

In general, care is similar to that described for laparoscopic cholecystectomy. The peritoneal drain can be removed in 24-48 hours if there is no evidence of bile leakage. The T-tube or cystic duct drain is left in place for approximately 2 weeks before obtaining a tube cholangiogram. If there are no retained stones, the tube can be removed in the outpatient office.

Complications

Retained common duct stones may require endoscopic removal. Injury to the common bile duct may occur during common bile duct exploration if it is not carefully done. Bile duct stricture can be a long-term complication.

Follow-Up

Patients should be followed short-term at intervals until tubes are removed and liver function tests are normal. Long-term follow-up is described under laparoscopic cholecystectomy.

Acknowledgment

The editors and author wish to acknowledge Kenric M. Murayama for contributing to the previous version of this chapter.

Chapter 10

Repair Common Bile Duct Injury: Open

Jonathan Fryer

Indications

Injuries to the common bile duct (CBD) or more proximal branches of the biliary tract occur most commonly as a consequence of surgical misadventure. The most common operation associated with biliary tract injury is a laparoscopic cholecystectomy. The diagnosis is typically made based on clinical findings (jaundice, abnormal liver function tests, bile leak) and defined by radiologic studies (magnetic resonance cholangiopancreatography [MRCP], computerized tomographic (CT) cholangiography, HIDA scan) and/or endoscopic retrograde cholangiopancreatography (ERCP). Some simple biliary tract injuries (i.e., cystic duct stump leak) are amenable to repair using interventional ERCP or percutaneous transhepatic cholangiography (PTC) techniques such as stenting and/or dilatation. Few common bile duct injuries are amenable to repair using laparoscopic surgical techniques. The indications for open surgical repair are: injury of the biliary tract with (i) suspicion of a thermal, ischemic or occlusive injury and (ii) greater than 50% circumferential disruption of the bile duct. Because the exact etiology of the injury is often not clear after a laparoscopic cholecystectomy or other surgical procedure, an open surgical exploration and common bile duct repair is frequently required.

10

Preop

Patients with suspected biliary tract injury must be assessed for hemodynamic instability due to sepsis and/or bleeding and resuscitated as needed. In addition to bile (if available) and blood cultures, CBC, INR, amylase, lipase, liver function tests (LFTs) should be obtained to support the diagnosis of biliary injury. Broad spectrum antibiotics with effective biliary penetration should be initiated. Evidence of bile leak should be sought by evaluating for bile in drains or signs of bile peritonitis and/or by reviewing imaging studies for intra-abdominal fluid collections consistent with bile leakage. Biliary imaging studies (MRCP, ERCP, PTC, CT cholangiography) are required to better define the nature and location of the injury to aid in planning the necessary intervention(s). Once the diagnosis of a bile leak is confirmed, it must be determined whether the bile collections have been drained adequately. If not they need to be adequately drained as soon as possible and contents sent for culture. Patients with prolonged bile leakage are at risk for malabsorption of fat-soluble vitamins, and therefore administration of vitamin K should be considered, especially if the INR is elevated. If surgery is planned, blood products should be made available for the OR. The patient's cardiopulmonary status and other risk factors for surgical procedure should be evaluated as patients with prohibitive co-morbidities may be best served by nonsurgical strategies.

Northwestern Handbook of Surgical Procedures, 2nd Edition, edited by Nathaniel J. Soper and Dixon B. Kaufman. ©2011 Landes Bioscience.

It is also important to evaluate the hepatic arterial system via a standard angiogram, magnetic resonance angiogram (MRA) or a CT angiogram (CTA) to determine whether an associated arterial injury was sustained at the time of the bile duct injury. This is important as concomitant vascular injuries are not uncommon (10%), and resultant ischemia may significantly impede healing of any bile duct repair required.

Procedure

Step 1. The patient should be placed in the supine position. Adequate venous access should be secured, and an NG tube and Foley catheter should be placed to facilitate management in the perioperative and postoperative periods. The abdomen should be prepped and draped in a standard fashion.

Step 2. The procedure should be approached either by a right subcostal incision or by an upper midline incision. Upon entry into the abdominal cavity, adequate exposure should be obtained using retraction.

Step 3. If the gallbladder has not yet been removed, it should be dissected free from the liver cautiously using a fundus-to-duct approach. If the cystic duct is identifiable, it may be useful for obtaining intraopertive cholangiographic studies which may aid in defining the residual biliary tract anatomy, but should otherwise be ligated and divided.

Step 4. The boundaries of the porta hepatis should be defined between the first part of the duodenum and the liver. The superficial layer of peritoneum overlying the porta hepatis should be divided left to right to provide exposure to the underlying structures. Care should be taken to identify and preserve the hepatic arteries as their typical orientation may be distorted by previous surgical efforts.

Step 5. The common bile duct should be identified by visual inspection and dissection using imaging studies as a guide. When identified, the common bile duct should be encircled carefully so as not to injure surrounding structures or the blood supply to the common bile duct.

Step 6. The exact location of the biliary tract injury, based on a preoperative radiologic finding, should be determined intraoperatively. The extent and nature of the common bile duct injury should be determined based on history, imaging studies and careful visual inspection of the biliary tree. If the injury is related to a sharp injury only, the status of the tissue proximal and distal to the injury should be carefully evaluated to determine if a primary closure is possible. If a thermal, ischemic, or occlusive injury is identified or suspected, all areas of the biliary tree believed to be involved in the injury should be resected. If the injury is related to ligation from a suture or clip, an attempt may be made to remove the clip or the suture. An extremely high level of suspicion should be held that the bile duct has been permanently injured at this location. Once the bile duct injury has been defined and all devitalized tissue has been dissected, a decision will need to be made about how to repair the bile duct.

Step 7. If a direct duct-to-duct anastomosis can be performed between two healthy ducts without tension, this method of closure could be considered. A duct-to-duct biliary anastomosis should be performed via microsurgical techniques, using 5-0 or 6-0 absorbable suture material in interrupted fashion.

Step 8. To prevent bleeding during the performance of anastomosis, the bile duct arteries should be carefully ligated with fine 6-0 Prolene. These Prolene sutures will be located at the 3 o'clock and 9 o'clock positions and can be used to align the bile ducts for anastomosis.

10

Step 9. The first suture of the anastomosis should be performed in the middle of the back wall with the suture being tied outside of the bile duct. The back wall will be completed by placing and tying two or three additional, interrupted "back wall" sutures on either side of the initial stitch, leaving the final sutures placed on either side long, to function as corner stitches.

Step 10. After completion of back wall stitches, the front wall stitches can be placed serially and held with rubber-shod forceps to keep them in alignment. After the interrupted front wall sutures have been placed, they can be tied serially, emphasizing the precision of their placement.

Step 11. Depending on the nature of the injury, the preference of the surgeon, and the underlying status of the patient, it may be preferred to place a T-tube in the common bile duct after completion of the bile duct repair. A T-tube should be placed, preferably at a site distal to the bile duct repair.

Step 12. A transverse incision for placement of the T-tube is preferred to avoid narrowing of the bile duct. The T-tube should be fashioned in such a way that the one arm of the "T"-d portion of the tube is passed proximally across the anastomosis and the other arm passed distally to the T-tube insertion site to facilitate drainage of bile during the healing of the bile duct injury. The bile duct should be repaired securely around the T-tube to prevent leakage. This repair should be performed using interrupted absorbable suture such as Maxon.

Step 13. After securing a T-tube in the bile duct, it should be brought out through a lateral percutaneous stab incision and secured at the skin level to avoid inadvertent dislodgement. The T-tube should be left in place for a minimum of 3 weeks, at which point a T-tube cholangiogram should be performed to confirm healing of the bile duct repair. The T-tube can be removed safely once a fibrous tract has formed around the tube thereby preventing leakage of bile from the T-tube insertional site into the peritoneal cavity.

Step 14. If after removing all devitalized tissue and demonstrating adequate vascularity of both ends of the bile duct, it appears that an end-to-end anastomosis cannot be performed safely without tension, a Roux-en-Y choledochojejunostomy should be performed.

Step 15. This is accomplished by identifying a segment of jejunum immediately distal to the ligament of Treitz that will reach without tension to the porta hepatis. The jejunum should be transected at this location and the distal segment of the transaction (i.e., Roux loop) brought to the porta. The proximal segment of the jejunal transaction should be anastomosed end-to-side to a site approximately 40 cm from the end of the distal segment of the jejunal transaction (i.e., entero-enterostomy).

Step 16. The proximal (i.e., liver side) segment of the biliary tree is then anastomosed end-to-side to the most proximal segment of the Roux loop. A small defect is created in the serosal surface of the proximal Roux loop. The mucosa is grasped with small forceps and cauterized to create a small mucosal defect in the same location.

Step 17. The proximal segment of biliary tree is then anastomosed to this transmural defect in the proximal Roux limb, using the same interrupted technique as described above for a duct-to-duct anastomosis.

Step 18. After completion of the bile duct repair, the abdominal cavity should be extensively irrigated to remove and dilute any spillage created from opening the bile duct. It is recommended that a Jackson-Pratt drain should be placed posterior to the bile duct repair and brought out through a lateral stab wound to facilitate early diagnosis of a bile leak should one occur.

10

Step 19. A heavy monofilament suture material is used to close the fascia. The skin can be closed with subcuticular stitches or staples.

Postop

Postoperatively, patients should be monitored for sepsis and bleeding. LFTs should be monitored to ensure no injury to the liver, obstruction of the bile duct repair, or leakage of bile. The JP should be closely inspected for evidence of bile leaks. Because of manipulation of the common bile duct, amylase and lipase should also be checked postoperatively. Patients requiring a direct common bile duct repair will likely be started on clear fluids within 24 hours of their operation and their diet quickly advanced. Patients who underwent a Roux-en-Y hepaticojejunostomy will need to be kept NPO for additional days to protect the entero-enterostomy. Patients should be covered perioperatively with antibiotics, although if concerns about sepsis are high due to massive spillage or inadequately-drained segments of the biliary tree, antibiotics may need to be continued for longer periods. Patients who are at immobile should be administered prophylactic anticoagulation to prevent DVT and/or should be placed in pneumatic compression devices. Providing the patient is stable, a Foley catheter should be removed at 24 hours of the operation. As mentioned previously, if a T-tube has been placed a T-tube cholangiogram should be performed approximately three weeks postoperatively. Based on the findings of the T-tube cholangiogram, plans can be made to remove the T-tube. If no T-tube is in place and the LFTs normalize, it may not be necessary to obtain additional imaging of the biliary tree. If there are abnormalities in the LFTs, an ERCP or MRCP is indicated to better define the anatomy of the biliary tree and intervene if necessary.

Complications

Potential complications specific to bile duct injury are sepsis, bleeding, bile duct leak, bile duct stricture, bile leak from the T-tube placement site, hepatic artery injury, portal vein injury and pancreatitis. Patients undergoing surgical repair of a bile duct injury are at risk of complications associated with any general surgery procedure including deep vein thrombosis, pulmonary embolus, pneumonia, wound infection and complications specific to administration general anesthesia.

Follow-Up

Patients who have undergone a common bile duct repair without a T-tube should be followed up postoperatively to monitor for wound infection, intra-abdominal abscess, biliary sepsis, bile leak, pancreatitis, jaundice or persistent LFT abnormalities. If concerns regarding the status of the biliary tree are prompted by these findings, it will be necessary to restudy the biliary tree with MRCP or ERCP.

Chapter 11

Hepaticojejunostomy: Roux-en-Y

David Bentrem and Eric Cheon

Indications

Roux-en-Y hepaticojejunostomy is indicated for reconstruction following resection for carcinoma of the proximal bile duct and hepatic duct bifurcation (Klatskin's tumors). When tumors of the distal third of the bile duct or head of the pancreas are not resectable, a bypass between the hepatic duct and a Roux-en-Y limb of jejunum may be done to relieve distal biliary obstruction. Benign indications for Roux-en-Y hepaticojejunostomy include: extrahepatic biliary stricture secondary to previous surgical injury, common bile duct obstruction secondary to recurrent stones, and distal common bile duct stricture secondary to chronic pancreatitis. Contraindications to resection and reconstruction for malignancy include distant metastases or extensive, bilateral liver involvement. Lymph node metastases outside the region of the porta hepatis are usually considered a contraindication to resection. Local extension of the tumor to include the main portal vein with thrombosis of the vein and encasement of the main common hepatic artery are also contraindications to resection. Unilateral involvement of the proximal bile duct, portal vein, or hepatic artery branch may necessitate combined hepatic resection with biliary resection and reconstruction.

Preop

In cancer patients, a triphasic helical CT scan of the liver and porta hepatis is done for assessment of local tumor extension. Percutaneous transhepatic cholangiography (PTC) and/or endoscopic retrograde cholangiopancreatography (ERCP) are done to demonstrate the extent and location of tumors or strictures of the biliary system and for preoperative relief of biliary obstruction. Placement of the percutaneous stent through the hepatic duct stricture or tumor into the duodenum will facilitate identification and dissection at surgery. Magnetic resonance imaging may be required to delineate the relation of a tumor to the main vascular structures within the porta hepatis. On the morning of surgery, patients receive preoperative antibiotics and an epidural catheter for intraoperative and postoperative analgesia. Deep vein thrombosis (DVT) prophylaxis with subcutaneous heparin and/or sequential compression devices is used according to the patient's risk factors for thromboembolus.

Procedure

Step 1. The patient is placed in the supine position. General anesthesia is induced and the abdomen is prepared for exploration. An upper midline incision is suitable for most patients. If a right subcostal incision was used for a previous operation, the abdomen may be entered through the prior incision.

Northwestern Handbook of Surgical Procedures, 2nd Edition, edited by Nathaniel J. Soper and Dixon B. Kaufman. ©2011 Landes Bioscience.

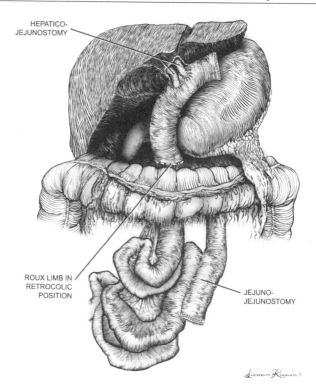

HEPATICO-
JEJUNOSTOMY

ROUX LIMB IN
RETROCOLIC
POSITION

JEJUNO-
JEJUNOSTOMY

Simon Kimm

Figure 11.1. Roux-en-Y hepaticojejunostomy.

Step 2. When operating for malignant strictures, exploration of the abdomen is done to rule out disseminated hepatic disease or peritoneal and omental metastases. In the absence of disseminated metastatic disease, exposure of the porta hepatis is begun.

Step 3. The falciform ligament is divided and the lateral suspensory ligaments of the liver are also divided to facilitate mobilization and retraction of the liver.

Step 4. A Kocher maneuver is done to facilitate mobilization and exposure of the distal common bile duct.

Step 5. Mechanical self-retaining retractors are placed in the upper abdomen, and the liver is retracted superiorly and laterally into the right upper quadrant. Manual countertraction is applied inferiorly and medially to provide exposure of the soft tissues in the porta hepatis

Step 6. If the gallbladder has not been previously removed, gentle traction is applied superiorly and anteriorly on the infundibulum of the gallbladder. The cystic duct is identified and dissected down to its junction with the common hepatic duct. The anterior peritoneum of the porta hepatis is completely divided across the bile duct and the hepatic artery. Circumferential dissection is then accomplished at the level of the cystic duct-common hepatic duct junction. A vessel loop is placed around the common bile duct, and the bile duct is then retracted laterally. The gallbladder is removed at this point or may be removed later en bloc with the specimen.

Step 7. If the gallbladder has been previously removed, identification of the bile duct is facilitated by the previously placed percutaneous biliary stents. By palpating the porta hepatis, the stent may be felt. Then the layer of peritoneum overlying the common hepatic duct can be divided. The peritoneum is divided medially to also expose the common hepatic artery. The bile duct is retracted laterally and the hepatic artery retracted medially. The soft tissues between the common hepatic artery and the common bile duct are dissected toward the main bile duct. The portal vein is identified posteriorly and dissected free of the bile duct. In cases of malignant strictures, the relationship of the proximal bile duct and tumor to the portal vein and hepatic artery branches is inspected. If the tumor has not extended into the main portal vein or has not encased the common hepatic artery, the dissection is continued. When there is local extension, the usual bile duct tumor will involve the right hepatic artery and potentially the right portal vein. In these situations, consideration should be given to right hepatic lobectomy or trisegmentectomy with biliary reconstruction to the left hepatic duct. In the cases of a benign biliary stricture, care must also be taken to identify the right hepatic artery and the bifurcation of the common hepatic artery as these areas may be densely adherent to the area of stricture and inflammation.

Step 8. Once the anatomic relationship of the biliary tree to the vascular structures in the porta hepatis has been identified and cleanly dissected, preparation is begun for resection and reconstruction of the biliary duct. The mobilized distal bile duct is retracted superiorly and divided distally above the entrance of the bile duct into the head of the pancreas. The distal common duct is closed with a running 4-0 suture. The proximal end is retracted anteriorly and superiorly, and dissection of the posterior wall of the common bile duct off of the anterior wall of the portal vein is carried to the level of the hepatic duct bifurcation.

Step 9. In cases of benign biliary strictures, the common hepatic duct may be divided at the level of the right and left ductal confluence if the ductal tissue is healthy at that point. It is at this point that the diameter of the common hepatic duct will be largest to facilitate subsequent biliary enteric anastomosis.

Step 10. In cases of malignant biliary obstruction, dissection into the hepatic parenchyma along the main right and left hepatic ducts may be required to obtain a clear proximal margin. In such cases bilateral hepaticojejunostomies may be required for reconstruction. Frozen-section examination of the proximal portions of the right and left ducts is done to determine if the tumor has been completely resected.

Step 11. In malignant cases, posterior lymph nodes in the porta hepatis are dissected completely off the portal vein with care taken to identify aberrant or anomalous right hepatic arteries. Dissection is carried proximally on the common hepatic artery back toward the level of the celiac trunk. No such lymphadenectomy is required in cases of benign biliary strictures.

Step 12. Reconstruction is begun with preparation of the Roux-en-Y limb. The proximal jejunum is divided with a linear stapling device approximately 15 cm distal to the ligament of Treitz. The mesentery is divided between clamps or with a tissue sealing device down to the level of the superior mesenteric artery.

Step 13. The distal end of the cut jejunum is brought through an opening in the transverse mesocolon to rest in a retrocolic position. If there is a large mass in the head of the pancreas or dissection is particularly difficult in this area, the limb may be put in an antecolic position. A primary biliary-to-enteric anastomosis between the end of the common hepatic duct and the antimesenteric border of the jejunum is done in a single layer with interrupted 4-0 or 5-0 absorbable sutures.

Step 14. Gastrointestinal tract continuity is then reestablished with a side-to-side anastomosis of the proximal jejunum to the descending limb of the Roux-en-Y segment with a linear stapling device or in hand sewn fashion. An area drain may placed in the right upper quadrant, posterior to the biliary anastomosis. The abdomen is closed in layers and the skin closed with staples.

Postop

Nasogastric decompression is maintained for <24 hours. Antibiotics are discontinued after the 24-hour perioperative period.

Complications

Early complications include trauma to the portal vein or hepatic artery during dissection of the porta hepatis and postoperative hemorrhage. Subsequent complications also include anastomotic leakage with persistent biliary fistula, subhepatic abscess formation, and small bowel obstruction secondary to angulation or torsion of the small bowel anastomosis. The most important long-term complication is anastomotic stricture secondary to ischemia or tumor recurrence.

Follow-Up

For patients with malignant disease, liver function tests and clinical status should be followed at regular intervals. A change in status or chemistries requires re-evaluation with imaging to rule out tumor recurrence. For patients operated for benign strictures, regular lifelong surveillance of liver chemistries is indicated because of a significant rate of recurrent stricture formation, which may be asymptomatic but lead to secondary biliary cirrhosis if not corrected.

Acknowledgment

The editors and authors wish to acknowledge Mark S. Talamonti for contributing to the previous version of this chapter.

Chapter 12

Pancreatic Cystogastrostomy

David M. Mahvi

Indications

Pancreatic cystogastrostomy is indicated for the treatment of large, persistent, and/or symptomatic pseudocysts of the pancreas. The pseudocyst should indent the posterior wall of the stomach by CT or other imaging.

Preop

Intraoperative ultrasound should be available. An epidural catheter for postoperative pain control is placed immediately prior to surgery. An intravenous antibiotic for surgical site infection prophylaxis (cefazolin 1.0 g IV) is given 30 minutes prior to incision. Pneumatic compression boots and/or subcutaneous heparin are used for deep venous thrombosis prophylaxis according to patient risk. A nasogastric (NG) tube and urinary catheter are placed after induction of anesthesia.

Procedure

Step 1. An upper midline abdominal incision is made. The abdomen is explored to the extent possible.

Step 2. The pseudocyst is palpated through the stomach. Using its location as a guide, a transverse incision about 5 cm long is made through the anterior gastric wall with electrocautery.

Step 3. Using narrow Deaver retractors or sutures for retraction of the edges of the stomach incision, the point where the pseudocyst bulges maximally into the posterior gastric wall is established by inspection and palpation.

Step 4. A 20-gauge needle attached to a 10 ml syringe is passed through the posterior gastric wall, aspirating continuously. Pseudocyst fluid should be encountered within 1-2 cm. If no fluid is encountered, aspiration in a different site or intraoperative ultrasound should be considered. The posterior wall of the stomach must not be opened unless fluid is encountered by aspiration within a short distance of the gastric wall.

Step 5. If fluid is encountered, an incision is made through the posterior wall of the stomach with electrocautery. Initially, a 2-3 cm incision is made to enter the pseudocyst. The incision may be enlarged depending on the size of the pseudocyst. The final incision should be at least 3-4 cm long. It is inadvisable to palpate the inside of the pseudocyst vigorously, as bleeding may develop which is difficult to control.

Step 6. Using the electrocautery, a thin ellipse of tissue is removed from the edge of the incision into the pseudocyst which encompasses some of the wall of the pseudocyst. The tissue must be sent to pathology for frozen section to rule out a cystic neoplasm of the pancreas, which will differ from a pseudocyst by possessing an epithelial lining.

Northwestern Handbook of Surgical Procedures, 2nd Edition, edited by Nathaniel J. Soper and Dixon B. Kaufman. ©2011 Landes Bioscience.

Step 7. The edge of the cystogastrostomy incision should be checked for bleeding, which should be controlled with cautery or suture ligatures. A running mono-filament suture should be run continuously around the cystgastrostomy as a method to control bleeding.

Step 8. Starting a separate suture at each corner, the anterior wall of the stomach is closed with an inner layer of continuous absorbable 3-0 suture, incorporating all layers of the stomach wall. The closure is completed with an outer layer of interrupted seromuscular 3-0 silk sutures.

Step 9. The midline fascia and skin are closed with suture material of choice.

Postop

The NG tube can be removed on postoperative day 1 and diet begun as tolerated.

Complications

The most significant procedure-specific complication is gastrointestinal bleeding from the cystogastrostomy. This occurs in approximately 10% of patients. Bleeding should be evaluated in the early postoperative period by angiography to rule out a pseudoaneurysm which can be embolized.

Follow-Up

After recovery from the procedure, patients should be followed at intervals to determine if any symptoms recur. Recurrence of symptoms should prompt imaging studies of the abdomen to rule out pseudocyst recurrence, which occurs in about 10% of cases, primarily in the first year after surgery.

Acknowledgment

The editors and author wish to acknowledge Richard H. Bell, Jr. for contributing to the previous version of this chapter.

Chapter 13

Pancreatic Necrosis: Debridement

David M. Mahvi

Indications

Debridement is indicated for infected pancreatic necrosis and associated sepsis following an attack of acute pancreatitis. Debridement for noninfected necrosis or viable pancreas with sepsis is not clearly indicated. Though percutaneous or trans-gastric drainage is critical in the management of infected pseudocysts, the necrotic pancreas cannot be removed without surgical access to the pancreas.

Preop

Retroperitoneal operative procedures in a patient with pancreatitis can result in significant blood loss. The patient thus should be typed and cross-matched for 2-4 units of blood. Preoperative antibiotics should be continued. An epidural catheter for postoperative pain control should be considered preoperatively even in these acutely ill patients. Deep venous thrombosis prophylaxis with sequential compression devices and/or subcutaneous heparin (depending on patient risk factors) should be instituted prior to the initiation of general anesthesia.

13

Procedure

Step 1. After prepping and draping the entire abdomen, a midline incision is made. An alternative approach is a bilateral subcostal incision. We prefer the midline incision for the ease of extension. If a percutaneous drain is in place, it should be prepped into the field as it may be the only means of finding the infected pancreas.

Step 2. The pancreas is approached by entering the lesser sac through the gastrocolic ligament. The gastrocolic omentum is opened just below the gastroepiploic vessels. If entry into the lesser sac is difficult due to the inflammation, it may be safer to enter the lesser sac immediately adjacent to the stomach to avoid injury to the transverse mesocolon and middle colic vessels.

Step 3. If the retroperitoneal inflammation prevents entry through the gastrocolic omentum, the pancreas may be approached through the base of the transverse mesocolon. An incision to the right of the middle colic vessels is used to debride the head of the gland while an incision to the left of the vessels is used to debride the body and tail. If a drain is in place carefully following the course of the drain may safely lead to the pancreas.

Step 4. The necrotic portions of the pancreas and peripancreatic tissue are removed manually. Laparotomy sponges may be useful. In addition, ring forceps are helpful to remove nonviable tissue. Sharp dissection should be used very sparingly. The goal of the procedure is to remove all nonviable tissue without injuring surrounding enteral or vascular structures.

Northwestern Handbook of Surgical Procedures, 2nd Edition, edited by Nathaniel J. Soper and Dixon B. Kaufman. ©2011 Landes Bioscience.

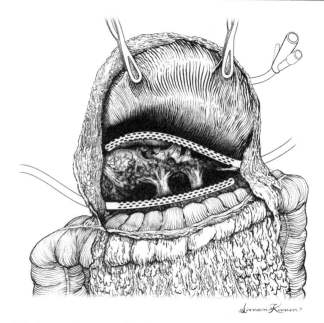

Figure 13.1. Pancreatic necrosis debridement.

Step 5. Bleeding is controlled with direct pressure or suture ligatures.

Step 6. At least two closed suction drains are left in the pancreatic bed. These should be placed over the head and the body and tail of the gland. They are brought through separate incisions in the abdominal wall.

Step 7. A feeding jejunostomy tube for nutritional support should be placed. A gastrostomy tube for gastric decompression should be considered as many of these patients may have a prolonged gastric ileus.

Step 8. The abdomen is closed. The skin should be left open as in any contaminated operative procedure.

Step 9. If the patient has extensive necrotic tissue in the retroperitoneum which cannot be debrided in one operation, the lesser sac can be marsupialized and packed with gauze and the abdominal wall left open, with the cavity dressings changed in the intensive care unit under sedation. Alternatively, the patient can be returned to the operating room several days in a row until the nonviable tissue has all been removed. In this case, temporary closure of the abdomen with Gore-Tex® or silicone sheeting sewn to the fascia around the periphery of the wound can be quite helpful. When returning to the operating room, the sheeting can be reopened in its midportion and reclosed without the necessity of disturbing the fascia.

Postop

These patients are often critically ill and require management in an intensive care unit. The drains are placed on low suction and irrigated each shift to prevent obstruction. The gastrostomy tube is placed on gravity drainage. Jejunostomy feeds can be started within 24-48 hours.

13

Complications

The most common causes of death in patients who have undergone pancreatic debridement are persistent retroperitoneal sepsis and bleeding. If the patient is not improving postoperatively, a CT scan should be performed to determine if undrained fluid collections are present or if additional nonviable tissue is present. Complications of pancreatic debdridement also include respiratory failure, renal failure, colonic ischemia, and gastrointestinal fistulae.

Follow-Up

Patients who survive the acute episode of necrosis need to be followed for the development of pancreatic endocrine and exocrine insufficiency. Patients who are treated by the open packing method will usually develop a ventral hernia which ultimately requires repair.

Acknowledgment

The editors and author wish to acknowledge Woody Denham for contributing to the previous version of this chapter.

13

Chapter 14

Transduodenal Sphincteroplasty

David Bentrem and Amanda Hayman

Indications

Transduodenal sphincteroplasty is indicated for the treatment of sphincter stenosis or dysfunction in selected cases. It is also indicated in the presence of multiple common duct stones in a nondilated system or to remove an impacted stone at the ampulla of Vater that is not removable by any other means. At present, many such patients are treated endoscopically yet expertise can be center-dependent.

Preop

A prophylactic antibiotic is administered 30 minutes prior to incision. Deep venous thrombosis prophylaxis with sequential compression devices and subcutaneous heparin is provided dependent on patient risk factors.

Procedure

Step 1. The abdomen is entered through a midline or right subcostal incision. A Kocher maneuver of the duodenum is performed.

Step 2. The common bile duct is exposed above the duodenum. Access to the common duct is required to aid in the performance of a sphincteroplasty. If the sphincteroplasty is being done in combination with a common bile duct exploration, access is through the choledochostomy. If the common duct does not need to be opened, access can be gained through the cystic duct stump.

Step 3. A #4 or #5 Fogarty catheter is passed through the cystic duct or choledochostomy into the duodenum. The balloon is inflated once the catheter tip is in the duodenum and the catheter pulled back snugly so that the balloon impacts the ampulla.

Step 4. A longitudinal incision is made in the duodenum over the ampulla. The incision is centered over the Fogarty balloon on the opposite wall of the duodenum.

Step 5. The ampulla of Vater is located and 3-0 silk stay sutures placed on either side of the ampulla to elevate it.

Step 6. The stay sutures are lifted forward, the balloon deflated, and the catheter partially withdrawn into the bile duct. A metal probe or grooved director is placed in the bile duct. The bile duct will be at the 11 o'clock position on the ampulla. The pancreatic duct is in the 4 o'clock position.

Step 7. With cautery, the ampulla is progressively incised over the probe, unroofing the bile duct 3-4 mm at a time. Each time a new 3-4 mm incision is made, interrupted 4-0 absorbable sutures should be placed from the bile duct mucosa to the duodenal mucosa on each side of the opening, being careful not to compromise the pancreatic duct opening.

14

Northwestern Handbook of Surgical Procedures, 2nd Edition, edited by Nathaniel J. Soper and Dixon B. Kaufman. ©2011 Landes Bioscience.

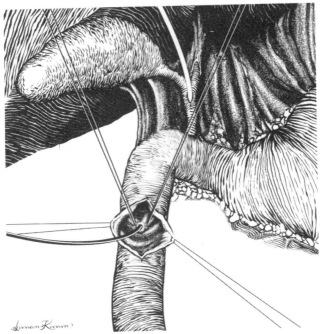

Figure 14.1. Transduodenal sphincteroplasty. Ampulla of Vater exposed and cannulated from the cystic duct.

14

Step 8. The opening in the bile duct should be extended until a dilator equal to the diameter of the common duct passes easily.

Step 9. A final 4-0 absorbable suture is placed at the apex.

Step 10. A pancreatic duct sphincteroplasty can be performed in the same manner except that the probe is placed in the pancreatic duct (4 o'clock position). If sphincteroplasties of the common bile duct and pancreatic duct are performed, the mucosa of the common wall between the two is sutured together with interrupted absorbable suture. The opposite wall of each duct is attached to the duodenal mucosa as previously described.

Step 11. The duodenum is closed transversely in two layers to prevent narrowing. If excess tension is encountered with transverse closure, longitudinal closure should be considered. Drain placement at the time of operation is based on surgeon preference.

Postop

Intravenous fluids and nothing by mouth should be continued until bowel function returns. Antibiotics are discontinued within 24 hours, except in patients with preexisting cholangitis. Pain is controlled by intravenous narcotics until the patient is tolerating a diet.

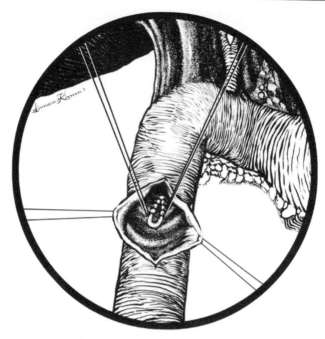

Figure 14.2. Transduodenal sphincteroplasty of ampulla of Vater complete.

Complications

Wound infection is a risk in the face of infected bile. Bleeding from the sphinctero-plasty may occur. Close suture placement will aid in hemostasis with well-vascularized duodenal mucosa. If the sphincteroplasty is carried too deep, the duodenal wall may be perforated and retroperitoneal leakage occur. The duodenal closure can also leak and is a dreaded complication.

Follow-Up

Patients should be followed until wounds are healed. If a T-tube was left in the common duct as part of the operation, a T-tube cholangiogram should be done as an outpatient and the T-tube removed about 2 weeks postoperatively.

Acknowledgment

The editors and authors wish to acknowledge Woody Denham for contributing to the previous version of this chapter.

14

Chapter 15

Longitudinal Pancreaticojejunostomy: Puestow Procedure

David M. Mahvi

Indications

Longitudinal pancreaticojejunostomy is indicated in patients with chronic pancreatitis with refractory pain, who have failed symptomatic therapy, whose entire pancreatic duct is dilated, and whose gland is generally atrophic.

Preop

Intraoperative ultrasound should be requested. The patient should be NPO for 8 hours preoperatively. An intravenous antibiotic for surgical site infection prophylaxis (cefazolin 1 g IV) is given 30 minutes before incision. A nasogastric (NG) tube and urinary catheter are placed after general anesthesia is induced. Sequential calf compression boots and/or low-dose subcutaneous heparin are administered for deep vein thrombosis (DVT) prophylaxis depending on patient risk factors.

Procedure

Step 1. A bilateral subcostal incision is made extending to the lateral edge of the rectus abdominis muscle on each side. A general exploration of the abdomen is performed. The NG tube position is confirmed.

Step 2. An upper abdominal retractor system is placed.

Step 3. The hepatic flexure of the colon is mobilized by incising the peritoneum above the flexure, and the base of the transverse mesocolon is swept inferiorly to expose the duodenum and head of the pancreas.

Step 4. After incising the peritoneum lateral to it, the duodenum is mobilized with a Kocher maneuver extending from the right edge of the hepatoduodenal ligament to the superior mesenteric vessels.

Step 5. The greater omentum is dissected upward and away from the transverse colon and the lesser sac thereby entered. The stomach is mobilized in a cephalad direction by blunt and sharp dissection and the stomach and omentum retracted upward to allow identification of the body and tail of the pancreas. The entire anterior surface of the body and tail of the gland should be cleaned and exposed.

Step 6. On the upper edge of the transverse mesocolon, the middle colic vein is followed down to its entrance into the superior mesenteric vein (SMV). The right gastroepiploic vein, which usually joins the middle colic vein to form the gastrocolic trunk before they enter the SMV, is identified. The right gastroepiploic vein is then ligated and divided as close to its entrance into the gastrocolic trunk or SMV as possible, taking care not to injure the middle colic vein. Because the right gastroepiploic vein crosses the neck of the pancreas from top to bottom to enter the SMV, it must be divided to provide adequate exposure of the junction between the head and body of the pancreas.

15

Northwestern Handbook of Surgical Procedures, 2nd Edition, edited by Nathaniel J. Soper and Dixon B. Kaufman. ©2011 Landes Bioscience.

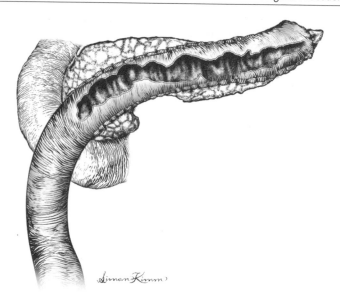

Figure 15.1. Longitudinal pancreaticojejunostomy.

15

Step 7. With the entire anterior surface of the pancreas well-exposed from the head to the tail, the ultrasound probe is applied to an area of the gland that is easily accessible and the pancreatic duct identified. A 20-gauge needle is passed into the duct, aspirating with a syringe to confirm the return of pancreatic juice. The needle is left in place once juice is returned, and the electrocautery used to cut down on the duct and enter it.

Step 8. The duct is opened as far as possible in both directions by passing a right-angle clamp into it and progressively dividing overlying parenchyma with electrocautery. To the patient's right, the duct should be opened to within 1-2 cm of the duodenum. To the left, the duct should be opened along the body and tail of the gland to within 3-4 cm of the tail of the gland. The gland should not be opened so far to the left that subsequent anastomosis will be overly difficult due to poor visibility.

Step 9. Once the pancreatic dissection is complete, an appropriate point in the proximal jejunum is identified to create a Roux-en-Y limb and the jejunum divided with a linear stapler. The mesentery of the jejunum is divided perpendicular to the axis of the bowel to allow mobilization of the limb. The distal end of the jejunum is oversewn with interrupted sutures of 3-0 silk. A small defect is created in the mesentery of the transverse colon and the distal end of the divided jejunum passed through the defect, bringing the end of the jejunum to the patient's left. The Roux limb should lie easily over the opened pancreas without tension.

Step 10. The Roux limb is sewn to the inferior border of the pancreas with interrupted sutures of 3-0 silk which are placed into the seromuscular layers of the bowel and into the pancreatic parenchyma. The small bowel is then opened with cautery, making the opening parallel to the pancreatic duct but keeping the opening in the bowel slightly shorter than the opening in the pancreas.

Step 11. Starting in the middle of the back row of the anastomosis, a double-armed suture of 3-0 absorbable monofilament material is placed and then a circumferential anastomosis of all bowel layers to the cut edge of the pancreatic defect is performed. The anastomosis is completed with an anterior row of interrupted 3-0 silk sutures.

Step 12. An end-to-side jejunojejunostomy is performed 40 cm below the pancreatic anastomosis to complete the Roux limb and the leaves of the mesentery sutured to prevent internal hernia. The edge of the defect in the transverse mesocolon is also sutured to the serosa of the Roux limb.

Step 13. Closed-suction drains are placed both above and below the anastomosis and brought out through the abdominal wall.

Step 14. The abdominal wall is closed in layers with suture material of choice.

Postop

The NG tube can be removed on postoperative day 1. Diet is begun as tolerated. The drains are removed after the patient takes a diet if no amylase-rich drainage is present.

Complications

The most important procedure-specific complication is leakage from the pancreaticojejunostomy, which occurs in fewer than 5% of patients. If a leak occurs, the patient should be converted to total parenteral nutrition, which may be continued at home if the leak is controlled. If the leak is associated with sepsis or bleeding from drains, reexploration may be required.

Follow-Up

After recovery, patients should be followed indefinitely at regular intervals (6-12 months) to determine if pain is relieved and to determine if additional complications of chronic pancreatitis have developed, such as diabetes, exocrine insufficiency, or left-sided portal hypertension.

Acknowledgment

The editors and author wish to acknowledge Richard H. Bell, Jr. for contributing to the previous version of this chapter.

15

Chapter 16

Duodenum-Preserving Subtotal Pancreatic Head Resection: Frey Procedure

David Bentrem

Indications

Duodenum-preserving subtotal resection of the pancreatic head is indicated in patients with chronic pancreatitis with refractory pain not responsive to nonsurgical therapy, whose pancreatic head is enlarged (greater than 7 cm in diameter), and whose distal pancreatic duct is dilated. The procedure is also applicable if patients have associated compression of the common bile duct or duodenum.

Preop

Intraoperative ultrasound should be requested. Bowel prep is not necessary. An intravenous antibiotic for prophylaxis of surgical site infection (cefazolin, 1 g IV) is given 30 minutes before incision. A nasogastric (NG) tube and urinary catheter are placed after general anesthesia is induced. Pneumatic calf compression boots and/or low-dose subcutaneous heparin are used for deep vein thrombosis (DVT) prophylaxis according to patient risk profile.

Procedure

Step 1. A bilateral subcostal incision is made extending through the rectus abdominis muscle on each side. The abdomen is explored. NG tube position is confirmed.

Step 2. An upper abdominal retractor system is placed.

Step 3. The hepatic flexure of the colon is moblized and the base of the transverse mesocolon swept inferiorly to expose the duodenum and head of the pancreas.

Step 4. By incising the peritoneum lateral to the duodenum, a Kocher maneuver is performed, mobilizing the duodenum from the lateral edge of the hepatoduodenal ligament to the superior mesenteric vessels.

Step 5. If possible, the anterior surface of the superior mesenteric vein (SMV) is identified at this point by incising the peritoneum and fatty tissue overlying the vein. The vein is followed to the point where it passes behind the neck of the pancreas. It is not necessary to dissect behind the neck of the pancreas. When there is significant peripancreatic inflammation present, dissection of the vein may not be possible at this time and the vein can be identified later.

Step 6. The greater omentum is dissected upward and away from the transverse colon and the lesser sac entered. The stomach is mobilized in a cephalad direction by blunt and sharp dissection and the stomach and omentum retracted upward. The body and tail of the pancreas are identified and the entire anterior surface exposed.

Northwestern Handbook of Surgical Procedures, 2nd Edition, edited by Nathaniel J. Soper and Dixon B. Kaufman. ©2011 Landes Bioscience.

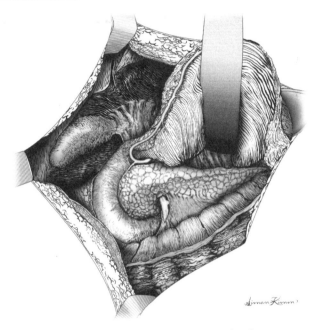

Figure 16.1. Duodenum-preserving subtotal pancreatic head resection. Pancreas exposure.

16

Step 7. On the upper surface of the transverse mesocolon, the middle colic vein is identified and followed down to its entrance into the SMV. If not done previously, the anterior surface of the SMV is cleaned by incising the peritoneum and fatty tissue overlying the vein. The vein is followed to the point where it passes behind the neck of the pancreas. The right gastroepiploic vein is identified, which usually joins the middle colic vein to form the gastrocolic trunk before the two veins enter the SMV. The right gastroepiploic vein is ligated as close as possible to its entrance into the gastrocolic trunk or SMV, taking care not to injure the middle colic vein. Because the right gastroepiploic vein crosses the neck of the pancreas from top to bottom to enter the SMV, it must be divided to provide adequate exposure of the junction between the head and body of the pancreas.

Step 8. After assuring that the entire anterior surface of the pancreas is well-exposed from the head to the tail, the ultrasound probe is applied to an area of the gland that is easily accessible and the pancreatic duct identified. A 20-gauge needle is passed into the duct, aspirating into a syringe to confirm the return of pancreatic juice. The needle is left in place once juice is returned and the electrocautery used to cut down on the duct and enter it.

Step 9. The pancreatic duct is opened as far as possible in both directions by passing a clamp into the duct and progressively dividing the overlying parenchyma with electrocautery. To the patient's right, the duct should be opened to within 1-2 cm of the duodenum. To the left, the duct should be opened along the body and tail of the gland to within 3-4 cm of the tail of the gland. The duct should not be opened so far to the left that subsequent anastomosis will be overly difficult due to poor visibility.

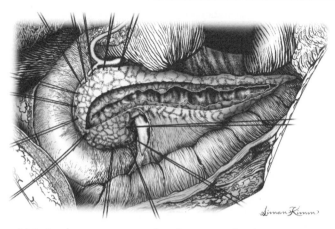

Figure 16.2. Duodenum-preserving subtotal pancreatic head resection. Pancreatic duct is open.

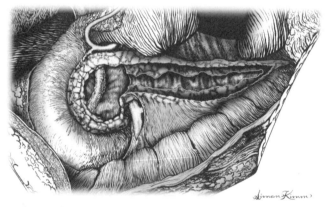

16

Figure 16.3. Duodenum-preserving subtotal pancreatic head resection. Partial resection of head of the pancreas.

Step 10. At this point, small portions of the head of the gland are resected by working from the opened duct outward. Pieces of the head are removed with a scalpel, obtaining hemostasis as needed with cautery or ligatures. The dissection can be quite bloody. The left hand is placed behind the head of the gland to guide the dissection. The objective is to leave a rim of the pancreatic head approximately 1 cm in thickness in all directions. If the bile duct is compressed, it is necessary to identify the duct within the head of the gland and decompress it from the point where it enters the gland to the point where it enters the ampulla. This dissection is easier if a metal probe (a Bakes dilator works well) is placed into the common duct through the cystic duct (necessitating cholecystectomy in patients who have not already had their gallbladder removed).

Step 11. Once the pancreatic dissection is complete, an appropriate point in the proximal jejunum is identified to create a Roux-en-Y limb and the jejunum divided with a linear stapler. The mesentery of the jejunum is divided perpendicular to the axis of the bowel to allow mobilization of the limb. A small defect is created in the mesentery of the transverse colon and the distal end of the divided jejunum passed through the defect, bringing the end of the jejunum to the patient's left. The Roux limb should lie easily over the opened pancreas without tension.

Step 12. The Roux limb is attached to the inferior border of the pancreas with interrupted sutures of 3-0 silk which are placed into the seromuscular layers of the bowel and into the pancreatic parenchyma. The small bowel is then opened, making the jejunotomy slightly smaller than the pancreatic defect.

Step 13. Starting in the middle of the back row of the anastomosis, a double-armed suture of continuous 3-0 nonabsorbable suture is placed, and then a continuous circumferential anastomosis of all bowel layers to the cut edge of the pancreatic defect is performed. The anastomosis is completed with a superior row of interrupted 3-0 silk sutures.

Step 14. A jejunojejunostomy is performed 40 cm below the pancreatic anastomosis to complete the Roux limb. The leaves of the mesentery are approximated to prevent internal hernia. The edge of the defect in the transverse mesocolon is likewise sewn to the serosa of the Roux limb.

Step 15. Closed-suction drains are placed both above and below the anastomosis.

Step 16. The abdominal wall is closed in layers with suture material of choice.

Postop

The NG tube can be removed on postoperative day 1 and diet instituted as tolerated. The drains are removed after the patient takes a diet if there is no amylase-rich drainage.

16

Complications

The most significant procedure-specific complication is leakage from the pancreaticojejunostomy, which occurs in under 5% of patients. If a leak occurs, the patient should be converted to total parenteral nutrition, which may be continued at home if the leak is controlled. If the leak is associated with sepsis or bleeding from drains, reexploration may be required.

Follow-Up

After recovery, patients should be followed indefinitely at regular intervals (6-12 months) to determine if pain is relieved and to determine if additional complications of chronic pancreatitis have developed, such as diabetes, exocrine insufficiency, or left-sided portal hypertension.

Acknowledgment

The editors and author wish to acknowledge Richard H. Bell, Jr. for contributing to the previous version of this chapter.

Chapter 17

Distal Pancreatectomy and Splenectomy

David M. Mahvi

Indications

Distal pancreatectomy is indicated for neoplasms of the body or tail of the pancreas, for the treatment of symptomatic chronic pancreatitis limited to the pancreatic body/tail, and for the treatment of chronic pancreatic fistula/pseudocyst arising from the distal pancreas. The spleen is generally removed en bloc with the distal pancreas, though recently as laparoscopic pancreatectomy has become more common several reports of splenic preservation have been presented. The goal of the procedure however must be the complete extirpation of the tumor which usually involves splenectomy.

Preop

Preoperative vaccination for encapsulated organisms should be performed preferably several weeks prior to the procedure. An epidural catheter for postoperative pain control should be placed preoperatively unless contraindicated. Deep venous thrombosis prophylaxis with sequential compression devices and/or subcutaneous heparin (depending on patient risk factors) should be instituted prior to the initiation of general anesthesia. A prophylactic perioperative antibiotic is administered intravenously 30 minutes prior to incision.

Procedure

Step 1. A bilateral subcostal incision is preferred. A midline incision is also acceptable.

Step 2. The lesser sac is entered inferior to the stomach with careful preservation of the gastroepiploic artery. The body and tail of the pancreas is exposed.

Step 3. The peritoneum along the inferior border of the pancreas is incised. The body and tail of the pancreas can be elevated by gentle dissection behind the gland. If the tumor extends posterior to the pancreas, a deeper plane on the kidney capsule may be necessary.

Step 4. The short gastric vessels between the spleen and stomach are ligated and divided starting at the midportion of the greater curve of the stomach and moving upward. If exposure is difficult the highest short gastric vessels are more easily ligated after the spleen has been mobilized (Step 5).

Step 5. The lateral peritoneal attachments of the spleen are divided and the spleen mobilized anteriorly. Splenic attachments to the colon and diaphragm must be divided. The attachments to the colon may require ligation and division due to their vascular nature. Any residual short gastric vessels spanning from the spleen to the stomach are divided so that the stomach can be completely retracted to the patient's right.

Northwestern Handbook of Surgical Procedures, 2nd Edition, edited by Nathaniel J. Soper and Dixon B. Kaufman. ©2011 Landes Bioscience.

Step 6. The pancreas and spleen are mobilized upwards and towards the patient's right by blunt dissection between the kidney and the tail of the pancreas. After some mobilization, the hand meets the retropancreatic space already created in Step 3.

Step 7. The inferior mesenteric vein, if encountered, is ligated and divided as it joins the splenic vein. Dissection of the peripancreatic soft tissues continues until the confluence of the splenic vein and superior mesenteric vein is visualized.

Step 8. Along the upper border of the pancreas, the splenic artery is identified near its origin from the celiac axis before it enters the pancreatic substance. It is doubly ligated proximally with 2-0 suture, ligated once distally, and divided.

Step 9. The splenic vein is divided either with a vascular stapling device or by clamping the vessel with vascular clamps and dividing between the clamps. The portal end of the vein is oversewn with continuous 5-0 polypropylene suture. The splenic end can be ligated with a 2-0 suture ligature if a stapling device was not utilized.

Step 10. The pancreatic parenchyma is divided. In patients with a neoplasm of the pancreas, a margin of at least 1 cm must be present.

Step 11. The treatment of the cut edge of the pancreas is controversial. The gland can be stapled and the pancreatic duct oversewn. The pancreas can be transected with a bipolar heating device. The pancreas can also be divided with cautery with subsequent ligation of the main pancreatic duct. Interrupted horizontal mattress sutures are then placed through the edge of the pancreatic parenchyma to close the cut edge.

Step 12. A soft, closed-suction drain is placed adjacent the cut edge of the pancreas and brought out through the left lateral abdominal wall.

Postop

A nasogastric tube is usually left for 24 hours, and then diet is begun gradually. The closed-suction drain is removed after oral intake is resumed if there is low output and the amylase level in the drain fluid is not elevated.

Complications

The most significant complications of distal pancreatectomy and splenectomy are pancreatic fistulae from the cut edge of the gland and/or left subphrenic abscess. Fistula is usually self-limiting, but may require prolonged drainage.

Follow-Up

In the early postoperative weeks, the platelet count should be monitored. If it rises above 1,000,000/mm^3, aspirin therapy should be initiated. Patients should also be followed long-term for the development of diabetes or steatorrhea. This is particularly true for patients who undergo distal pancreatectomy for chronic pancreatitis. In this group, up to 33% develop postoperative diabetes and 20% may develop steatorrhea.

Acknowledgment

The editors and author wish to acknowledge Woody Denham for contributing to the previous version of this chapter.

17

Chapter 18

Pancreaticoduodenectomy: Whipple Procedure

David Bentrem and Seth Krantz

Indications

Patients undergo pancreaticoduodenectomy for solid and cystic lesions of the pancreatic head and distal common bile duct. Ductal adenocarcinoma is the most common tumor type. Other primary tumor types for which pancreaticoduodenectomy is indicated include carcinoma of the ampulla, duodenal carcinoma, carcinoma of the distal common bile duct, islet cell carcinoma, and mucinous cystic neoplasms. Absolute contraindications include distant metastases to the liver, peritoneum, omentum, or lungs. Lymph node metastases outside the region of the head of the pancreas are usually considered a contraindication to resection. Other contraindications include portal/superior mesenteric vein thrombosis and encasement of the superior mesenteric or hepatic artery.

Preop

Preoperative radiologic staging is done by triphasic helical CT scan of the pancreas with 1-2 mm sections through the head of the gland. In patients with profound jaundice (bilirubin greater than 20 mg/dl) and malnutrition, consideration should be given to placement of an endoscopic biliary stent with a period of nutritional supplementation prior to the procedure. On the morning of surgery, patients receive a preoperative prophylactic antibiotic. Patients are offered an epidural catheter for postoperative analgesia. Deep vein thrombosis prophylaxis with subcutaneous heparin and sequential compression boots should be considered.

Procedure

Step 1. The patient is placed in the supine position, general anesthesia is induced, and the abdomen is prepared for exploration. In the setting of known pancreatic or biliary malignancy, staging laparoscopy is considered. A relatively thin patient may be explored through a midline incision. For patients with previous mid or lower abdominal surgery or with a wide costal margin, a bilateral subcostal incision is employed.

Step 2. Exploration of the entire abdomen is done to rule out hepatic, peritoneal, or omental metastases. In the absence of disseminated metastatic disease, exposure of the pancreas is begun.

Step 3. The colon and transverse mesocolon is retracted inferiorly and the omentum and the stomach are retracted superiorly and anteriorly. The lesser sac is entered through the greater omentum. The middle colic vein is identified in the transverse mesocolon and dissected down to its junction with the superior mesenteric vein.

Step 4. The inferior neck of the pancreas is inspected for evidence of tumor extension from the inferior edge of the pancreas on to the superior and anterior surfaces of the superior mesenteric vein and into the root of the small bowel mesentery. In the absence of these findings, dissection is continued.

Step 5. The third portion of the duodenum and the uncinate process are retracted superiorly and the hepatic flexure and transverse mesocolon are retracted inferiorly. Attachments between these organs are divided with a combination of sharp dissection and electrocautery. Inspection is carried out for evidence of locally advanced extension into the root of the small bowel or into the retroperitoneum.

Step 6. An extended Kocher maneuver is then performed, mobilizing the duodenum and the head of the pancreas off of the retroperitoneal structures. The inferior vena cava is identified, and the posterior pancreatic soft tissues including the posterior pancreatic lymph nodes are completely dissected off of the inferior vena cava and the aorta. Extension of the Kocher maneuver up to the porta hepatis is then performed. En bloc dissection of the retroperitoneal soft tissues is accomplished up to the level of the left renal vein.

Step 7. The gallbladder is retracted superiorly and anteriorly. The cystic duct is identified and dissected down to its junction with the common hepatic duct. The anterior peritoneum of the porta hepatis is completely divided across the bile duct and the hepatic artery. Circumferential dissection is then accomplished at the level of the cystic duct and common hepatic duct confluence. A vessel loop is placed around the common bile duct, and the bile duct is retracted laterally.

Step 8. The gallbladder is dissected off of the inferior surface of the liver . The cystic artery originating from the right hepatic artery is identified, ligated, and divided. The common hepatic artery is identified and the take-off of the gastroduodenal artery is exposed. The right gastric artery is identified, ligated, and divided if there is no plan to perform a pylorus-preserving procedure. The right gastric artery should be preserved if a pylorus-preserving pancreaticoduodenectomy is planned. Dissection is then carried proximally on the common hepatic artery back toward the level of the celiac trunk. The common hepatic artery lymph nodes are dissected off of the hepatic artery and sent for frozen section. If biopsies at the level of the hepatic artery and celiac trunk fail to reveal evidence of metastatic cancer, the procedure is continued.

Step 9. The gastroduodenal artery is identified and ligated just distal to its origin from the common hepatic artery with care taken to avoid narrowing of the common hepatic artery. The gastroduodenal artery is oversewn with a 4-0 polypropylene suture ligature.

Step 10. The vessel loop around the common bile duct is retracted laterally, and the soft tissues between the common hepatic artery and the common bile duct are then dissected toward the main specimen. The anterior surface of the portal vein is identified and dissected down to the superior neck of the pancreas. This area is inspected to determine if there is tumor extension onto the portal vein. In the absence of any locally advanced extension into the common hepatic artery, or circumferentially around the portal vein, the dissection is continued.

Step 11. The posterior lymph nodes in the porta hepatis are then dissected completely off of the portal vein. Care is taken to identify any aberrant right hepatic arteries, which generally lie posterolateral to the vein.

Step 12. The common hepatic duct is divided generally proximal to the cystic duct junction.

18

Step 13. If a pylorus-preserving Whipple procedure is being done, the subpyloric lymph nodes are dissected off of the inferior portion of the duodenum and the duodenum is divided approximately 2-3 cm distal to the pyloric valve with a linear stapling device. In the standard Whipple procedure, which includes resection of the distal stomach, the greater omentum is dissected off of the transverse mesocolon up to the level of the confluence of the left gastroepiploic and right gastroepiploic arteries on the greater curvature of the stomach. The bare area between this confluence is chosen as the line of division on the greater curvature. On the lesser curvature, the branches of the left gastric artery are identified and at approximately the level of the second or third branch, the lesser omentum is dissected off the lesser curvature of the stomach, and then the stomach is divided generally using a linear stapling device. The proximal portion of the staple line is oversewn with a running 3-0 polypropylene Lembert suture. The distal stomach is retracted to the right of the porta hepatis.

Step 14. The proximal jejunum is then divided approximately 10-15 cm distal to the ligament of Treitz with a linear stapling device. Mesenteric branches to the proximal jejunum and distal duodenum are divided between clamps and oversewn or divided with a tissue sealing device in close proximity to the distal duodenum. Dissection is carried back to the level of the superior mesenteric artery and uncinate process.

Step 15. The fourth portion of the duodenum and the first portion of the jejunum are delivered through the retroperitoneum from left to right beneath the mesenteric vessels.

Step 16. The neck of the pancreas is carefully dissected off the anterior surface of the superior mesenteric vein and the anterior surface of the portal vein. In preparation for the division of the neck of the pancreas, four separate 3-0 polypropylene figure-of-eight stitches are placed on either side of the inferior and superior neck of the pancreas to control the inferior and superior pancreatic arterial arcade.

Step 17. The pancreas is divided with electrocautery. Care is taken to identify the pancreatic duct as it is divided.

Step 18. The pancreatic neck is dissected off the right lateral surface of the portal vein. The superior uncinate vein and the inferior uncinate vein are identified and controlled with small vascular clamps and divided. The veins are ligated on the mesenteric vein side and the portal vein side with 5-0 polypropylene sutures with care taken to avoid narrowing of the superior mesenteric vein or the portal vein.

Step 19. The uncinate process is retracted to the right, the portal vein retracted to the left, and the superior mesenteric artery identified by visualization and palpation. Small branches of the superior mesenteric artery into the uncinate process are identified and sequentially divided between fine clamps. The uncinate process is completely dissected off the lateral surface of the superior mesenteric artery, and the branches of the superior mesenteric artery are controlled with 4-0 and 3-0 polypropylene suture ligatures. Alternatively, a linear stapling device with a vascular staple load may be used to divide the uncinate process attachments.

Step 20. The specimen is removed and sent to pathology for frozen sections on the bile duct margin, the uncinate process margin, and the pancreatic neck margin. Titanium clips are placed along the retroperitoneal margin on the lateral edge of the superior mesenteric artery for postoperative localization by radiation oncology. The cut end of the left pancreas is then retracted superiorly and the posterior branches from the superior mesenteric vein and the splenic vein are divided between clamps and oversewn with 5-0 polypropylene suture ligatures. Approximately 1 inch of the proximal body of the pancreas is mobilized for preparation of the anastomosis.

18

Step 21. The proximal end of the jejunum is then brought through an opening in the transverse mesocolon to the right of the middle colic vessels. In this retrocolic position, an end-to-side pancreatic jejunal anastomosis is done. The end of the pancreas is sewn to the antimesenteric border of the jejunum in two layers. A posterior layer of interrupted 4-0 polypropylene sutures are placed . When possible, separate sutures are placed in the corners of the pancreatic duct and then stitched to the opening in the small bowel in a full-thickness layer. This helps maintain patency of the pancreatic duct and minimizes anastomotic leaks. A pancreatic stent is considered to assist with suture placement. An anterior row of interrupted sutures is then placed.

Step 22. The next anastomosis performed is the hepaticojejunostomy. This is also performed in an end-to-side fashion between the end of the common hepatic duct and the antimesenteric border of the jejunum. This anastomosis is placed approximately 10-15 cm distal to the pancreatic anastomosis and is performed in a single layer with interrupted 4-0 absorbable monofilament sutures. The opening in the transverse mesocolon is reapproximated with interrupted 3-0 absorbable sutures.

Step 23. An end-to-side gastrojejunal anastomosis is then placed approximately 30-35 cm distal to the biliary anastomosis. This may be placed in a retrocolic or antecolic position, but the antecolic position is preferred as it may minimize postoperative radiation injury to the anastomosis or subsequent blockage from local recurrent cancer.

Step 24. Two 10 mm flat closed suction drains are placed in the right upper quadrant. An inferior drain is placed posterior to the biliary anastomosis, and a superior drain is placed between the pancreatic anastomosis and the incision.

Step 25. A feeding jejunostomy may be placed approximately 30 cm distal to the gastrojejunal anastomosis. The abdomen is irrigated, hemostasis is assured, the abdominal wall is closed in layers.

Postop

Nasogastric decompression is maintained for 24 hours. Drain outputs are monitored for volume and consistency. Glucose levels are monitored for postoperative insulin requirements. Erythromycin 500 mg intravenously every 8 hours may be started on postoperative day 3 as a motility agonist. Deep venous thrombosis prophylaxis with subcutaneous heparin and sequential compression devices is essential. Antibiotics are discontinued after the 24-hour perioperative period.

Complications

Early complications include postoperative hemorrhage, which may be from the gastric anastomosis, the cut edge of the pancreas, or from branches of the superior mesenteric artery or superior mesenteric vein. Subsequent complications include leakage from the pancreaticojejunal anastomosis or the hepaticojejunal anastomosis, delayed gastric emptying, and postoperative marginal ulcer formation.

Follow-Up

Patients should be seen frequently in the weeks following surgery to assess healing and return of functional status. Insulin and/or pancreatic enzyme replacement may be required. Adjuvant chemoradiation therapy is ordinarily given unless done preoperatively. Given the high likelihood of recurrence, patients should be seen regularly and repeat imaging performed if dictated by a deterioration in clinical status.

Acknowledgment

The editors and authors wish to acknowledge Mark S. Talamonti for contributing to the previous version of this chapter.

18

Chapter 19

Splenectomy: Open

Mark Toyama

Indications

Indications for splenectomy are similar for both laparoscopic and open approaches and include blood dyscrasias refractory to medical management (i.e., autoimmune anemias, hereditary spherocytosis, immune thrombocytopenic purpura, and thrombotic thrombocytopenic purpura), and primary splenic diseases such as cysts or abscess. Traumatic splenic injuries that need surgery are usually done in an open manner.

Preop

In patients undergoing elective splenectomy, immunizations should be given 1-2 weeks preoperatively against pneumococcus, *H. influenzae*, and *N. meningitides*, all of which can cause the rare, but potentially significant complication of overwhelming postsplenectomy sepsis. On the day of surgery, a preoperative prophylactic antibiotic is given. Postoperative antibiotics are usually not indicated. A nasogastric tube is placed. Deep vein thrombosis (DVT) prophylaxis should be provided by the use of sequential compression devices and/or subcutaneous heparin as appropriate for the patient's risk factors.

Procedure

Step 1. Either a midline or a left subcostal incision can be used.

Step 2. The lesser sac is entered through the left part of the gastrocolic omentum.

Step 3. Moving upward along the greater curvature of the stomach, the short gastric veins are divided with clamps and ties or divided with the aid of the harmonic scalpel. If the highest short gastric vessels are difficult to visualize, they can be ligated and divided after the spleen is mobilized.

Step 4. The splenocolic ligament between the spleen and the splenic flexure of the colon is divided, reflecting the colon downward.

Step 5. Manually retracting the spleen towards the patient's midline, the peritoneum lateral to the spleen is opened and the splenorenal and splenophrenic ligaments are divided. The surgeon's hand is passed behind the spleen and tail of the pancreas, mobilizing the spleen anteriorly and medially by blunt dissection until the spleen is brought up anteriorly into the operative field.

Step 6. Any remaining short gastric vessels between the spleen and stomach are divided, retracting the stomach to the patient's right.

Step 7. The splenic artery is isolated just distal to the tail of the pancreas and ligated. Next the splenic vein is separately ligated and the spleen is removed. Vessels can also be divided with a vascular staple load.

Step 8. The retroperitoneum up into the left subphrenic space is thoroughly inspected for hemostasis.

Northwestern Handbook of Surgical Procedures, 2nd Edition, edited by Nathaniel J. Soper and Dixon B. Kaufman. ©2011 Landes Bioscience.

19

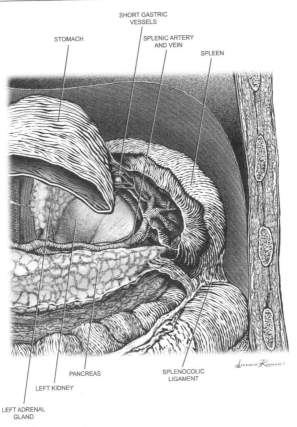

Figure 19.1. Splenectomy. Open.

Postop

The NG tube can be left in place overnight and is usually removed the first postoperative day. Diet is advanced as tolerated. If the procedure was done for thrombocytopenia, coagulation status and platelet counts should be rechecked postoperatively. Discharge is possible when the patient has tolerated a diet without problems. If the procedure was done emergently, immunizations are often given before the patient leaves the hospital.

Complications

Immediate complications include postoperative hemorrhage, a pancreatic leak from damage to the tail of the pancreas, or necrosis of a portion of the greater curvature of the stomach. Postoperative subphrenic abscesses or fluid collections may occur. An arteriovenous fistula may develop between the splenic artery and vein stumps. Postsplenectomy sepsis occurs in approximately 1% of patients. Arterial thrombosis and stroke have been reported secondary to the postoperative thrombocytosis.

Follow-Up

If the patient was operated for a hematologic indication, the response to therapy must be followed. In immune thrombocytopenic purpura, for example, there is a relapse rate of 15-20%. Postsplenectomy patients should generally be treated with prophylactic oral antibiotics when undergoing dental work or similar procedures that entail a risk of bacteremia.

Acknowledgment

The editors and author wish to acknowledge Malcolm M. Bilimoria for contributing to the previous version of this chapter.

19

Chapter 20

Splenectomy: Laparoscopic

Mark Toyama

Indications

Indications for splenectomy include blood dyscrasias refractory to medical management (i.e., autoimmune anemias, hereditary spherocytosis, immune thrombocytopenic purpura, and thrombotic thrombocytopenic purpura) and primary splenic diseases such as cysts or abscess.

Preop

The patient should receive immunizations against pneumococcus, meningococcus, and *H. influenzae* 1-2 weeks preoperatively. Routine preoperative antibiotics are administered 30 minutes prior to incision. A nasogastric (NG) tube is placed to decompress the stomach during the procedure. Sequential compression stockings or subcutaneous heparin should be administered for patients at risk for deep vein thrombosis/pulmonary embolism.

Procedure

Step 1. The patient is placed in a 45° right lateral decubitus position.

Step 2. Obtain peritoneal access at or above the umbilicus, depending on the patient's body habitus and insufflate in the standard fashion to a pressure of 15 mm Hg. Inspect the abdomen with a 30° or 45° laparoscope.

Step 3. Three or four additional ports are placed along the left costal margin. A combination of 5, 10 and 12 mm ports may be used depending on instrument preference. One of the lateral ports is usually a 12 mm port to accommodate the laparoscopic stapling device. An additional port is sometimes needed to retract the left lobe of the liver. A hand assist port can also be utilized in the epigastric position, particularly with large spleens.

Step 4. Mobilize the splenic flexure of the colon with the harmonic shears by carefully retracting it down and to the right and dividing the tissues between the colon and spleen.

Step 5. Enter the lesser sac by dividing the greater omentum between the stomach and spleen.

Step 6. Continue with the posterior dissection of the spleen by dividing the splenophrenic and splenorenal ligaments. Some of the posterior attachments can be divided early or left until after vascular division. This allows the decubitus position to suspend the spleen in the field, often allowing better visualization of the hilar vessels.

Step 7. Moving towards the top of the spleen, sequentially divide the short gastric vessels with the harmonic shears. Careful retraction of the gastric fundus and upper pole of the spleen will assist in dividing the highest short gastric vessels.

20

Northwestern Handbook of Surgical Procedures, 2nd Edition, edited by Nathaniel J. Soper and Dixon B. Kaufman. ©2011 Landes Bioscience.

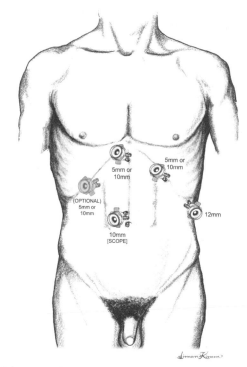

Figure 26.1. Splenectomy. Laparoscopic.

Step 8. Identify the hilum of the spleen and carefully dissect free the splenic artery and vein.

Step 9. If possible, divide the splenic artery first by passing the laparoscopic stapling device with a vascular staple load through a lateral port and positioning it perpendicular to the vessel.

Step 10. Divide the splenic vein with the stapling device in a similar fashion.

Step 11. Multiple firings of the stapling device may be necessary to divide all of the vascular attachments to the spleen.

Step 12. Divide any remaining posterior attachments of the spleen with the harmonic shears.

Step 13. Place the spleen into a large specimen bag and remove it by either morcellization through a large port site or by extending one port site incision and removing the spleen intact. The spleen can also be removed through the hand port site, if one has been used.

Step 14. Carefully inspect the splenic hilar area and omentum for any evidence of an accessory spleen(s).

Step 15. Irrigate the operative field and suction. Drain placement is not required unless there is concern for a pancreatic injury.

Step 16. Close the fascia at all port sites 10 mm in size or larger. Close the skin at all port sites.

Postop

Depending on surgeon preference, the NG tube can be removed in the operating room or left in place overnight. If the procedure was done for thrombocytopenia, coagulation status and platelet counts should be rechecked postoperatively. Discharge is usually possible within 24-48 hours.

Complications

Immediate complications include postoperative hemorrhage, a pancreatic leak from damage to the tail of the pancreas, or necrosis of a portion of the greater curvature of the stomach. Postoperative subphrenic abscess may occur. An arteriovenous fistula may develop between the splenic artery and vein stumps. Splenic vein thrombosis has also been observed. Postsplenectomy sepsis occurs in approximately 1% of patients.

Follow-Up

Patients are usually seen 2-3 weeks after their procedure in routine postoperative follow-up. If the patient was operated for a hematologic condition, long-term follow-up with their hematologist may be necessary for monitoring and any further medical therapy.

Acknowledgment

The editors and author wish to acknowledge Malcolm M. Bilimoria for contributing to the previous version of this chapter.

20

Chapter 21

Splenorrhaphy: Open

Michael B. Shapiro

Indications

Splenorrhaphy is indicated when the grade of splenic injury requires repair, but does not require splenectomy. Minor injuries (e.g., Grade I capsular tear) should be treated with local measures or topical hemostasis and do not require splenorrhaphy. In general, Grade II and III spleen injuries would have the greatest likelihood of benefiting from splenorrhaphy. If possible, it is always desirable to preserve spleen mass and function to minimize the risk of late infection or immune defects.

Preop

Prior to performing exploratory laparotomy the patient should have appropriate venous access and should (if possible) be well-resuscitated. Blood for type and cross-match should be sent. It is advantageous to place a Foley catheter prior to abdominal exploration. A nasogastric (NG) suction catheter should be placed preoperatively. When performing exploratory laparotomy for trauma, the surgeon should be sure that there are two suction devices and carefully position, prep, and drape the patient such that the chest and/or mediastinum can be accessed intraoperatively. Antibiotic prophylaxis should be instituted prior to making the incision. Choice of agent should be based on the pathogens likely to be encountered. A second-generation cephalosporin or other agents that cover aerobic and anaerobic enteric pathogens are frequently used.

Procedure

Step 1. Exploratory laparotomy is performed through a midline incision.

Step 2. A large body wall retractor is placed in the left upper quadrant to permit examination of the spleen. Initially examine the spleen in situ, looking for evidence of deep lacerations, active bleeding, or injury to hilar structures. It is not necessary to mobilize the spleen in the absence of splenic injury. Mobilization of the spleen may exacerbate bleeding from minor injuries. If there is evidence of significant splenic injury and ongoing hemorrhage, the spleen should be completely mobilized.

Step 3. Splenic mobilization is performed by retracting the spleen anteriorly and medially using the surgeon's nondominant hand. This maneuver places the spleno-renal ligament "on stretch." This ligament can then be incised using a scissors or cautery, beginning at the inferior aspect of the spleen and continuing in a cephalad direction. In some instances this maneuver must be performed blindly, by "feel." The spleen is mobilized medially using predominantly blunt dissection in the plane posterior to the pancreas.

Northwestern Handbook of Surgical Procedures, 2nd Edition, edited by Nathaniel J. Soper and Dixon B. Kaufman. ©2011 Landes Bioscience.

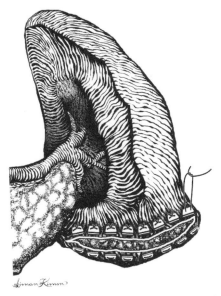

Figure 21.1. Splenorrhaphy. Horizontal mattress sutures with pledgets to control hemorrhage.

Step 4. The short gastric vessels in the splenogastric ligament should be divided. In many instances, the side of the vessel on the greater curve of the stomach should be suture-ligated to avoid dislodgment from postoperative gastric distension. The spleno-colic ligament is likewise divided and the splenic flexure of the colon swept inferiorly. Occasionally this structure will contain significant vessels that require clamping. The mobilized spleen is rotated upward into the midline abdominal incision. This may be facilitated with placement of several laparotomy pads in the splenic fossa.

Step 5. Hemorrhage can be temporarily controlled (if needed) by application of digital pressure to the hilum of the mobilized spleen. Debridement should be performed very sparingly, if at all.

Step 6. Horizontal mattress sutures are employed to control hemorrhage and reapproximate isolated lacerations of the splenic parenchyma. Some type of pledget or "bolster" material is usually required to prevent tearing the splenic parenchyma with exacerbation of bleeding. Teflon felt, polyglycolic acid mesh, and omentum have been employed for this purpose. Absorbable sutures (2-0) on a large taper or blunt needle are carefully placed in an interlocking horizontal mattress fashion. In most instances, the sutures are best placed so that the knots are on the diaphragmatic surface.

Step 7. In instances where multiple or deeper lacerations are found, the spleen can sometimes be salvaged by performing a wrap using polyglycolic acid mesh. The splenic wrap is performed using a large sheet of absorbable mesh material. A "keyhole" slit to accommodate the hilar vessels is cut on one side of the mesh. The mesh is passed behind the spleen and the mesh folded to envelop the organ. The free edges of the mesh are approximated with a running 2-0 absorbable suture. It is desirable to fashion the mesh and secure the closure to achieve a slight degree of compression. However, care must be taken to avoid vascular compromise or splenic infarction.

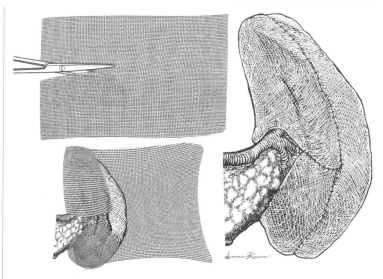

Figure 21.2. Splenorrhaphy. A splenic wrap with polyglycolic acid mesh.

Step 8. Anatomic resection may be employed if injury is localized to one pole of the spleen. The appropriate hilar vessel is ligated and horizontal mattress sutures are placed proximal to the line of vascular demarcation using the technique described in Step 6. After the sutures have been secured, the nonviable splenic parenchyma is resected distal to the suture line.

Step 9. The splenic bed is carefully inspected to ensure that hemostasis is complete prior to closure. Laparotomy sponges in the splenic fossa can be rolled anteriorly to assist the surgeon to visualize this area. Drains are not used.

Step 10. Abdominal fascial and skin closure are performed as described for exploratory laparotomy.

Additional Comments

Complete abdominal exploration should be performed prior to attempting splenorrhaphy. This is important to determine the presence of other injuries and to appropriately prioritize the operative treatment(s).

Postop

Careful postoperative hemodynamic management and fluid management are indicated following splenorrhaphy. Hypertension should be avoided. Routine postoperative CT scanning confers little benefit.

Complications

The most common complication of splenorrhaphy is rebleeding. Splenectomy should be performed if reoperation is required for ongoing bleeding. Infectious complications may occur in the left subphrenic area. Most fluid collections or abscesses can be drained percutaneously. Many surgeons feel that avoiding Teflon pledgets minimizes infectious complications after splenorrhaphy.

Follow-Up

No specific long-term follow-up is required. If splenic preservation was successful, there is no need to administer polyvalent vaccines against encapsulated bacteria.

Acknowledgment

The editors and author wish to acknowledge Michael A. West for contributing to the previous version of this chapter.

21

Chapter 22

Antireflux Procedure: Laparoscopic (Nissen)

Eric Hungness

Indications

Nissen fundoplication is indicated in operable patients with objectively documented gastroesophageal reflux disease (GERD) whose symptoms/signs have not been controlled by medical management with proton pump inhibitor (PPI) therapy and other measures.

Preop

Prior to surgery, patients should be fully evaluated by esophagogastroduodenoscopy (EGD) and distal esophageal biopsy, esophageal manometry, 24-hour pH study (if no objective evidence of reflux or if symptoms are atypical), and barium swallow if hiatus hernia is large (>3-4 cm). Immediately prior to surgery, patients should have appropriate deep vein thrombosis (DVT) prophylaxis based on risk factors.

Procedure

Step 1. The patient is positioned supine in modified lithotomy with the bed rotated 20° clockwise. The surgeon stands between the patient's legs and the assistant stands at the left side.

Step 2. After prepping and draping, CO_2 pneumoperitoneum is established and five 5-10 mm laparoscopic ports are placed in the upper abdomen: one in the supra-umbilical region to the left of midline and four in the subcostal region (one at each anterior axillary line and one at each lateral rectus).

Step 3. A 5-10 mm liver retractor is placed through the far right port in the anterior axillary line, elevating the left hepatic lobe, thereby exposing the proximal stomach and esophageal hiatus. The retractor is secured to a table-mounted retractor holder.

Step 4. An orogastric (OG) tube is positioned in the midstomach for decompression.

Step 5. The gastrocolic and gastrosplenic omenta are divided along the greater curvature from midstomach up to the angle of His with an ultrasonic scalpel, mobilizing the fundus off the left half of the diaphragmatic crus and opening the retroperitoneum behind and at the left side of the gastric cardia.

Step 6. The upper portion of the gastrohepatic omentum is divided and the phrenoesophageal membrane opened over the esophagogastric junction (EGJ), identifying the anterior vagus nerve.

Step 7. The retroperitoneum medial to the right half of the diaphragmatic crus is entered and the retroesophageal space behind the posterior vagus nerve opened, and this dissection connected to the left subphrenic space, going behind the esophagus.

Step 8. A half-inch Penrose drain is passed through the retroesophageal space to surround the EGJ. It is secured with a locking Allis laparoscopic grasping forceps, placing gentle downward traction on the EGJ.

Northwestern Handbook of Surgical Procedures, 2nd Edition, edited by Nathaniel J. Soper and Dixon B. Kaufman. ©2011 Landes Bioscience.

Step 9. The distal esophagus is mobilized out of the posterior mediastinum and hiatal canal and any hiatus hernia reduced. At this point, the anesthesiologist should remove the OG tube and gently pass a 50-52 F esophageal dilator down the esophagus through the EGJ into the stomach.

Step 10. An enlarged hiatal orifice behind the esophagus is eliminated by reapproximating the left and right halves of the crus with interrupted 0 or 2-0 braided polyester sutures over PTFE pledgets.

Step 11. The gastric fundus is passed from left to right behind the EGJ and the stomach circumferentially plicated around the EGJ with at least two anteriorly placed interrupted 2-0 braided polyester sutures, grasping the esophagus to the right of the anterior vagus nerve between bites of fundus. The length of the wrap should be approximately 2 cm.

Step 12. The esophageal dilator is removed. The OG tube is replaced in the mid body of stomach only if the stomach appears distended with air or fluid.

Step 13. The abdomen is irrigated with warm saline, inspecting the subphrenic space and spleen for bleeding, and then aspirating all saline irrigation.

Step 14. The liver retractor, the table-mounted retractor holder, and all laparoscopic ports are removed. The fascia at sites of 10 mm ports is closed. Skin incisions are closed with subcuticular sutures.

Additional Comments

If inadequate esophageal length is obtained after reducing the hiatus hernia in Step 9 and the EGJ cannot be returned to an intra-abdominal position, a Collis gastroplasty may be indicated.

Postop

A Foley catheter is necessary only in patients having a long procedure or in whom there is indication to monitor urine output closely. An NG or OG tube is usually not necessary. Oral intake of liquids and soft foods is begun as soon as the patient has no nausea, usually on the day of surgery. Red meat and dry bread are avoided for 2-4 weeks. Pain is controlled with injectable narcotics for 6-12 hours and then with oral analgesics.

Complications

Complications of Nissen fundoplication include vagal injury, esophageal perforation, gastric perforation, splenic laceration and hemorrhage, gastroparesis, gastric bloating, esophageal dysmotility (dysphagia), excess flatulence, and fundoplication dehiscence with recurrent GERD.

Long-Term Follow-Up

A barium swallow and/or esophagogastroduodenoscopy (EGD) should be performed in patients with persistent dysphagia after 4-8 weeks. A postsurgical EGD should be done in 3-6 months in patients operated for erosive esophagitis. In patients with Barrett's esophagus, a follow-up EGD should be done in 6-12 months and then every 6 months to 3 years depending on the presence or absence of dysphagia. Recurrent GERD symptoms require a complete evaluation including a barium swallow, esophageal manometry, EGD, and possibly a 24-hour pH study.

Acknowledgment

The editors and author wish to acknowledge Raymond J. Joehl for contributing to the previous version of this chapter.

Chapter 23

Repair of Paraesophageal Hernia: Open

Eric Hungness

Indications

Most elective paraesophageal hernia repairs are now performed laparoscopically. Open paraesophageal hernia repair is indicated when converting from a laparoscopic approach due to complication or technical difficulty or more usually in the setting of an emergent operation for acute incarceration. Symptoms usually manifest as postprandial chest pain or dysphagia.

Preop

Prior to elective surgery, patients should be fully evaluated by esophagogastroduodenoscopy (EGD), esophageal manometry, and barium swallow. Immediately prior to surgery, patients should have appropriate antibiotic and deep vein thrombosis (DVT) prophylaxis based on risk factors. Patients needing emergent operations usually have a CT scan as part of their workup confirming the diagnosis.

Procedure

Step 1. The patient is positioned supine.

Step 2. After prepping and draping, an upper midline incision is made from the xiphoid to the umbilicus. Extension laterally along the left side of the xiphoid is usually needed for adequate exposure.

Step 3. An Omni or Bookwalter retractor is used for costal retraction.

Step 4. If possible, an orogastric (OG) tube is positioned in the midstomach for decompression, although the hernia may make this difficult. EGD is sometimes required to adequately decompress the stomach and can decompress a acutely incarcerated stomach.

Step 5. The left lobe of the liver is mobilized medially by dissecting the triangular ligament for better hiatus exposure.

Step 6. The hernia contents are carefully manually reduced as much as possible. If gastric necrosis is suspected, or if the incarcerated stomach is not able to be reduced, the left crura and diaphragm may need to be incised.

Step 7. The upper portion of the gastrohepatic omentum is divided and the hernia sac is opened along the medial aspect of the right crus. The leading edge of hernia sac is retracted inferiorly and the entire hernia sac is carefully reduced with blunt and electrocautery dissection dividing the hernia sac anteriorly until the left crus.

Step 8. The esophageal hiatus is then meticulously dissected, identifying the posterior vagus nerve

Step 9. The distal esophagus is mobilized out of the posterior mediastinum and hiatal canal in an orad direction as far a possible.

Northwestern Handbook of Surgical Procedures, 2nd Edition, edited by Nathaniel J. Soper and Dixon B. Kaufman. ©2011 Landes Bioscience.

Step 10. At this point, any necrotic stomach is resected and reconstructed appropriately. Also, a decision needs to be made as to whether to proceed with fundoplication or G-tube/gastropexy alone. If gastropexy is chosen (usually for emergent cases), a gastrostomy tube is placed in the mid body to keep the stomach adequately reduced to prevent an acute recurrence. Additional 2-O tacking sutures from the stomach to the abdominal wall also placed.

Step 11. If proceeding with fundoplication, the gastrocolic and gastrosplenic ligaments are divided along the greater curvature from midstomach up to the angle of His, mobilizing the fundus off the left half of the diaphragmatic crus and opening the retroperitoneum behind and at the left side of the gastric cardia.

Step 12. The gastric fundus is passed from left to right behind the EGJ and used to retract the esophagus and expose the enlarged hiatal defect. A half-inch Penrose drain passed through the retroesophageal space may also be used for retraction.

Step 13. The crura are reapproximated with interrupted 0 or 2-0 braided polyester sutures. Large crural defect closures should be buttressed with a bioabsorbable mesh either carefully sutured or "glued" in place with fibrin sealant.

Step 14. At this point, the anesthesiologist should remove the OG tube and gently pass a 50 F followed by a 60 F esophageal dilator down the esophagus through the EGJ into the stomach.

Step 15. The gastric fundus is checked for rotational tension and torsion and the stomach circumferentially plicated around the EGJ with at least two anteriorly placed interrupted 2-0 braided polyester sutures, grasping the esophagus to the right of the anterior vagus nerve between bites of fundus. The length of the "short, floppy" fundoplication should be approximately 2 cm. Patients with severe esophageal dysmotility should have a partial fundoplication.

Step 16. The esophageal dilator is removed. The OG tube is replaced in the mid body of stomach only if the stomach appears distended with air or fluid.

Step 17. The abdomen is irrigated with warm saline, inspecting the subphrenic space and spleen for bleeding, and then aspirating all saline irrigation.

Step 16. The retractor is removed and the fascia is closed in a running fashion.

Additional Comments

If inadequate esophageal length is obtained after reducing the hiatus hernia in Step 9 and the EGJ cannot be returned to an intra-abdominal position, a Collis gastroplasty may be indicated.

23

Postop

A Foley catheter is necessary only in patients having a long procedure or in whom there is indication to monitor urine output closely. An NG or OG tube is usually not necessary. Oral intake of liquids and soft foods is begun as soon as the patient has no nausea, usually on the day of surgery. Red meat and dry bread are avoided for 4 weeks. Pain and nausea are controlled with injectable narcotics, ketorolac and ondansetron for 6-12 hours and then with oral analgesics.

Complications

Complications of open paraesophageal hernia repair include vagal injury, esophageal perforation, gastric perforation, splenic laceration and hemorrhage, gastroparesis, gastric bloating, esophageal dysmotility (dysphagia), excess flatulence, and fundoplication dehiscence with recurrent GERD.

Long-Term Follow-Up

A barium swallow and/or esophagogastroduodenoscopy (EGD) should be performed in patients with persistent dysphagia after 4-8 weeks. A follow-up barium swallow is suggested at 6 months and if symptoms recur. If a G-tube was placed for gastropexy, it can usually be removed 3-4 weeks postop.

23

Chapter 24

Repair of Paraesophageal Hernia: Laparoscopic

Eric Hungness

Indications

Laparoscopic paraesophageal hernia repair is indicated in operable patients with a symptomatic type 2, 3 or 4 paraesophageal hernia. Symptoms usually manifest as postprandial chest pain or dysphagia.

Preop

Prior to surgery, patients should be fully evaluated by esophagogastroduodenoscopy (EGD), esophageal manometry, and barium swallow. Immediately prior to surgery, patients should have appropriate deep vein thrombosis (DVT) prophylaxis based on risk factors.

Procedure

Step 1. The patient is positioned supine in modified lithotomy or split legs with the bed rotated 20° clockwise. The surgeon stands between the patient's legs and the assistant stands at the left side.

Step 2. After prepping and draping, CO_2 pneumoperitoneum is established with placement of a 10 mm port 12 cm below the xiphoid process just to the left of midline. A 30° angled laparoscope is inserted, and the patient is placed in steep heads-up position. The surgeon's 10 mm right hand port is placed 10 cm from the xiphoid 2 fingerbreadths below the left costal margin. A 5 mm liver retractor port is placed at least 15 cm from the xiphoid process 2 fingerbreadths below the right costal margin. The assistant's 5 mm port is then placed halfway between these two ports.

Step 3. A 5 mm "snake" liver retractor is placed through the right lateral port, elevating the left hepatic lobe, thereby exposing the proximal stomach and esophageal hiatus. The retractor is secured to a table-mounted retractor holder. The surgeon's 5 mm left hand port is then placed to the right of midline just below the inferior edge of the retracted left lobe of the liver. A Veress needle is used to "sound out" potential sites for this port.

Step 4. An orogastric (OG) tube is positioned in the midstomach for decompression although the hernia may make this difficult. EGD is sometimes required to adequately decompress the stomach.

Step 5. The hernia contents are carefully manually reduced as much as possible.

Step 6. The upper portion of the gastrohepatic omentum is divided and the hernia sac is opened along the medial aspect of the right crus. The leading edge of hernia sac is retracted inferiorly and the entire hernia sac is carefully reduced with blunt and ultrasonic shear dissection dividing the hernia sac anteriorly until the left crus.

Step 7. The esophageal hiatus is then meticulously dissected, identifying the posterior vagus nerve

24

Northwestern Handbook of Surgical Procedures, 2nd Edition, edited by Nathaniel J. Soper and Dixon B. Kaufman. ©2011 Landes Bioscience.

Step 8. The distal esophagus is mobilized out of the posterior mediastinum and hiatal canal in an orad direction as far a possible.

Step 9. The gastrocolic and gastrosplenic omenta are divided along the greater curvature from midstomach up to the angle of His with an ultrasonic scalpel, mobilizing the fundus off the left half of the diaphragmatic crus and opening the retroperitoneum behind and at the left side of the gastric cardia.

Step 10. The gastric fundus is passed from left to right behind the EGJ and used to retract the esophagus and expose the enlarged hiatal defect. A half-inch Penrose drain passed through the retroesophageal space may also be used for retraction.

Step 11. The crura are reapproximated with interrupted 0 or 2-0 braided polyester sutures. Large crural defect closures should be buttressed with a bioabsorbable mesh either carefully sutured or "glued" in place with fibrin sealant.

Step 12. At this point, the anesthesiologist should remove the OG tube and gently pass a 50 F followed by a 60 F esophageal dilator down the esophagus through the EGJ into the stomach under direct laparoscopic visualization to avoid inadvertent esophageal perforation.

Step 13. The gastric fundus is checked for rotational tension and torsion and the stomach circumferentially plicated around the EGJ with at least two anteriorly placed interrupted 2-0 braided polyester sutures, grasping the esophagus to the right of the anterior vagus nerve between bites of fundus. The length of the "short, floppy" fundoplication should be approximately 2 cm. Patients with severe esophageal dysmotility should have a partial fundoplication.

Step 14. The esophageal dilator is removed. The OG tube is replaced in the mid body of stomach only if the stomach appears distended with air or fluid.

Step 15. The abdomen is irrigated with warm saline, inspecting the subphrenic space and spleen for bleeding, and then aspirating all saline irrigation.

Step 16. The liver retractor, the table-mounted retractor holder, and all laparoscopic ports are removed. The fascia at sites of 10 mm ports is closed. Skin incisions are closed with subcuticular sutures.

Additional Comments

If inadequate esophageal length is obtained after reducing the hiatus hernia in Step 9 and the EGJ cannot be returned to an intra-abdominal position, a Collis gastroplasty may be indicated.

Postop

A Foley catheter is necessary only in patients having a long procedure or in whom there is indication to monitor urine output closely. An NG or OG tube is usually not necessary. Oral intake of liquids and soft foods is begun as soon as the patient has no nausea, usually on the day of surgery. Red meat and dry bread are avoided for 4 weeks. Pain and nausea are controlled with injectable narcotics, ketorolac and ondansetron for 6-12 hours and then with oral analgesics.

Complications

Complications of laparoscopic paraesophageal hernia repair include vagal injury, esophageal perforation, gastric perforation, splenic laceration and hemorrhage, gastroparesis, gastric bloating, esophageal dysmotility (dysphagia), excess flatulence, and fundoplication dehiscence with recurrent GERD.

Long-Term Follow-Up

A barium swallow and/or esophagogastroduodenoscopy (EGD) should be performed in patients with persistent dysphagia after 4-8 weeks. A follow-up barium swallow is suggested at 6 months and if symptoms recur.

24

Chapter 25

Thoracic Esophageal Perforation Repair

Alberto de Hoyos and Malcolm DeCamp

Indications

1. Boerhaave's esophagus—violent retching against closed glottis causes esophageal blowout usually on left side just above gastroesophageal (GE) junction.
2. Iatrogenic—following endoscopy/dilatation/biopsy/foreign body extraction/ transesophageal echocardiogram.

Preop

Time from diagnosis to operative treatment is critical for successful primary repair and avoidance of life-threatening sepsis. Diagnosis is established by chest X-ray, upper gastrointestinal study (gastrograffin or thin barium) or computed tomography (CT) scan. The location of the perforation will dictate the side for transthoracic repair. Preoperative resuscitation includes IV hydration and broad-spectrum IV antibiotics.

Procedure

Step 1. A posterolateral thoracotomy is performed on the side of pleural soilage. If there is no pleural effusion, upper to mid esophageal tears should be approached through a right 5th intercostal space thoracotomy. Lower esophageal tears should be approached through a left 7th to 8th intercostal space thoracotomy. Harvest a pedicled intercostal muscle flap with a generous pleural paddle at the time of the incision for later use as coverage of the repair. The ribs are spread gently with a mechanical retractor taking care to protect the muscle pedicle.

Step 2. Mobilize the lung superiorly and divide the inferior pulmonary ligament to the level of the inferior pulmonary vein. Open the mediastinal pleura widely to expose the esophagus.

Step 3. Copiously irrigate the mediastinum and pleural space.

Step 4. Debride nonviable mediastinal and esophageal tissue.

Step 5. Identify the mucosal limits of the esophageal tear. This may require a longitudinal myotomy to locate the mucosal edges.

Step 6. If tissues appears healthy (usually the case within 24-36 hours of the insult), proceed with primary repair.

Step 7. Have the anesthesiologist pass a nasogastric tube into the stomach beyond the defect under the surgeon's direct vision. Perform a two-layer closure. Close the mucosa with interrupted 4-0 polyglactan sutures followed by closure of the muscular layers with 3-0 interrupted silk sutures.

Step 8. Buttress the repair with the intercostal muscle flap. If this is not available, consider coverage with a parietal pleural flap or pedicled pericardial fat pad to provide additional support.

Northwestern Handbook of Surgical Procedures, 2nd Edition, edited by Nathaniel J. Soper and Dixon B. Kaufman. ©2011 Landes Bioscience.

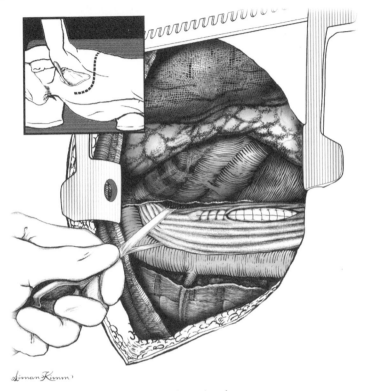

Figure 25.1. Repair of mid-thoracic esophageal perforation.

Step 9. Insert two chest tubes anterior and posterior to the repair and close the thoracotomy.

Step 10. Drain opposite pleural space if an effusion is present.

Addsteps. If a primary repair of the esophageal injury is not feasible and the patient is hemodynamically stable, esophageal resection and reconstruction with a gastric pull-up can be considered. For less stable patients, esophageal exclusion with delayed reconstruction is necessary. This is generally best accomplished by esophagectomy with construction of an end cervical esophagostomy. A decompressing gastrostomy and a feeding jejunostomy are also performed. Patients with underlying pathology causing distal obstruction (cancer, achalasia, radiation or reflux-induced stricture) need to have the obstruction relieved by either esophagectomy, myotomy or esophageal dilation if successful repair and/or recovery is to be achieved.

Postop

Broad-spectrum antibiotics and fluid resuscitation are required for the first several days, Enteral nutrition through the jejunostomy tube is started 2-3 days following the repair. An upper gastrointestinal contrast study should be planned for 5-7 days postoperatively.

Complications

Complications specific to esophageal repair are primarily recurrent leak and mediastinal sepsis.

Follow-Up

Long-term follow-up is geared towards treating the underlying condition (stricture, acid-reflux, achalasia, cancer, alcoholism, etc.) and vigilance for fibrotic stricture development at the repair site.

Acknowledgment

The editors and authors wish to acknowledge Michael J. Liptay for contributing to the previous version of this chapter.

25

Chapter 26

Heller Myotomy: Laparoscopic

Nathaniel J. Soper

Indications

Laparoscopic Heller myotomy is indicated in operable patients who are candidates for laparoscopy and who have had diagnostic studies indicating a swallowing disorder consistent with achalasia. Tests that should be performed include a barium swallow (showing the characteristic bird-beak appearance of the narrowed esophagogastric junction [EGJ]), an upper endoscopy to rule out the possibility of a neoplasm (pseudoachalasia), and esophageal manometry that shows absence of peristalsis in the proximal and midesophagus and absence of swallow-induced relaxation of the lower esophageal sphincter (LES), usually with high basal pressure >20 mm Hg.

Preop

Preoperative studies should include barium swallow, esophagogastroduodenoscopy (EGD), and esophageal manometry. The patient should refrain from eating solids for 48 hours preoperatively.

Procedure

Step 1. The patient is placed supine on a 'split leg' OR table. The body is stabilized with a bean bag mattress. The surgeon stands between the patient's legs and assistant stands at the right side.

Step 2. After prepping and draping, CO_2 pneumoperitoneum is established and five 5-10 mm laparoscopic ports are placed in the upper abdomen: one in the supra-umbilical region to the left of midline, and four in the subcostal region (one at each anterior axillary line and one at each lateral rectus).

Step 3. A 5-10 mm liver retractor is placed through the far right port in the anterior axillary line, elevating the left hepatic lobe and exposing the proximal stomach and esophageal hiatus. The retractor is secured to a table-mounted retractor holder. The patient is placed in a steep reverse Trendelenburg (head-up) position.

Step 4. The gastrosplenic omentum is divided from the fundus of the stomach to the angle of His with an ultrasonic scalpel, mobilizing the fundus off the left half of the diaphragmatic crus and incising the retroperitoneum behind and at the left side of gastric cardia.

Step 5. The upper portion of the gastrohepatic omentum and the phrenoesophageal membrane over the EGJ are opened, identifying the anterior vagus nerve.

Step 6. In the retroperitoneum medial to the right half of the diaphragmatic crus, the retroesophageal space behind the posterior vagus nerve is opened. The dissection is then connected to the left subphrenic space and the esophagus is mobilized well into the mediastinum. Downward traction is placed on the proximal stomach by grasping the epiphrenic fat pad or by using a Penrose drain sling.

26

Northwestern Handbook of Surgical Procedures, 2nd Edition, edited by Nathaniel J. Soper and Dixon B. Kaufman. ©2011 Landes Bioscience.

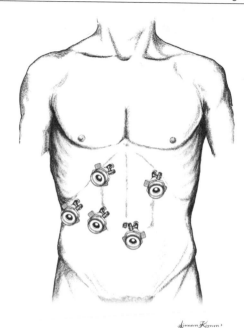

Figure 26.1. Heller myotomy, laparoscopic. Port sites.

Step 7. The distal esophagus is separated from the apex of crural diaphragm and the anterior vagus nerve is mobilized away from the underlying esophagus.

Step 8. The myotomy is marked out by the application of low-wattage cautery to the midline of the distal esophagus and carried down on to the proximal stomach.

Step 9. The myotomy is begun on the distal esophagus by grasping and elevating the longitudinal muscle on each side of the cautery mark using atraumatic graspers; the muscles are then gently teased apart to enter the submucosal plane. Using the L-hook cautery, the myotomy is extended proximally 6-8 cm and then distally onto the stomach for 1-2 cm.

Step 10. Endoscopy is performed again to assess myotomy completeness by positioning just proximal to the squamocolumnar junction and showing a widely patent GE junction. The endoscope also assesses for perforations.

Step 11. When the myotomy is judged to be complete by endoscopy, all fluid and air are suctioned from the stomach. A partial fundoplication is then performed. We perform a posterior (Toupet) fundoplication primarily, suturing the fundus to each side of the myotomy. However, if the posterior wrap angulates the GE junction, or a perforation has been identified and repaired, a Dor anterior fundoplasty can be performed by folding the fundus of the stomach over the myotomy surface and suturing it to the right edge of the myotomy and to the right half of the diaphragmatic crus using nonabsorbable 2-0 suture.

Step 12. The upper abdomen and subphrenic space are irrigated with warm saline, inspecting the subphrenic space and spleen for bleeding. The fascia is closed at the sites of 10 mm ports.

26

Figure 26.2. Heller myotomy, laparoscopic.

Additional Comments
- If a hiatal hernia is noted, the hiatus may be narrowed with suture(s) placed to approximate the left and right halves of the crus behind the EGJ.
- If esophageal perforation occurs during the myotomy, repair the perforation with interrupted 3-0 Vicryl suture using intracorporeal knot-tying.

Postop
It is not necessary to use a nasogastric tube. Oral intake of liquids and soft foods (with no red meat or dry bread) is begun as soon as the patient has no nausea, usually on the day of surgery. Pain is controlled with injectable ketorolac +/- narcotics for 6-12 hours and then oral pain medicine. If intraoperative esophageal perforation occurred (5% chance), the patient is kept NPO and a water-soluble-contrast X-ray swallow is performed the following morning; if there are no signs of leakage, PO fluids are begun. If leakage is noted, the patient should be monitored closely for clinical signs of esophageal perforation (fever, vital sign changes suggesting sepsis, breathing difficulties, chest pain, subcutaneous emphysema) and a thoracic surgeon consulted. If there are no clinical signs of esophageal perforation, the patient is kept NPO and a repeat water-soluble-contrast X-ray swallow performed in 5-7 days.

Complications
Vagal injury, esophageal perforation, gastric perforation, splenic laceration and hemorrhage, gastroesophageal reflux disease (GERD).

Long-Term Follow-Up
Proton pump inhibitor (PPI) therapy is given for one month, then discontinued; if symptoms of GERD develop, then lifelong daily PPI therapy may be indicated. A timed barium swallow is done in 6 months. At one year, a timed barium swallow, EGD, and manometry are performed. If patients develop persistent dysphagia at any time after Heller myotomy, a full evaluation with barium swallow, esophageal manometry, and EGD is indicated.

Acknowledgment
The editors and author wish to acknowledge Raymond J. Joehl for contributing to the previous version of this chapter.

26

Chapter 27

Esophageal Diverticulectomy: Zenker's

Malcolm DeCamp and Matthew G. Blum

Indications

A Zenker's (pharyngoesophageal) diverticulum is an acquired pulsion diverticulum that develops secondary to dysfunction of the pharyngoesophageal junction. It consists of a mucosal outpouching and hence it is a pseudodiverticulum. The diverticulum, which typically enlarges with time, and typically occurs in the 6th to the 8th decades of life, arises from a triangular weakening (Killian's triangle) in the posterior midline of the lower pharynx between the transverse (cricopharyngeous) and oblique (thyropharyngeal) muscle fibers of the lower pharyngeal constrictor. Operative intervention should be considered in symptomatic patients (dysphagia, food regurgitation, aspiration pneumonia).

Preop

A barium swallow is essential to confirm the diagnosis of Zenker's diverticulum. On radiologic evaluation, most diverticula are larger than 2 cm (76%). Upper endoscopy is not routinely recommended and must be performed with extreme caution to prevent perforation. Esophageal manometry is reserved for patients suspected of having associated foregut disorders. Patients with associated gastroesophageal reflux disease (25%) should be treated adequately preoperatively. Full evaluation of the esophagus (complete barium swallow and endoscopy) is completed after repair of the diverticulum, when easy transit from pharynx to esophagus has resumed. General anesthesia is performed with single-lumen endotracheal intubation. Patients with large diverticula should be kept on clear fluids 2-3 days prior to surgery to decrease the risk of aspiration during induction of anesthesia. Prophylactic antibiotics and deep vein thrombosis prophylaxis are routinely used.

Procedure

Step 1. Place the patient in supine position with arms at the sides and insert a shoulder roll to extend the neck. Rotate the head slightly to the right.

Step 2. Place a nasogastric (NG) tube into the stomach using an esophagogastroscope to prevent perforation of the diverticulum from a "blind" insertion The NGT to aid in the identification and dissection of the esophagus. Alternatively, once the esophagus is exposed, the NG may be advanced guiding it manually.

Step 3. Make a longitudinal incision along the anterior border of the left sternocleidomastoid. Incise the platysma and divide the omohyoid. Gently retract the carotid artery, internal jugular vein, and vagus nerve laterally allowing the retroesophageal area to be developed with blunt dissection. If necessary, divide the middle thyroid vein. The inferior thyroidal artery may also be divided if necessary. Vascular ligation should be done away from the tracheoesophageal groove to prevent injury to the recurrent laryngeal nerve.

Northwestern Handbook of Surgical Procedures, 2nd Edition, edited by Nathaniel J. Soper and Dixon B. Kaufman. ©2011 Landes Bioscience.

27

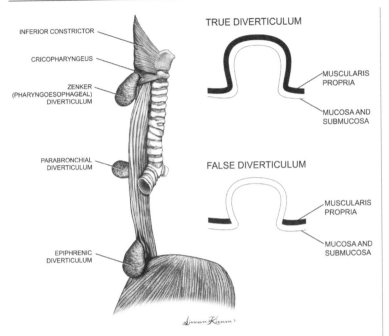

INFERIOR CONSTRICTOR

CRICOPHARYNGEUS

ZENKER
(PHARYNGOESOPHAGEAL)
DIVERTICULUM

PARABRONCHIAL
DIVERTICULUM

EPIPHRENIC
DIVERTICULUM

TRUE DIVERTICULUM

MUSCULARIS
PROPRIA

MUCOSA AND
SUBMUCOSA

FALSE DIVERTICULUM

MUSCULARIS
PROPRIA

MUCOSA AND
SUBMUCOSA

Figure 27.1. Esophageal diverticula.

Step 4. Dissect the esophagus from the prevertebral fascia. Avoid dissection in the tracheoesophageal groove.

Step 5. Locate and dissect the diverticulum posteriorly, just inferior to the cricoid cartilage. Using blunt and sharp dissection, identify the neck of the diverticulum. Grasp the diverticulum gently with a Babcock forcep and elevate it. Remove the nasogastric tube and place a Maloney bougie (48-54 F) within the esophageal lumen

Step 6. The cricopharyngeus forms the inferior muscular border of the diverticulum and division is the critical step in success of the operation. With the bougie serving as an intraesophageal stent, the myotomy is started on the cervical esophagus, progresses through the cricopharyngeus and is extended cranially 2 to 3 cm on the hypopharynx. A right angle clamp is used to elevate the muscularis off of the mucosa for safe division.

Step 7. Excise the diverticulum transversely using a mechanical stapler device or by standard suture technique. Alternatively, suspend the diverticulum from the posterior wall of the pharynx by placing four to five silk sutures taking care to avoid penetrating the mucosa to prevent leaks (diverticulopexy).

Step 8. Remove the bougie and gently place an esophagogastroscope into the proximal esophagus. Insufflate the esophagus under water to identify any unrecognized perforation or a staple line leak that requires repair. The myotomy may be left open or reapproximated on top of the staple line

Step 9. Place a Jackson-Pratt drain in the retroesophageal space. Reapproximate the deep cervical fascia with interrupted absorbable suture. Close the subcutaneous tissue and skin with a continuous absorbable suture.

Transoral stapling of the diverticulum is gaining popularity as an alternative to open repair.

Postop

Keep the patient NPO and perform a gastrograffin or dilute barium swallow 1-2 days postoperatively to rule out a leak from the myotomy or diverticulectomy site. If no leak is detected, start the patient on clear fluids and advance diet over 2-3 days. If a JP was placed, remove it after the swallow study has demonstrated no leak.

Complications

Esophageal leak, inadequate myotomy, esophageal stricture, and recurrent laryngeal nerve injury.

Follow-Up

Dietary counseling and treatment of associated gastroesophageal reflux disease.

Acknowledgment

The editors and authors wish to acknowledge Sean C. Grondin for contributing to the previous version of this chapter.

27

Chapter 28

Gastrostomy: Open

Eric Hungness

Indications

The indication for open gastrostomy is primarily the inability of the patient to be nourished by the oral route due to obstruction or dysfunction of the oral cavity, pharynx, or esophagus. With the advent of percutaneous endoscopic gastrostomy (PEG), open gastrostomy as an isolated procedure is typically limited to those in whom PEG placement is not possible or is contraindicated. Open gastrostomy tube placement may also be performed as an adjunct to other abdominal procedures in patients with poor respiratory reserve who may not tolerate a nasogastric tube or in patients who are expected to have a long period before they are able to resume oral feedings.

Preop

Patients in need of open gastrostomy should have fluid and electrolyte balance optimized to the extent possible. Coagulation factors may need to be brought to acceptable levels since these patients are often nutritionally depleted by their underlying disease.

Procedure

Step 1. The patient is placed in the supine postion and prepped from above the xiphoid to the pubis, laterally to each side of the abdominal wall. Drapes are placed at the xiphoid, umbilicus, right rectus edge, and left edge of the anterior abdominal wall.

Step 2. An incision is made in the midline below the xiphoid, large enough to admit approximately four fingerbreadths and to visualize the anterior gastric wall.

Step 3. The skin site for the gastrostomy is chosen that is away from the costal margin and overlies the stomach so that the gastric wall can be easily brought up and attached to the abdominal wall.

Step 4. A 0.5-1.0 cm incision is made in the skin and a 6 inch clamp placed through the full thickness into the peritoneal cavity, grasping an 8 inch clamp and bringing the latter back outside the peritoneal cavity.

Step 5. The tip of a 24 F Foley catheter with a 5 ml balloon is then grasped by the 8" clamp and brought back inside the abdominal cavity.

Step 6. Two Babcock clamps grasp the gastric wall on each side of the gastrostomy site and electrocautery is used to make a small opening in the muscular layers. The submucosa and mucosa are grasped with mosquito clamps and the lumen entered, tying the tissue with 4-0 absorbable suture.

Step 7. The opening is enlarged and the Foley tip is inserted. Two concentric pursestring sutures of 2-0 or 3-0 silk are placed sequentially, tying the inner one to invert the gastric edge, then tying the outer one afterward.

28

Northwestern Handbook of Surgical Procedures, 2nd Edition, edited by Nathaniel J. Soper and Dixon B. Kaufman. ©2011 Landes Bioscience.

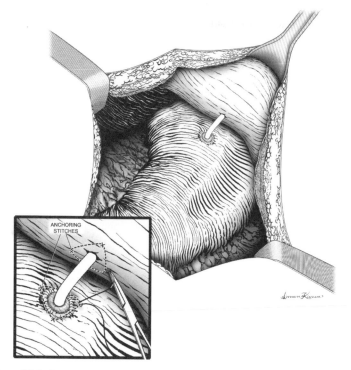

Figure 28.1. Open gastrostomy.

Step 8. Three or four 2-0 silk sutures are then placed adjacent to the openings in the peritoneum and stomach, starting at the point most distant from the operator, and then closer.

Step 9. The Foley balloon is inflated and the stomach is pulled up to the abdominal wall by traction on the tube. The sutures are then tied in the order in which they were placed.

Step 10. The tube is fixed to the skin of the abdominal wall with a 2-0 silk or nylon suture.

Step 11. The abdominal incision is closed with 0 nylon on the fascia, 3-0 absorbable suture on Scarpa's layer, and 4-0 subcuticular sutures, staples, or steri-strips. The gastrostomy tube is attached to a drainage bag, and a tape "mesentery" is attached to the tube to protect it from being pulled out.

28

Postop

The gastrostomy tube is left open to drainage for 24 hours and may then be used for feeding, if that was its intent, or to continued drainage. Topical povidone may be used daily to prevent skin inflammation from the holding suture. Maintaining a tape mesentery on the tube itself protects against inadvertent pulling on the suture holding the tube in place. If the tube is used for feeding, it may be coiled up on the abdominal wall and taped in place when not in use.

Complications

Complications associated with this procedure include bleeding from the gastric mucosa, dislodgement of the gastrostomy tube, leakage of gastric contents with excoriation of the protective skin surface, and infection at the tube site. Gastric obstruction may occur if the tube is placed too proximal on the gastric wall causing tenting of the stomach or if the tube migrates distally and the balloon occludes the pylorus.

Follow-Up

Long-term follow-up for open gastrostomies involves maintaining of the tube lumen and preventing infection at the tube site. Changing the tube is easily accomplished by deflating the balloon, removing the tube after cutting the suture, and replacing it immediately. If the tube comes out inadvertently at home, immediate return for replacement is mandatory; in certain cases, the patient can be taught to replace the tube.

Acknowledgment

The editors and author wish to acknowledge Richard S. Berk for contributing to the previous version of this chapter.

28

Chapter 29

Gastrectomy: Subtotal or Partial

Jeffrey D. Wayne

Indications

Subtotal gastrectomy is indicated for the treatment of gastric adenocarcinoma in the absence of distant metastases. In a subtotal gastrectomy for cancer, it is typical to remove about 85% of the stomach, leaving only a small fundic remnant attached proximally to the esophagus. Subtotal or partial gastrectomy may be indicated for other less common gastric neoplasms, such as gastrointestinal stromal tumors (GIST).

Partial gastrectomy is also indicated for the treatment of gastric ulcer. In this case, the goal is removal of the gastric antrum only, and approximately 40% of the stomach is typically removed. No vagotomy is performed.

Finally, partial gastrectomy (antrectomy) with truncal vagotomy is occasionally indicated in the treatment of severe or recurrent duodenal ulcer, although the operation is performed much less commonly than a few decades ago.

In all cases of partial or subtotal gastrectomy, the distal margin of resection should be in the duodenum, approximately 1-2 cm beyond the pylorus.

Preop

An intravenous dose of a first-generation cephalosporin should be given prior to making an incision. Antiembolism prophylaxis should be employed, using subcutaneous heparin and/or sequential compression boots depending on individual risk factors.

Procedure

Step 1. Under general anesthesia, a midline incision is made from the xiphoid to the umbilicus. A bilateral subcostal incision may be used if the patient has a shallow costal angle. The subsequent course of the operation varies depending on the indication (cancer or benign ulcer), so the steps are described separately below.

Subtotal Gastrectomy for Cancer

Step 2. The omentum should be dissected away from the transverse colon along its entire length and mobilized superiorly en bloc with the stomach.

Step 3. A Kocher maneuver is performed to mobilize the duodenum.

Step 4. The right gastroepiploic vessels are divided beyond the pylorus. The vessels should be ligated as far away from the gastric wall as possible so that any accompanying lymph nodes can be swept toward the stomach and included in the specimen.

Step 5. The right gastric vessels should be divided at the left edge of the porta hepatis as the right gastric artery arises from the proper hepatic artery.

Step 6. The duodenum is circumferentially dissected about 2 cm distal to the pylorus and divided with a linear stapler. The staple line on the distal duodenal stump is imbricated beneath 3-0 Lembert sutures.

Northwestern Handbook of Surgical Procedures, 2nd Edition, edited by Nathaniel J. Soper and Dixon B. Kaufman. ©2011 Landes Bioscience.

Step 7. Holding the stomach up and to the patient's left, the lesser omentum is divided close to the liver, ultimately exposing the left gastric artery as it sweeps up onto the lesser curvature of the stomach.

Step 8. The left gastric artery is divided and ligated at its base as it arises from the celiac axis. A suture ligature should be used on the stump of the artery. The accompanying vein is also divided and ligated. All soft tissue and nodes running along the left gastric vessels are included in the specimen.

Step 9. The short gastric vessels on the greater curvature of the stomach are ligated and divided up to the level of proposed transection of the stomach.

Step 10. The stomach is divided using a linear stapler. The first stapler is applied at approximate right angles to the greater curvature, and the length of the first cut is designed to be the length of the planned gastrojejunal anastomosis. The second firing of the stapler should angle up to the top of the lesser curvature of the stomach, within 1-2 cm of the esophagus. With the second firing of the stapler, the specimen is removed. The second staple line is then oversewn with running or interrupted suture.

Step 11. The proximal jejunum is brought up to the stomach in either an antecolic or retrocolic position. A back row of interrupted sutures is placed between the jejunum and the posterior stomach. Once the posterior wall is complete, the first gastric staple line is excised and a matching or slightly smaller opening is made in the jejunum. The inner layer of the anastomosis is then performed using a running suture, beginning on the back wall and coming around the corners to the front. The anastomosis is completed with an anterior row of interrupted Lembert sutures.

Step 12. If the jejunum was passed in a retrocolic position, the defect in the transverse mesocolon should be closed to prevent an internal hernia. The abdominal wall is then closed with nonabsorbable sutures on the fascia and staples on the skin.

Operation for Ulcer Disease

Step 1. If an antrectomy is being performed for gastric ulcer or a vagotomy and antrectomy is being performed for duodenal ulcer, the dissection is usually begun proximally and extends distally towards the duodenum. The reason for this difference is that the duodenum may be quite scarred in an ulcer operation, and dividing the stomach proximally allows the distal stomach to be lifted up and manipulated, providing better circumferential exposure.

Step 2. If a truncal vagotomy is to be performed, the left triangular ligament of the liver is divided and the left lateral liver segment retracted to the right to expose the gastroesophageal junction. The peritoneum overlying the gastroesophageal junction is incised.

Step 3. The distal esophagus is encircled with a finger, working from left to right behind the esophagus. By pulling down on the esophagus, the left (anterior) vagus nerve is stretched and can be felt easily on the anterior surface of the esophagus. It is picked up with a nerve hook and a 2 cm length of the nerve cleaned. Two medium clips are placed on the nerve and a 1 cm section of nerve between the clips excised.

Step 4. A Penrose drain is placed around the esophagus and used to retract the gastroesophageal junction inferiorly and slightly to the patient's left. The right index finger is passed behind the esophagus and the right (posterior) vagus nerve identified by palpation as a thick band in the tissue between the aorta and the esophagus. The finger is used to push the right nerve up into view, where it can be grasped with a nerve hook, and a 1 cm section of nerve is excised between staples as described for the left vagus. The sections of both left and right nerve should be sent for frozen-section examination to confirm nerve tissue in both specimens.

29

Step 5. The proximal extent of the antrum on the lesser curvature of the stomach is estimated by looking for the point where the anterior nerve of Latarjet fans out from the lesser omentum over the anterior surface of the stomach (the so-called crow's foot). The lesser curve should be cleaned 1-2 cm proximal to this point. A point on the greater curvature is also cleared that is approximately halfway up the stomach, usually at the lower end of the short gastric vessels. The stomach is then divided with two fires of a linear cutting stapler,. The second staple line is oversewn.

Step 6. The distal stomach is grasped and used as a handle to facilitate further dissection. Both the greater curve and the lesser curve of the stomach are then skeletonized until the dissection reaches a point about 2 cm beyond the pylorus. In contradistinction to the cancer operation, all of this dissection may be right along the gastric wall. There is no need to include omentum or nodal tissue. The duodenum is then divided with another firing of the linear cutting stapler and the specimen removed.

Step 7. Reconstruction of the gastrointestinal tract can either be done by sewing the gastric remnant to the duodenum (Billroth I) or to the jejunum (Billroth II).

Postop

A nasogastric tube is typically placed after surgery and removed when there are signs of bowel activity.

Complications

Splenic injury may occur because of inadvertent traction during the dissection of the stomach or the performance of the vagotomy. The injury may not be recognized during surgery. Signs of significant blood loss should prompt reexploration. The duodenal "stump" may leak after a Billroth II reconstruction and require reoperation. Internal hernias may occur through the mesocolon after retrocolic reconstruction. Occasionally, patients may develop bile reflux gastritis or afferent loop syndrome. Recurrent ulcer is very rare. Finally, an aberrant left hepatic artery arising from the left gastric artery may be divided if it is not recognized when dividing the lesser omentum. This may cause ischemia or necrosis of the left hepatic lobe.

Follow-Up

After an operation for cancer, patients should be followed regularly because the incidence of recurrent cancer is high. In addition, referral to a medical oncologist is made, as combined, 5-FU-based chemo-radiation has now been shown to provide a survival advantage when compared to surgery alone. A change in symptoms may be a harbinger of recurrence and should prompt radiologic imaging. Initial follow-up is at 4-month intervals.

Patients operated for ulcer disease do not require antiulcer medications. They do not need endoscopic follow-up unless they develop recurrent symptoms.

Acknowledgment

The editors and author wish to acknowledge Richard H. Bell, Jr. for contributing to the previous version of this chapter.

29

Chapter 30

Gastrectomy: Total

Jeffrey D. Wayne

Indications

Total gastrectomy is indicated in patients with biopsy-proven cancer of the proximal stomach. Regional lymph node involvement which can be encompassed in the resection is acceptable but distant metastases are a contraindication to total gastrectomy.

Preop

Patients should be preoperatively staged with a chest X-ray, CT scan of the abdomen and pelvis, and esophagogastroduodenoscopy, often with endoscopic ultrasound. Standard prophylactic antibiotics, typically consisting of a first generation cephalosporin are administered preoperatively. A thoracic epidural catheter is often placed for postoperative pain management. Sequential compression boots and subcutaneous heparin are generally indicated for prophylaxis and should be started preoperatively.

Procedure

Step 1. The patient is placed in the supine position. Typically, a diagnostic laparoscopy is undertaken to rule out sub-radiographic metastases to the peritoneum or hepatic surface. If the laparoscopy is negative, open abdominal exploration is undertaken. Either a midline or bilateral subcostal incision can be used. A self-retaining retractor should be inserted to provide good exposure of the esophageal hiatus and celiac lymph nodes.

Step 2. The omentum is completely mobilized off the transverse colon so that it can be removed en bloc with the stomach. The posterior stomach is examined through the lesser sac. If the primary tumor invades the short gastric vessels or the splenic capsule, a splenectomy will be necessary.

Step 3. A Kocher maneuver is performed. The subpyloric nodes are separated from the pancreatic head, and the pyloric vein is ligated at its junction with the middle colic vein. The right gastric artery and veins are ligated just distal to the pylorus. The duodenum just distal to the pylorus is transected using a linear stapler. The stapled duodenal stump is oversewn with 3-0 sutures in a Lembert fashion.

Step 4. With the greater omentum retracted in a cephalad direction, the dissection is continued along the greater curvature of the stomach up to the short gastric vessels. If a splenectomy is not necessary, the short gastric vessels are ligated up to the gastroesophageal junction. If a splenectomy is required, the splenic artery is ligated near its origin from the celiac axis, and the splenic artery lymph nodes are taken in continuity with the spleen and stomach. The splenic vein is ligated and divided in the splenic hilum.

30

Northwestern Handbook of Surgical Procedures, 2nd Edition, edited by Nathaniel J. Soper and Dixon B. Kaufman. ©2011 Landes Bioscience.

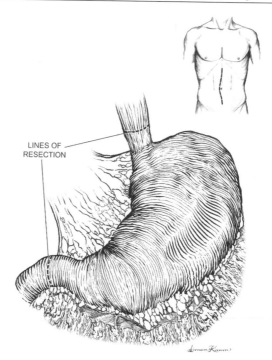

LINES OF
RESECTION

Figure 30.1. Total gastrectomy. Boundary of resection.

Step 5. The stomach is then retracted cephalad to expose the left gastric artery, the celiac lymph nodes, and the aorta. The left gastric artery is ligated at its junction with the celiac axis. The common hepatic artery lymph nodes are dissected off the artery and retracted medially into the left gastric artery lymph node bundle. If a splenectomy was not done, the splenic artery lymph nodes are dissected away from the artery and retracted in the left gastric artery lymph node bundle. The para-aortic lymph nodes are separated from the aorta and retracted up into the left gastric artery lymph node bundle, which is attached to the gastrectomy specimen. This maneuver completes the D2 lymph node dissection.

Step 6. An automatic purse string applier is then placed across the esophagus approximately 2 cm above the proximal margin of gross tumor or the gastroesophageal junction, whichever is more proximal. The esophagus is transected, and the gastrectomy specimen is sent to pathology for immediate analysis of the esophageal margin.

Step 7. Approximately 10 cm distal to the ligament of Treitz, the jejunum is mobilized and transected at that location using a linear stapler. The mesentery is divided vertically as necessary to gain mobility. The distal portion of the jejunum is brought through the bare area of the transverse mesocolon to the left of the midline.

Step 8. An end-to-side esophagojejunal anastomosis should be at least 25 mm in diameter to reduce the risk of postoperative stricture. The circular stapler anvil is positioned in the esophagus, and the purse string suture is tied down around it. The circular stapler is then placed into the end of the jejunal loop by removing the staple

30

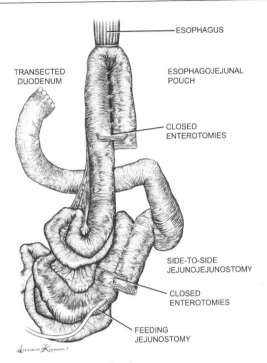

Figure 30.2. Total gastrectomy. Completed reconstruction.

line. The trocar of the stapler is advanced through the jejunal wall at a convenient location and is attached to the post of the anvil. The handle of the circular stapling device is rotated to bring the stapling rings together and the device is fired. The circular staple line is inspected to establish that the tissues are approximated. Defects in the suture line may be reinforced with interrupted sutures. The open end of the jejunum through which the stapler has been inserted is reclosed using a linear stapler or a two-layer hand-sewn closure. The nasogastric tube is repositioned so that it passes through the esophagojejunal anastomosis.

Alternative Method for Esophagojejunostomy: Hunt-Lawrence Pouch

Approximately 10 cm of the jejunum is doubled back onto itself to create an inverted "J" configuration. Two small punctures are made side-by-side in the two adjacent limbs of the "J," and a linear cutter is used to create the pouch. Once this is done, the surgeon puts the index finger into the pouch as far as possible through the common enterotomy and loops the finger around the remaining web of common jejunal wall that still exists in the proximal pouch. A second firing of the linear cutter transects the residual web. The circular stapler is then placed into the pouch through the distal enterotomy and the esophagojejunostomy performed as above. While there is no convincing long-term data to support the routine use of a jejunal pouch, many authors believe that early enteral capacity is improved, especially in thin females.

30

Step 10. 50-60 cm below the esophagojejunostomy, a side-to-side jejunojejunostomy or functional stapled end-to-side jejunojejunostomy is constructed to re-establish intestinal continuity.

Step 11. A feeding jejunostomy is often placed for postoperative nutrition.

Step 12. The abdominal fascia is closed with a monofilament suture, and the skin is closed with staples.

Postop

Antibiotics are discontinued after three doses. Subcutaneous heparin is continued while the patient is hospitalized. A gastrograffin upper gastrointestinal study may be done about 5 days after the operation, but is often omitted is the patient is clinically well. Otherwise, once there are signs of bowel function, the nasogastric tube may be removed.

Complications

A leak at the esophagojejunostomy may manifest clinically with tachycardia, fever, pain, and/or an elevated white blood cell count. The most common time for a leak to occur is approximately five to seven days postoperatively.

Follow-Up

If the patient received neoadjuvant chemotherapy (most commonly epirubicin, cisplatin, and 5-fluorouracil), this regimen is often reinstituted after recovery from surgery. If no neoadjuvant therapy was employed, postoperative 5-FU based chemoradiation should be considered. Patients should be followed postoperatively with careful attention to symptoms that might suggest recurrence. There is no advantage to routine imaging in patients who are asymptomatic.

Acknowledgment

The editors and author wish to acknowledge Stephen F. Sener and Malcolm M. Bilimoria for contributing to the previous version of this chapter.

30

Chapter 31

Perforated Duodenal Ulcer Repair: Omental Patch

Carla Pugh

Indications
Omental patch repair is indicated for perforated duodenal ulcers.

Preop
Patients with perforated ulcers may be dehydrated, so intravenous fluid resuscitation should be undertaken. Prophylactic intravenous antibiotics are indicated. Patients are generally begun on H2 receptor antagonists or proton pump inhibitors.

Procedure
Step 1. The patient is placed in the supine position; an upper midline incision is made.

Step 2. Intra-abdominal fluid should be collected in a syringe for culture and sensitivity, then any remaining fluid aspirated.

Step 3. Place 2-3 seromuscular sutures of 2-0 or 3-0 silk so that the sutures cross the perforation but are placed in relatively healthy tissue. Do not tie the sutures yet.

Step 4. Cover the perforation with a tongue of omental fat and tie the previously placed sutures over the omentum.

Step 5. Irrigate copiously and close fascia. Consider the use of retention sutures for elderly patients, poorly nourished patients, immunocompromised patients, etc.

Postop
Remove the nasogastric tube when bowel activity resumes. For complicated repairs a gastrograffin swallow is indicated. Advance diet as tolerated. Switch from IV to oral antiulcer medications before discharge.

Complications
Wound infection, abscess; intra-abdominal abscess; suture line leak; duodenal fistula.

Follow-Up
Continue antiulcer medications for approximately 6 weeks. At that point, patients should be tested for *H. pylori* and treated if positive. Upper endoscopy should also be scheduled.

Chapter 32

Truncal Vagotomy and Pyloroplasty

John J. Coyle

Indications

Truncal vagotomy and pyloroplasty is indicated for the treatment of complicated duodenal ulcer disease. It is the standard operation for bleeding duodenal ulcer. It is less often used for perforation or obstruction. It is also an operative option in patients with ulcers refractory to medical treatment.

Preop

The operation should rarely if ever be done without documentation of a duodenal ulcer by endoscopy or barium studies. In cases of bleeding, large-bore IVs should be placed and aggressive volume resuscitation begun. A nasogastric tube should be placed. A Foley catheter should be placed for monitoring volume status in emergency cases. A prophylactic antibiotic should be given IV 30 minutes prior to incision. Sequential compression devices should be used for deep venous thrombosis prophylaxis.

Procedure

Step 1. A midline abdominal incision is made from the xiphoid process to the umbilicus or slightly below the umbilicus. The incision should be carried up along the side of the xiphoid process to allow maximal exposure of the gastroesophageal (GE) junction area.

Step 2. After entering the abdomen, the left triangular ligament of the liver is incised and the left lateral segment retracted to the patient's right to reveal the GE junction. The peritoneum over the GE junction is opened just below the esophageal hiatus of the diaphragm.

Step 3. The distal esophagus is mobilized circumferentially by blunt dissection and the esophagus surrounded with a Penrose drain.

Step 4. Using the Penrose drain to provide downward traction on the stomach, the anterior and posterior vagal trunks are identified. Caudad retraction of the stomach is often helpful in identifying the nerve trunks by palpation. The anterior nerve crosses the anterior surface of the GE junction. The posterior nerve lies to the right posterior of the esophagus and is a short distance away from the esophagus.

Step 5. Each nerve trunk is isolated in turn with a nerve hook and lifted into view.

Step 6. Metal clips are applied twice to each nerve and a portion of each nerve 5-10 mm long is resected between the clips. Pathologic evaluation of the specimens by frozen section should be done to confirm that they contain nervous tissue.

Step 7. The duodenum is then mobilized liberally with a Kocher maneuver.

Northwestern Handbook of Surgical Procedures, 2nd Edition, edited by Nathaniel J. Soper and Dixon B. Kaufman. ©2011 Landes Bioscience.

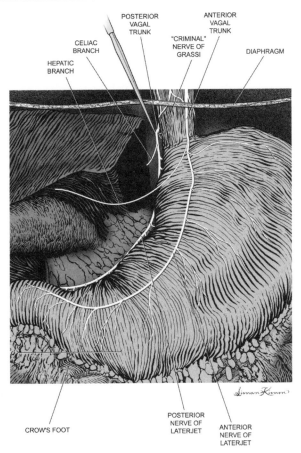

Figure 32.1. Truncal vagotomy.

Step 8. Incise the anterior stomach and duodenum at least 2 cm proximally and distally from the pylorus along the long axis of the bowel. Include perforation in the incision when present.

Step 9. Close the resultant defect along transverse axis of the bowel.

Postop

Nasogastric suction until evidence of bowel function.

Complications

Subhepatic abscess, wound infection, gastric atony, dehiscence of pyloroplasty, esophageal perforation.

Follow-Up

Assessment of gastric function, ulcer recurrence.

32

Chapter 33

Gastric Bypass: Roux-en-Y: Open

Jay B. Prystyowsky

Indications

Gastric bypass is an operation for morbid obesity and is indicated in patients with a body mass index >40 kg/m^2 or >35 kg/m^2 with significant obesity-related illnesses. The patient should be an appropriate surgical risk and understand the life-style changes that the operation requires.

Preop

The patient should be evaluated by a multidisciplinary team and understand the dietary changes that will occur postoperatively. On the day of surgery, antibiotic prophylaxis should be provided and antithrombotic measures instituted, including sequential compression devices and subcutaneous heparin.

Procedure

Step 1. With the patient supine, the abdomen is entered through an upper midline incision; a self-retaining retractor system is placed. The patient is then placed in reverse Trendelenburg position.

Step 2. The left triangular ligament of the liver is divided to allow retraction of the left lateral segment of the liver to the patient's right. At this point the gastroesophageal (GE) junction is mobilized and the distal esophagus elevated anteriorly with a Penrose drain.

Step 3. The stomach is transected with a linear stapler, beginning about 6 cm distal to the GE junction along the lesser curvature and concluding about 2 cm from the GE junction along the greater curvature of the stomach.

Step 4. The small bowel is divided about 40-50 cm below the ligament of Treitz with a linear stapler. A 75-150 cm Roux-en-Y limb is created by performing a hand-sewn end-to side or a stapled functional end-to-side small bowel anastomosis and closure of the mesenteric defect.

Step 5. The Roux limb is brought into the left upper quadrant in either an antecolic or retrocolic position.

Step 6. An end-to-side gastrojejunostomy with a 21 mm EEA stapler or a hand-sewn gastrojejunostomy is performed.

Step 7. The small bowel is temporarily occluded beyond the gastric anastomosis to test for anastomotic leaks with instillation of dye or air into the stomach; endoscopy may also be used.

Step 8. Drain placement around the anastomosis is optional.

Step 9. Any defect in the transverse mesocolon is closed if the retrocolic approach was used for the jejunal limb.

Step 10. The midline fascia and skin are closed.

Northwestern Handbook of Surgical Procedures, 2nd Edition, edited by Nathaniel J. Soper and Dixon B. Kaufman. ©2011 Landes Bioscience.

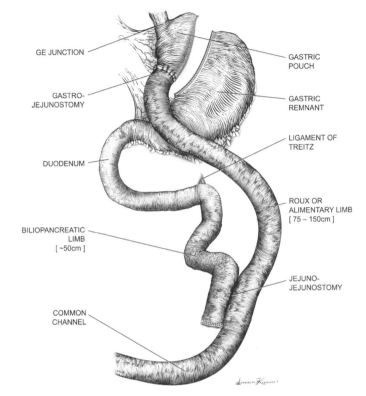

Figure 33.1. Roux-en-Y gastric bypass.

Postop

The use of a nasogastric tube is optional. On postoperative day 1, the patient is given 30 ml water per hour. On postoperative day 2, a gastrograffin swallow may be performed to rule out a leak; if the study is negative, then the patient is advanced to a sugar-free clear liquid diet.

Patients should ambulate early and can ordinarily be discharged home on postop day 2-4.

Complications

Complications associated with the procedure include anastomotic leak, intra-abdominal abscess, splenic injury, anastomotic stricture, wound infection, and wound dehiscence.

Follow-Up

The patient should have frequent follow-up in the first 3-6 months to aid in adjustment to the new diet, then less frequently to assess nutritional status; expected weight loss is about two-thirds of excess weight (defined as preoperative weight—ideal body weight).

Chapter 34

Gastric Bypass: Roux-en-Y: Laparoscopic

Alexander P. Nagle

Indications

Severe obesity has been notoriously refractory to virtually every method of non-surgical treatment. The failure rate of diet and behavior modification treatment at 2 years in the morbidly obese approaches 100%. Surgery has been shown to be the most effective treatment for achieving sustained weight loss with subsequent control of obesity-related co-morbidities. The 1991 National Institutes of Health (NIH) Consensus Development Conference has helped to form the basis of the indications for the surgical treatment of obesity that includes:

1. Body Mass Index (BMI) greater than 40 kg/m² or greater than 35 kg/m² with associated obesity-related illnesses.
2. Failure of sustained weight loss on supervised dietary and/or medical regimens.
3. Patient shows understanding of the risks and benefits of surgery and understands the lifestyle changes subsequent to the operation.
4. Acceptable operative risk.

Preop

Preoperative evaluation of patients for bariatric surgery is a process that involves multiple healthcare providers. It is important both for the purpose of patient selection and patient education. A thorough history and physical examination should be supplemented by routine blood tests, chest radiographs, and electrocardiogram. Pulmonary function tests and cardiac stress tests are frequently useful for accurate risk stratification. Polysomnography to detect sleep apnea is indicated when the diagnosis is suspected. As for any major operation, other consultations and/or tests may be needed for optimal preoperative preparation and perioperative care. A thorough discussion of the operations and subsequent lifestyle changes by the surgeon and healthcare team are mandatory for each patient. A complete explanation of the risks of surgery including mortality and major morbidity is indicated. Patients also meet with a dietitian and psychologist preoperatively. Behavioral modification is critical to the long-term success of most bariatric surgical procedures.

Procedure

Step 1. All patients should receive both preoperative antibiotics and preoperative chemoprophylaxis for DVT/PE.

Step 2. Peritoneal access and trocar placement. The initial trocar is typically placed approximately 15-16 cm below the xyphoid and in the midline. The initial trocar can be placed using an open (Hassan) or Veress needle technique. Alternatively, an optical-viewing trocar can be utilized. Five additional trocars are placed—two for the surgeons right and left hand, two for the assistants right and left hand, and one to retract the left lateral segments of the liver to exposure the area of the GE junction.

Northwestern Handbook of Surgical Procedures, 2nd Edition, edited by Nathaniel J. Soper and Dixon B. Kaufman. ©2011 Landes Bioscience.

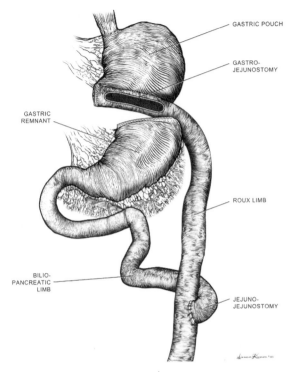

GASTRIC POUCH

GASTRO-
JEJUNOSTOMY

GASTRIC
REMNANT

ROUX LIMB

BILIO-
PANCREATIC
LIMB

JEJUNO-
JEJUNOSTOMY

34

Figure 34.1. Laparoscopic Roux-en-Y gastric bypass.

Step 3. Creation of the Roux limb. The ligament of Treitz is identified by elevating the transverse colon mesentery. The jejunum is then followed distally for 40 cm and divided at that point with a linear GIA stapler. A point 150 cm distal to the transected jejunum is identified and marked. At that point the jejunum is anastomosed to the distal end of the jejunal segment that is in continuity with the ligament of Treitz—thereby creating a 40 cm biliopancreatic limb and a 150-cm Roux limb (Fig. 34. 1). The jejunojestostmy is performed using a linear stapler and the closure of the enteroenterostomy can be performed with either a hand-sewn or linear GIA stapler technique. The mesenteric defect should also be closed with suture to minimize the risk of an internal hernia.

Step 4. Creation of the gastric pouch. The angle of His is identified and mobilized by dissecting posteriorly on the left crus of the diagram. The gastrohepatic ligament (pars flaccida) is opened to expose the lesser sac and the left gastric vascular pedicle. The gastrohepatic ligament is divided with linear GIA stapler just inferior to the left gastric pedicle—thereby preserving the left gastric as the primary blood supply to the gastric pouch. The stomach is transected transversely 4 cm below the GE junction using a 60 mm linear GIA stapler. The stomach is then transected vertically (using linear GIA staplers) aiming directly for the angle of His. The volume of the gastric pouch should be 20-30 ml. Care must be taken to prevent incorporating the fundus as a part of the gastric pouch as this will create a large gastric pouch.

Step 5. Creation of the gastrojejunostomy. The proximal end of the Roux limb is brought to the upper abdomen either via an antecolic (over the transverse colon) or retrocolic (under the transverse colon and through the mesentery) approach. We prefer an antecolic approach; however, in rare cases, if there is tension in bringing the Roux limb to the upper abdomen, a retrocolic approach is necessary. The gastrojejunostomy can be performed using either a handsewn, linear GIA stapler or circular EEA stapler technique. The specific technique that is utilized should be the one in which the surgeon is most comfortable performing and has the most experience. The size of the gastrojejunostomy should be 10-12 mm in diameter. The gastrojejunostomy should be tested intraoperatively for leak. This can be performed by inject air or blue dye via an oral-gastric tube. We prefer to utilize a flexible endoscope to gently insufflate the gastric pouch while the anastomosis is held under saline irrigation.

Postop

Most patients are discharged on POD#2. NG tube is not necessary postoperatively. On POD#1 patients are started on sips of water and then advanced to sugar-free liquids. Most patients are discharged on a pureed diet and are advanced to a soft diet after 7-10 days. During the first week postoperatively, emphasis is on acquiring enough water intake to prevent dehydration. The Foley catheter is removed on POD#1 and patients are encouraged to ambulate to minimize the risk of DVT/PE. Intermittent pneumatic compression devices and DVT/PE chemoprophylaxis should be continued while in the hospital. The routine use of extended DVT/PE chemoprophylaxis beyond the in-hospital stay is less clear—but in select patients extended DVT/PE prophylaxis is recommended. Routine postoperative upper GI contrast study is not necessary.

Complications

Complications after RYGB are typically divided into early (within 30 days of surgery) and late. Additionally, operative complications are further divided into major and minor complications. Major operative complications include death, anastomotic leak, pulmonary embolus, intestinal obstruction, and hemorrhage. Overall mortally from RYGB is low—around 0.2% to 0.5%. The most common cause of death after RYGB is related to pulmonary embolus or anastomotic/staple line leak leading to sepsis. The incidence of an anastomotic leak following a Roux-en-Y gastric bypass is reported to be about 1-2%. Gastrointestinal leaks can be difficult to recognize after RYGB because fever and abdominal tenderness may be absent during the first 48 hours after a leak has occurred. Persistent tachycardia and progressive tachypnea are the most common early signs. Because failure to recognize a leak can result in death, exploratory laparotomy should be empirically performed in patients suspected of leak. Late complications after RYGB include anastomotic ulcer/strictures at the gastrojejunostomy (~10%), symptomatic gallstones, and internal hernia. An internal hernia can also be difficult to diagnose, and, given the potential devastating consequences, early exploration is indicated if an internal hernia is suspected.

Follow-Up

Long-term follow-up after RYGB is critically important for both the bariatric surgery program and the patient. The success of any bariatric surgery is largely dependent on proper patient selection with particular attention to commitment and compliance. Bariatric surgery patients are at risk for potential nutritional complications; such as protein malnutrition and certain vitamin deficiencies. The most common micronutrient deficiencies following RYGB are iron, folate, B12, thiamine, calcium and vitamin D. Lifelong vitamin supplementation and monitoring is required. Weight regain after bariatric surgery is another long-term challenge; significant weight regain may occur in up to 20% of patients following RYGB. This weight regain after bariatric surgery supports the concept that obesity is a chronic and progressive disease. Weight regain occurs probably due to anatomical and physiological adaptations that occurs over time; which may permit the patient to return to bad old habits. The consumption of highly caloric liquids and the presence of abnormal eating patterns, such as binge-eating and snack-eating patterns, as well as a progressive increase in the diet's total calorie intake, seem to be related to weight regain. A comprehensive team concept is most appropriate to help support the patient in making the necessary lifestyle changes.

34

Chapter 35

Laparoscopic Gastric Banding for Obesity

Eric Hungness

Indications

All morbidly obese patients interested in weight loss surgery (BMI >40, or 35 with significant comorbidities) should be evaluated in a multidisciplinary fashion (internist, dietitian, psychologist or psychiatrist, and surgeon) and counseled regarding surgical options. Gastric banding results in slower weight loss but has a safer complication profile compared with other bariatric operations. Prior bariatric surgery is not an absolute contraindication for gastric band placement. Patients with gastroesophageal reflux should be evaluated for hiatal hernia, as the presence of a large hiatal hernia is a contraindication for adjustable gastric banding.

Preop

A first or second generation cephalosporin or an antibiotic of equivalent coverage is given 30 minutes prior to surgery. In addition, a 5000 unit subcutaneous dose of heparin is given in preop holding for deep venous thrombosis chemoprophylaxis.

Procedure

Step 1. Using a #11 blade, a 1 cm skin incision is made 2 fingerbreadths below the costal margin in the midclavicular line. The abdominal wall is elevated and the Veress needle is inserted into the peritoneal cavity. The needle should be hooked to a syringe partially filled with saline. Aspiration of bloody or bile tinged fluid should be negative and the saline should easily flow into the abdomen. Carbon dioxide pneumoperitoneum is established to 12-15 mm Hg pressure. An 11 mm trocar is inserted in standard fashion.

Step 2. A 10 mm 45° angled laparoscope is inserted and two additional 5 mm ports are placed under direct visualization (far right lateral below the liver edge and far left lateral just below the costal margin).

Step 3. A 15 mm trocar is inserted through a 1 inch horizontal incision 8 cm below the xiphoid just to the left of the midline. This trocar is also positioned under direct visualization and has an oblique angle through the fascia to allow for the band tubing to gently enter the peritoneum.

Step 4. The patient is placed in steep reverse Trendelenburg and a Nathanson retractor is placed under direct visualization in the subxiphoid region left of the falciform ligament to retract the left lobe of the liver, exposing the diaphragmatic hiatus.

Step 5. Dissection starts by carefully mobilizing the angle of His with cautery and blunt dissection paying attention to the nearby spleen.

Step 6. The pars flaccida is then opened, the lesser sac is entered, and the base of the right crus is identified.

Northwestern Handbook of Surgical Procedures, 2nd Edition, edited by Nathaniel J. Soper and Dixon B. Kaufman. ©2011 Landes Bioscience.

Step 7. The peritoneum overlying the base of the right crus is opened and a blunt 5 mm grasper is then passed anterior to the right and left crus toward the angle of His, thereby creating a retrogastric window. The camera driver should start at the base of the right crus and move to the angle of His to directly visualize the passage of this instrument. There should be minimal resistance during this maneuver. Increased resistance may indicate that the grasper is being improperly passed and may lead to gastric or esophageal perforation. If suspected, upper endoscopy should be performed

Step 8. An appropriate sized adjustable gastric band is introduced via the 15 mm trocar. Patients with a large amount of mesenteric fat may need a larger sized band to prevent it from being too tight. The tubing and band are then passed behind the stomach through the retrogastric window. Again, anything more than minimal resistance during this maneuver should prompt reassessment.

Step 9. The tubing is then placed through the buckle of the band, and the band is locked into position just beneath the GE junction. It is inspected to ensure that it is not overly tight.

Step 10. The gastric fundus is then secured over the band to the gastric pouch utilizing 2-3 interrupted 2-0 permanent sutures. These are best placed with intracorporeal suturing to ensure adequate tissue purchase. The "buckle" of the band should not be covered. Care must be taken not to damage the band or tubing with the needle which will lead to a saline leak. Patients with large amounts of mesenteric fat may have a very large GE fat pad that may need to be excised to properly place these sutures.

Step 11. The band is inspected to ensure that it is not too tight (spins easily and able to pass a 5 mm instrument under it). The buckle of the band should remain exposed.

Step 12. The tubing is exteriorized via the 15 mm trocar site and all trocars and retractors are removed under direct visualization.

Step 13. The patient is positioned supine and a suprafascial pocket through the 15 mm trocar incision is created to accommodate the subcutaneous port. Excess tubing is cut and the tubing is connected to the subcutaneous port. The port is secured to the fascia with four 0 Prolene interrupted sutures. The tubing is inspected to ensure no kinking or bending of the tubing as it entered the peritoneal cavity.

Step 14. This port incision is closed in two layers and covered with a Tegaderm™ dressing. All other skin incisions are closed with 4-0 Maxon and dressed appropriately.

Postop

Patients are started on clear liquids the night of surgery and should be discharged on a liquid diet for at least 7 days. Early postoperative nausea should be aggressively prevented to avoid band slippage.

Complications

Major complications include bleeding, splenic injury, gastric or esophageal perforation.

Follow-Up

The patient should be seen in 1 week to examine wounds and for postop dietary counseling.

Chapter 36

Small Bowel Resection and Anastomosis (Enterectomy): Open

Carla Pugh

Indications

Enterectomy with anastomosis is indicated in a variety of conditions, including congenital atresia or stenosis of the small bowel, blunt and penetrating injuries of the small bowel, benign and malignant small bowel tumors, bleeding Meckel's diverticulum, inflammatory bowel disease, intestinal fistula, intestinal gangrene, intussusception in adults, some cases of meconium ileus or intestinal duplication, mesenteric tumors that in the course of removal could produce vascular compromise to adjacent small bowel, and when small bowel is adherent to intra-abdominal tumors arising in other organs.

Preop

Antibiotic coverage should be started 30 minutes before the operation if elective. Antibiotics may be continued postoperatively if there is significant spillage of intestinal contents during the resection, gangrene, or infection. Decompression of preoperative bowel obstruction with nasogastric intubation when possible is very important. General anesthesia is usually employed. Patients should receive deep venous thrombosis prophylaxis appropriate to their level of risk.

Procedure

Step 1. The patient lies supine on the operating table, and the abdomen can be explored through a variety of incisions, depending on the pathology.

Step 2. A thorough systematic exploration of the abdomen should be done in most cases. There are times when this is not feasible, but one must guard against partial examinations that can miss important pathology.

Step 3. The area of small bowel containing the pathology should be isolated along with the appropriate section of mesentery. Towels can be used to wall off the area from the general abdominal cavity. Noncrushing intestinal clamps can be applied at some distance from the proposed resection lines to minimize spillage of the intestinal contents.

Step 4. A small opening in the mesentery immediately adjacent to the bowel wall should be made at the proposed proximal and distal resection sites, being careful to adequately expose the serosa of the bowel wall at the proposed resection sites.

Step 5. The peritoneum overlying the mesenteric leaf should be scored with scissors to outline the proposed line of resection of the mesentery. This resection line should extend from the small aperture adjacent to the bowel wall in a U-shaped fashion toward the other proposed resection margin. During this step it is important to transilluminate and/or palpate the mesenteric vessels so that only that portion of the vascular supply that is supplying the area of the proposed resected bowel is removed.

Northwestern Handbook of Surgical Procedures, 2nd Edition, edited by Nathaniel J. Soper and Dixon B. Kaufman. ©2011 Landes Bioscience.

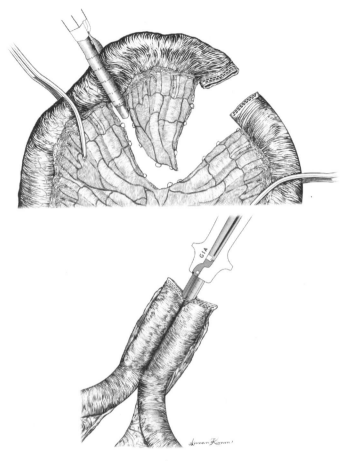

Figure 36.1. Small bowel resection and anastomosis.

Step 6. Starting with the space adjacent to the bowel at the proximal resection margin, the mesentery is divided between fine clamps along the U-shaped area to separate the mesentery of the resected bowel from the mesentery to be left behind. Mass ligatures should be avoided, and precise clamping and division of the individual vessels is preferred. Fine silk ligatures (4-0) are usually used depending on the size of the vessels. The extent of the mesenteric resection will depend on the amount of bowel to be removed. If a malignancy is present or suspected, an appropriate en bloc resection of mesenteric nodal tissue is required.

Step 7. The segment of the bowel to be removed is transected at each end with a linear stapler, which places two staggered rows of staples across the patient side and two more staggered rows on the specimen side of the bowel. The bowel is flattened and emptied, then the device is placed in a scissor-like fashion around the bowel wall and the staples fired. The built-in knife is then used to transect the bowel.

Step 8. The two ends of the bowel are placed adjacent to one another, and two sero-muscular sutures are placed to approximate the bowel loops. Care must be taken not to twist the mesentery of the bowel. A small section of the antemesenteric border of the staple line closure of both bowel ends is excised. The bowel openings should be large enough to introduce one anvil of the linear stapler into each limb of bowel. One fork of the instrument is placed into each side of the bowel, and the bowel ends aligned evenly on the forks. The instrument is closed and the staples are fired. Then the knife is used to divide the tissue between the staple line which produces the stoma between the bowel loops. After the instrument is removed, it is very important to inspect the staple line for any bleeding.

36

Step 9. The common bowel opening is closed using a 55 mm double-row stapling instrument. One traction suture should be placed at each end of the bowel to hold the entire area in the staple instrument until the instrument is fired. Care must be taken to overlap the ends of the previous staple lines. One must avoid direct apposition of the anastomotic staple lines.

Step 10. The lumen is checked for patency and any leaks. The adequacy of the blood supply to the bowel at the suture lines is checked. The mesenteric opening is carefully closed with absorbable 4-0 suture, being careful not to injure the vessels to the bowel at the area of the anastomosis.

Additional Considerations

1. Gentle handling of the tissues is imperative with anastomosis of grossly healthy tissue only. Open anastomosis may be used and the risk of infection and later anastomotic leaks minimized by the use of appropriate antibiotics and limiting spillage of the intestinal contents at the time of anastomosis.
2. Adequate resection of the involved segment beyond the area of pathology is important in some diseases, including a 5-6 cm margin of tissue especially for malignant tumors.
3. It is good practice to cut the bowel obliquely, removing a greater portion of tissue on the antemesenteric side of the bowel. This is particularly important if the mesenteric circulation is borderline.
4. One must be very cautious when attempting an anastomosis in segments of bowel that are dilated. An increased incidence of leakage may occur depending upon the degree of dilatation.

Postop

The patients are not allowed to eat until they pass gas or otherwise show signs of return of bowel function. Nasogastric suction is used in the early postoperative period until bowel function is evident.

Complications

A variety of complications can occur which depend on the patient's disease and state of health. Among the most serious problems is a leak at the anastomosis, which typically would occur at about postoperative day 5-7. Postoperative ileus or bowel obstruction may occur and must be differentiated. Bleeding and wound infection also occur early.

Follow-Up

Long-term follow-up is dependent on the nature of the disease. Resection of short segments of small intestine has few if any long-term sequelae.

Acknowledgment

The editors and author wish to acknowledge Joseph A. Caprini for contributing to the previous version of this chapter.

Chapter 37

Enterolysis for Small Bowel Obstruction: Open

Carla Pugh

Indications

Enterolysis is ordinarily performed for small bowel obstruction in a patient with adhesions from previous abdominal surgery, past intraperitoneal infections, or congenital adhesive bands. Occasionally, although rarely, it is indicated for chronic or recurrent signs of intestinal obstruction including bloating, vomiting, and severe abdominal cramps if radiological studies support a diagnosis of mechanical obstruction.

Preop

Patients with intestinal obstruction may be volume depleted and require significant fluid resuscitation, although this should be done expeditiously. Decompression of preoperative bowel obstruction with nasogastric intubation as much as possible is very important. Antibiotic coverage should be started before the operation and may be continued postoperatively depending upon whether or not an inadvertent injury to the bowel occurred during division of adhesions. Patients should receive deep venous thrombosis prophylaxis appropriate to their level of risk.

Procedure

Step 1. The patient lies supine on the operating table. The abdomen may be explored through the previous incision, but it is often wise to slightly extend the incision so that one can enter the peritoneal cavity in normal tissue. In these patients, adhesions between the bowel and the abdominal wall are frequently present, and bowel injury may occur when the peritoneum is opened.

Step 2. A thorough, systematic exploration of the abdomen should be done as a routine. There are times when this is not feasible, but one must guard against partial examinations that can miss important pathology. The surgeon should look for the transition point between dilated and decompressed small bowel which marks the point of obstruction.

Step 3. Traction and countertraction in a gentle fashion are used to stretch out adhesions and allow sharp dissection. Gentle handling of the tissues is imperative, along with avoiding excessive tension when dividing adhesions. Adhesions should be divided in avascular planes. The use of sharp dissection is generally preferred for longstanding adhesions. Blunt tissue dissection can tear adjacent structures including the bowel. When extensive adhesions are present, one must be very cautious to work from areas of known anatomy into the unknown scarred areas. When working in the pelvis, one should be particularly careful to avoid the ureter and iliac veins. Bowel loops should be protected with saline-soaked gauze pads when they are temporarily lifted out of the abdominal cavity. Saline should be used at body temperature. Cold saline can cause hypothermia, and the peritoneum and bowel loops can be damaged by excessively warm saline.

Northwestern Handbook of Surgical Procedures, 2nd Edition, edited by Nathaniel J. Soper and Dixon B. Kaufman. ©2011 Landes Bioscience.

Step 4. Examination of the entire small bowel is desirable, if possible, but is not absolutely necessary if the point of obstruction is obvious and dissection of the remaining bowel would be dangerous or excessively difficult. In patients with extensive adhesions, it is not necessary or desirable to take down all adhesions as long as it is clear that the adhesions causing the obstruction have been dealt with adequately. Areas of damage to the bowel serosa should be repaired with fine silk sutures. Once all of the adhesions have been divided as necessary to relieve the problem, the bowel is gently replaced in the abdominal cavity in as natural a position as possible.

Step 5. The abdominal cavity is irrigated with body-temperature saline, followed by closure of the abdominal wall with a large monofilament suture. An attempt should be made to cover the intestines with any available omentum to prevent bowel loops from coming into contact with the incision.

Postop

Nasogastric suction is used in the early postoperative period until bowel function is evident. The patients are allowed to initiate a diet once they pass flatus or otherwise show signs of return of bowel function.

Complications

Postoperative ileus and/or recurrent bowel obstruction may occur and must be differentiated. Wound dehiscence and wound infections are relatively common complications.

Follow-Up

The patient should be followed as an outpatient until wounds heal. Patients with a small bowel obstruction are at significantly increased risk of a recurrent obstruction.

Acknowledgment

The editors and author wish to acknowledge Joseph A. Caprini for contributing to the previous version of this chapter.

Chapter 38

Ileostomy: Open Loop

Steven J. Stryker

Indications

The most common indication for loop ileostomy is to protect a low pelvic anastomosis such as a coloproctostomy, coloanal anastomosis, or ileoanal anastomosis. Other indications include proximal diversion for an unresectable, obstructing malignancy, for a radiation stricture, for pelvic sepsis from an anastomotic leak or trauma, or for severe anorectal Crohn's disease.

38

Preop

For elective resectional surgery that incorporates a complementary loop ileostomy, a standard mechanical bowel prep along with perioperative intravenous antibiotics should be employed. The patient should meet with an enterostomal therapist to begin ileostomy instruction and to have the stoma site selected and marked.

Procedure

Step 1. The patient is positioned on the operating table as dictated by the access required for any associated procedure. (i.e., supine vs. synchronous combined lithotomy). The abdomen is prepped and draped and the stoma site mark is scratched with a sterile needle to facilitate subsequent identification.

Step 2. If the operation requires a bowel resection or drainage of pelvic sepsis, this is carried out first.

Step 3. Just prior to closing the abdominal incision, the segment of ileum to be exteriorized is chosen. Typically, this is 10-15 cm proximal to the ileocecal junction. When the loop ileostomy is used in conjunction with proctocolectomy and ileoanal anastomosis, the surgeon should identify the most distal segment of ileum which has sufficient mesenteric laxity to comfortably reach the surface of the abdomen at the previously marked stoma site.

Step 4. An oval of skin and the underlying subcutaneous fat is excised at the stoma site. The anterior rectus fascia is incised in a cruciate fashion, and the loop of ileum is drawn through the abdominal wall, passing through the rectus abdominis muscle. Care is taken to maintain the proper orientation of the loop and avoid twisting or angulation when exteriorizing the bowel.

Step 5. The loop is supported at the skin surface with a plastic rod slipped through a small opening in the mesentery adjacent to the bowel wall.

Step 6. The primary abdominal incision is closed completely. Then the loop ileostomy is matured by incising the distal limb transversely for about three-quarters of its circumference and folding it back over the proximal limb. The full thickness of the opened edge of bowel wall is then sewn to the dermis circumferentially with absorbable suture.

Northwestern Handbook of Surgical Procedures, 2nd Edition, edited by Nathaniel J. Soper and Dixon B. Kaufman. ©2011 Landes Bioscience.

Step 7. A stoma faceplate is cut to the corresponding size and placed over the newly fashioned stoma.

Postop

The patient is fed upon resumption of bowel activity. When the ileostomy begins to pass stool and flatus, the enterostomal therapist begins instruction in the care of the stoma. The peristomal skin is monitored for signs of irritation, which can be due to a poorly fitting stomal appliance. The skin margin adjacent to the distal limb opening is particularly prone to irritation with a loop ileostomy. The plastic rod supporting the loop ileostomy is typically removed prior to discharge from the hospital.

Complications

Early complications of loop ileostomy include excessive ileostomy output and dehydration, bleeding from the mucocutaneous junction, peristomal wound infection, and stomal retraction. Later complications include prolapse, peristomal hernia, fistula, or internal hernia.

Follow-Up

If the loop ileostomy is formed to protect a distal anastomosis, it is best to wait about 10 weeks before closure. Prior to closing the stoma, a contrast radiograph of the distal anastomosis is obtained to assess the adequacy of healing. If the ileostomy was created because of intra-abdominal or pelvic pathology, the stoma closure is deferred until the primary process has resolved.

Chapter 39

Open Feeding Jejunostomy

Carla Pugh

Indications

Enteral or parenteral feeding is advised when a patient is unable to eat for 7-14 days or longer. In the setting of a functional gut, enteral feeding is preferred. When the need for enteral feeding is anticipated to be 20 days or shorter, feedings through a nasogastric tube or a more distal nasoenteric tube are usually appropriate. However, because such tubes are associated with considerable discomfort and sinusitis and epistaxis are common complications, direct enteral access is preferred when feeding needs extend beyond 20 days. A feeding jejunostomy may also be indicated at the time of an abdominal operation for patients with known preoperative nutritional deficiencies and expected upper GI dysfunction.

Preop

Hydration and prophylactic intravenous antibiotics are indicated.

Procedure

Step 1. The patient is placed in supine position on the operating table. Under local or general anesthesia, the abdomen is entered through a 6-8 cm midline or paramedian incision. When using the paramedian incision, the anterior rectus sheath is incised and the rectus muscle is retracted laterally. The posterior sheath and peritoneum are then incised.

Step 2. The operating table is placed in reverse Trendelenberg and the table rotated to the right to facilitate exposure. A point on the jejunum 30-40 cm from the ligament of Treitz is identified. A 3-0 pursestring suture is placed, creating a circular pattern with at least a 2.5 cm diameter.

Step 3. A stab wound is placed in the center of the pursestring, and a feeding jejunostomy tube (8-16 French) is passed distally into the bowel. The pursestring suture is secured around the tube. The catheter is then tunneled along the bowel wall by placing four to six, 4-0 silk vertical mattress sutures. The first stitch is placed near the entrance site of the tube. Subsequent stitches are placed proximally, keeping the tube flat against the bowel wall. The tails of the last stitch are wrapped around the tube and tied to secure the tube in place.

Step 4. A stab wound is made in the left upper quadrant of the abdomen, and the feeding tube is pulled through with the assistance of a Kelly clamp. The jejunum is sutured to the anterior abdominal wall with three 3-0 silk sutures. One suture is placed near the tube exit site, and the others are placed proximally and distally. This is performed to prevent torsion.

Step 5. The abdominal incision is closed with an 0 or 1-0 suture. The tube is secured to the abdominal wall with a 4-0 nylon and flushed with saline to ensure patency. Low volume tube feeds may be started at once and advanced as tolerated.

Northwestern Handbook of Surgical Procedures, 2nd Edition, edited by Nathaniel J. Soper and Dixon B. Kaufman. ©2011 Landes Bioscience.

Postop

When the feeding jejunostomy is performed as part of larger abdominal operation, nasogastric suction is used in the early postoperative period. Patients must be monitored for signs of obstruction or torsion. Postoperative ileus may delay the start of tube feeds. Local wound care is indicated.

Complications

Torsion is one of the most common complications. Radiographic imaging will show the feeding tube facing in the wrong direction (proximally) and possibly a partial small bowel obstruction. If the patient becomes obstructed, tube removal and reoperation may be necessary. Other complications relate to the tube feeds. A small percentage of patients develop ischemic necrosis of the small bowel just distal to the tube. In this case tube feeds must be discontinued and reoperation is mandatory.

Follow-Up

The patient should be followed as an inpatient until tube feeds are at goal. Outpatient follow-up may follow the course of that planned for the patient's primary diagnosis.

39

Chapter 40

Appendectomy: Open

Alexander P. Nagle

Indications

Appendectomy is indicated in patients for acute appendicitis. The patient's history and physical examination remain the most important aspects of diagnosing acute appendicitis. However, the common signs and symptoms such as fever, anorexia, and right lower quadrant pain are often absent; up to 33% of patients have an atypical presentation. Diagnostic accuracy varies by sex and age. The diagnosis of acute appendicitis may be difficult in women of childbearing age because symptoms of acute gynecologic conditions may manifest similarly. In selected clinical settings, a CT scan can improve the diagnostic accuracy up to 95%.

Preop

Correction of electrolytes, fluid resuscitation, and a Foley catheter are standard practice. Broad-spectrum antibiotics including adequate anaerobic coverage are administered. Both single- and multiple-agent therapy have been shown to be effective. Lower extremity sequential compression devices are used for deep venous thrombosis prophylaxis.

Procedure

Step 1. A right lower quadrant skin incision is made in the direction of the skin lines. The incision should be centered over McBurney's point (two-thirds of the way along a line drawn from the umbilicus to the anterior superior iliac spine).

Step 2. The external oblique fascia is divided. The three muscle layers of the abdominal wall are split in the direction of their fibers and retracted. The peritoneum is opened transversely.

Step 3. As the peritoneal cavity is entered, any fluid should be noted, aspirated, and, if purulent, sent for a gram stain and culture. Any remaining fluid is then suctioned from the field.

Step 4. The cecum is identified, grasped, and exteriorized through the wound using a gauze sponge or blunt bowel graspers. The base of the appendix is identified at the confluence of the teniae coli. Often a finger can be used to sweep the appendix into the operative field. If the appendix is retrocecal, the cecum should be mobilized medially by incising the lateral peritoneal attachments of the cecum.

Step 5. Using a curved hemostat, a mesenteric window is created in an avascular area of the mesoappendix near the base of the appendix. The mesentery of the appendix is then divided between hemostats. It is important to correctly identify the appendiceal artery and assure that it is properly ligated. Any attachments at the appendiceal base are dissected in order to clearly identify the site of transection of the appendix.

Northwestern Handbook of Surgical Procedures, 2nd Edition, edited by Nathaniel J. Soper and Dixon B. Kaufman. ©2011 Landes Bioscience.

Step 6. The appendix is divided approximately 5 mm from the cecal wall using a suture ligature. A clamp is placed distal to the suture ligature. Using a sharp scalpel, the appendix is transected between the suture ligature and the clamp. The mucosa of the appendiceal stump is lightly touched with electrocautery.

Step 7. In some cases of severe acute necrotizing appendicitis, the base of the appendix may not be suitable for transection and it may be necessary to perform a partial cecectomy.

Step 8. It is important to inspect the cut edge of the mesoappendix and appendiceal stump for integrity and hemostasis.

Step 9. The lower abdominal cavity is irrigated thoroughly with normal saline. No drains are required for cases of acute appendicitis.

Step 10. Closure of the fascial layers of the wound is carried out using running or interrupted absorbable suture. The skin and subcutaneous tissue are usually closed in acute appendicitis. In gangrenous and perforated appendicitis, the wound may be left open.

Special Considerations

The debate continues regarding the management of a normal-appearing appendix found at the time of surgery for presumed appendicitis. There are three possible situations: (1) normal-appearing appendix and no other pathology, (2) normal-appearing appendix and medically treated pathology (e.g., inflammatory bowel disease or pelvic inflammatory disease), and (3) normal-appearing appendix and surgically treated pathology (e.g., acute cholecystitis or perforated duodenal ulcer). The appendix should be removed in all but the last circumstance. This eliminates confusion should the patient return with similar symptoms.

Postop

Uncomplicated acute appendicitis requires minimal postoperative care. A significant decrease in pain and the return of bowel function can be anticipated by the day following operation. Diet is instituted as tolerated. Most patients can be discharged on postoperative day one. Oral opiates are used for pain control. For patients with a perforated appendix, antibiotics are generally continued for at least 5 days.

Complications

Acute infectious complications include abdominal abscess formation, cecal fistulas, and wound infections. The most important determinant of wound infection is the severity of contamination at the time of surgery. Wound infection rates of 1-4% occur with negative exploration. The risk increases to 5-12% with gangrenous appendicitis and 20-50% with perforated appendicitis. Antibiotic coverage for anaerobic organisms and a policy of open wound management after severe contamination reduce the risk of wound infection substantially. The mortality of acute appendicitis is contingent upon the state of the appendix at exploration and the age of the patient. Nonperforated appendicitis carries a 0.1% mortality rate and most deaths are due to intercurrent illness. Perforated appendicitis has a mortality rate of approximately 1% overall, usually secondary to infectious complications. In the elderly population, high rates of perforation and intercurrent illness contribute to a mortality rate in excess of 5%.

Follow-Up

The patient should return to see their surgeon 10-14 days after their operation. At that time, the pathology report should be reviewed; the presence of malignancy and carcinoid should be excluded. If no cancer is found within the specimen, no further follow-up is needed.

Chapter 41

Appendectomy: Laparoscopic

Mark Toyama

Indications

Laparoscopic appendectomy is indicated for the treatment of acute appendicitis or for interval appendectomy following nonoperative treatment of complicated appendicitis.

Preop

Standard intravenous antibiotic prophylaxis against wound infection is given preoperatively. The patient is placed in the supine position. For female patients, lithotomy position may be chosen, which allows manipulation of pelvic organs. A Foley catheter is inserted in the urinary bladder after the induction of anesthesia. Arms are tucked at the patient's side. The surgeon stands on the patient's left side.

Procedure

Step 1. The peritoneal cavity is accessed at the umbilicus (using either a Veress needle or a Hasson cannula), and a 5 mm or 10 mm trocar and laparoscope are placed.

Step 2. A 12 mm trocar is placed in the midline above the symphysis pubis. The incision can be hidden in the pubic hairline.

Step 3. A 2 mm or 5 mm trocar is placed in the right upper quadrant.

Step 4. An additional trocar can be placed as needed to improve exposure.

Step 5. The patient is then placed in Trendelenburg position and rotated towards the left to help expose the cecum.

Step 6. Instruments are placed and the appendix inspected.

Step 7. If the appendix is normal, a search for other sources of right lower quadrant pain should be undertaken, including adnexa in females, distal small bowel, gallbladder, etc.

Step 8. If the appendix is to be removed, it is elevated with an instrument and dissected free of any adhesions. It is often helpful to start at the appendiceal base.

Step 9. The appendix is elevated with a grasper or a pre-tied ligature to expose the mesoappendix.

Step 10. The mesoappendix is isolated with a dissector by making a window between it and the appendix.

Step 11. The mesoappendix is divided with clips, harmonic shears, or a laparoscopic stapling device.

Step 12. The appendix is divided at its base with a laparoscopic linear stapling device or with pre-tied ligatures.

Step 13. The appendix is placed in a specimen bag and removed through the largest port site.

Northwestern Handbook of Surgical Procedures, 2nd Edition, edited by Nathaniel J. Soper and Dixon B. Kaufman. ©2011 Landes Bioscience.

41

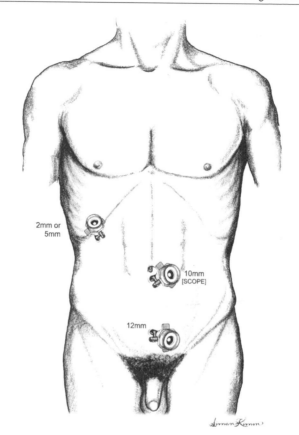

Figure 41.1. Laparoscopic appendectomy. Abdominal trocar placement.

Step 14. The abdomen is irrigated and suctioned. The appendiceal base and the mesoappendix should be inspected for hemostasis. The fascia of all port sites 10 mm in size or larger is closed.

Postop
Diet may be started when the patient is awake and alert if not nauseated. Discharge is often possible in less than 24 hours.

Complications
Postoperative complications include hemorrhage, periappendiceal abscess formation, wound infection, appendiceal stump leak with fistula formation. Incomplete appendectomy may become manifest much later with recurrent symptoms.

Follow-Up
Follow-up should occur in 1 to 2 weeks after discharge to inspect the incisions and review the pathology report.

Chapter 42

Hemicolectomy (Right): Open

Anne-Marie Boller

Indications

Open right hemicolectomy is indicated for treatment of selected benign and malignant diseases of the ascending colon and appendix.

Preop

Preoperative colonoscopy to evaluate the entire colon and tattoo small lesions is advised. Preoperative marking of potential stoma sites is prudent in the instance that creation of a stoma is needed. Mechanical bowel preparation is a matter of surgeon preference but is no longer an absolute mandate in colon surgery. A nasogastric tube need not be placed unless the patient presents with an obstruction but an orogastric tube should be placed once the patient is asleep. Sequential compression devices are applied to all patients and subcutaneous heparin administered. Prophylactic antibiotics should be given 30 minutes prior to the incision. A Foley catheter should be inserted.

Procedure

Step 1. The patient is placed in the supine position. After induction of general anesthesia, a midline incision is made that extends above and below the umbilicus.

Step 2. The abdomen is explored. Under elective circumstances, when operating for colon cancer, perform an intraoperative ultrasound of the liver.

Step 3. Place a self-retaining retractor.

Step 4. Incise the peritoneal reflection of the right colon with electrocautery. Continue the mobilization distally to include the hepatic flexure. It may be necessary to ligate a few blood vessels at the hepatic flexure.

Step 5. Mobilizing the colon anteriorly, expose the retroperitoneum and identify the right ureter and duodenum. Avoid using electrocautery near these structures.

Step 6. Mobilize the ileum distally as it joins the cecum.

Step 7. Transect the ileum approximately 10 cm proximal to the ileocecal valve using an automated linear cutting stapler.

Step 8. Transect the colon just distal to the hepatic flexure using a second load on the stapling device.

Step 9. Score the peritoneum covering the mesocolon between the two points of transection. This may be done with electrocautery. This should be carried down to the origin of the right colic artery if the operation is being performed for cancer. For benign conditions, a lymphadenectomy is not necessary and may increase the chance of injuring a retroperitoneal structure such as the ureter.

Northwestern Handbook of Surgical Procedures, 2nd Edition, edited by Nathaniel J. Soper and Dixon B. Kaufman. ©2011 Landes Bioscience.

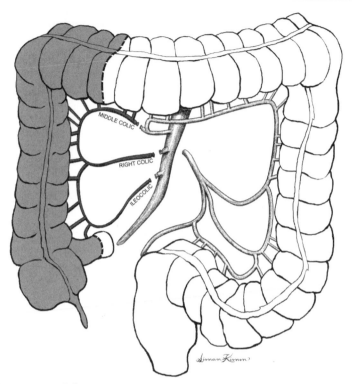

Figure 42.1. Right hemicolectomy.

Step 10. Clamp, ligate, and divide the mesenteric vessels with 2-0 silk suture or an appropriate device for sealing and dividing the vascular pedicle such as the EnSeal® or LigaSure™ device.

Step 11. Remove the specimen from the field and provide orientation for the pathologist. Open the specimen to inspect the pathology and to be sure the suspected lesion has been removed with an adequate margin.

Step 12. Create a functional end-to-end anastomosis by firing a third load of the automated stapling device between the ileum and transverse colon. This is accomplished by excising the corners of each stapled end and stapling the two pieces of bowel together along the antimesenteric border. Make certain that the bowel is not kinked or twisted.

Step 13. Inspect the intraluminal staple line to confirm hemostasis. If necessary, control bleeding with a 3-0 silk gastrointestinal (GI) suture.

Step 14. Close the ileocolotomy using a last reload of the stapling device. Prior to closing, line up the edges of the bowel using Allis clamps to ensure that the stapling device will encompass the entire opening and will reapproximate all layers of the bowel. Oversew the staple line with interrupted 3-0 Vicryl sutures.

Step 15. Place a 3-0 silk Vicryl suture at the distal corner of the anastomosis to relieve any tension.

Step 16. Irrigate the abdominal cavity with 2-4 L of warm saline. Confirm hemostasis. Confirm laparotomy pad counts, needle counts, and instrument counts.

Step 17. Close the abdominal wall fascia with two running looped O Maxon sutures. Close the skin with 4-0 monocryl subcuticular sutures or skin staples.

Step 18. The Foley catheter and orogastric tube may be removed prior to awakening the patient.

Postop

Intravenous antibiotics are continued for 24 hours. Deep venous thrombosis prophylaxis should be continued until the patient is ambulating. Intravenous fluid is continued until adequate oral intake has been established. The diet should be advanced according to the patient's appetite, lack of nausea and emesis, presence of bowel sounds, and bowel activity. The presence of nausea, hiccups, bloating, and anorexia should alert the physician that the patient should remain NPO. Most patients will be able to start clear liquids on the first day after surgery and advanced thereafter as tolerated. If there are skin staples, they may be removed on postoperative day 7-10. This may occur in the clinic, depending upon the pace of the recovery.

42

Complications

The main complications associated with this procedure include wound infections; pelvic, subphrenic, or subhepatic abscess; anastomotic dehiscence; abdominal wall dehiscence; and prolonged postoperative ileus.

Follow-Up

For benign conditions (adenomatous polyps), accepted screening measures should be followed. For malignancy, the patient should undergo colonoscopy one year following operation and enter lifelong follow-up because of the increased risk of a second colon cancer.

Chapter 43

Hemicolectomy (Right): Laparoscopic

Anne-Marie Boller

Indications

Laparoscopic right colectomy is indicated when removal of the entire right colon is required for benign or malignant diseases of the right colon or appendix. Large adenomatous polyp which cannot be removed endoscopically are adequately addressed with laparoscopic removal. Recent randomized, controlled clinical trials have proven laparoscopic techniques equivalent to open right hemicolectomies in the setting of malignant diseases.

Preop

Preoperative colonoscopy to evaluate the entire colon and tattoo small lesions is advised. Preoperative marking of potential stoma sites is prudent if a stoma is needed. Mechanical bowel preparation is a matter of surgeon preference but is no longer an absolute mandate in colon surgery. A nasogastric tube should be placed once the patient is asleep. Sequential compression devices are applied to all patients and subcutaneous heparin administered. Prophylactic antibiotics should be given 30 minutes prior to the incision. A Foley catheter should be inserted.

Procedure

Step 1. The patient should be positioned supine with both hands tucked at the sides. A Foley catheter and a naso/orogastric tube are placed in all patients. Standard skin cleansing is performed.

Step 2. A 10 mm laparoscope is used with two 5 mm trocars and one 10 mm trocar.

Step 3. Access to the peritoneal cavity is obtained for insufflation at the umbilicus. This can be accomplished via either closed (Veress needle) or open (Hasson cannula) technique. If the closed technique is used, then a "saline drop test" should be performed to ensure that the needle is in the peritoneal cavity. Subsequent trocars are placed under direct visualization. One trocar is placed in the epigastric region or the left lower quadrant, and the other is placed in the suprapubic region.

Step 4. The abdomen is explored using two atraumatic bowel graspers (Babcock-type) to manipulate bowel. A special effort should be made to examine the surfaces of the liver.

Step 5. The patient is placed in severe Trendelenberg with their right side elevated. The appendix is grasped with the atraumatic bowel clamp placed through the left lower quadrant port and retracted cephalad toward the left upper quadrant. The ultrasonic shears placed through the suprapubic trocar are used to mobilize the cecum by dividing the lateral peritoneal attachments starting at the base of the cecum. This dissection is carried cephalad along the white line of Toldt. The peritoneum overlying the mesentery of the terminal ileum is divided and blunt dissection is used to mobilize

Northwestern Handbook of Surgical Procedures, 2nd Edition, edited by Nathaniel J. Soper and Dixon B. Kaufman. ©2011 Landes Bioscience.

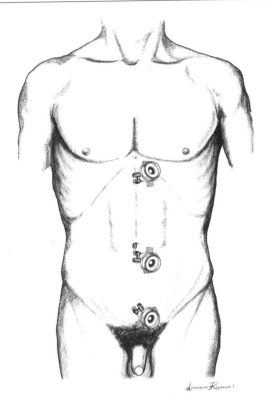

43

Figure 43.1. Laparoscopic right hemicolectomy. Trocar placement. The epigastric port may also be placed in the left lower quadrant.

the mesentery of the terminal ileum. The terminal ileum should be mobilized for at least 10-15 cm. During this dissection, special effort is made to identify and preserve the right ureter and gonadal vessels.

Step 6. The patient's position is changed to reverse Trendelenberg. The hepatic flexure is grasped with the atraumatic bowel clamp placed through the suprapubic port and retracted caudad. The ultrasonic shears placed through the left lower quadrant port are used to mobilize the hepatic flexure. The hepatocolic attachments are divided using the ultrasonic shears. Surgical clips may be necessary to control bleeding of large veins. Special care should be taken to avoid injury to the right kidney, duodenum, and right adrenal gland. The right half of the gastrocolic attachments is similarly divided, again taking special care to avoid injury to the duodenum and the middle colic vessels. Adequate mobilization of the hepatic flexure and the right transverse colon is essential to permit extraction of the colon for extracorporeal resection and anastomosis.

Step 7. The right colon is mobilized to the level of the medial border of the right psoas muscle or vena cava. At this point, the mesentery can be divided using laparoscopic linear stapling devices or an electronic device such as EnSeal® or a LigaSure™ device. Alternatively the mesentery can be divided extracorporeally.

124

Northwestern Handbook of Surgical Procedures

Step 8. The 10 mm incision at the level of the umbilicus is elongated to approximately 5 cm. This is carried down through the fascial and muscle layers using a muscle-splitting technique (as in an appendectomy). The peritoneum is opened, and a wound retractor is placed to facilitate the delivery of the right colon through the incision.

Step 9. The proximal and distal resection margins are divided using linear staplers in the same fashion as in the open procedure. If the mesentery was not divided intracorporeally, it can be divided in continuity between clamps and the proximal vessels can be ligated.

Step 10. A side-to-side (functional end-to-end) anastomosis is performed using a linear cutting stapler. The open end of the anastomosis can be closed with a linear cutting stapler, a non-cutting stapler, or a handsewn technique. The mesenteric defect does not require closure. The bowel is returned to the peritoneal cavity. Special care should be taken to be certain the bowel or the anastomosis is not traumatized while returning it to the peritoneal cavity.

Step 11. The fascia is reapproximated using O-Maxon in a running fashion. All layers are irrigated prior to closure to minimize the risk of wound infection.

Step 12. The abdomen is reinsufflated and explored using the laparoscope to examine for hemostasis. The anastomosis is examined, and the proper lay of the mesentery is ensured. Any problems can be dealt with laparoscopically.

Step 13. Skin incisions are closed with an absorbable, monofilament subcuticular suture closure.

Postop

The Foley catheter and naso/orogastric tubes are removed in the OR. Standard intravenous fluids are administered. Clear liquids are started on postoperative day one. If clear liquids are tolerated, diet is advanced the next day to a general diet and the patient can be discharged.

Complications

Anastomotic leak usually manifests on postoperative days 7-10 by either peritonitis (large leak) or occult fever or failure to progress (small leak with possible abscess). Wound infection is uncommon; if it occurs, it is generally in the transverse incision through which the colon was removed. Intraluminal bleeding from the anastomosis can sometimes be managed conservatively but may require reoperation.

Follow-Up

Follow-up is dictated by the underlying condition for which the colon resection was performed.

43

Chapter 44

Sigmoid Colectomy: Open

Steven J. Stryker

Indications

Indications for sigmoid colon resection include cancer, sessile polyps, diverticular disease, localized Crohn's colitis, volvulus, ischemic colitis, and as an adjunct to suture rectopexy for rectal prolapse.

Preop

When resection of diseased sigmoid is contemplated, it is desirable to rule out significant pathology in the remainder of the colon. In the absence of obstruction or acute inflammation, this is best accomplished by colonoscopy. A mechanical bowel prep includes a clear liquid diet for 48 hours preoperatively, in conjunction with magnesium citrate or a polyethylene-glycol (PEG) prep. Oral antibiotics are of no proven added advantage when perioperative intravenous antibiotics are utilized. Pneumatic compression sleeves for the lower extremities are used to minimize the risk of deep venous thrombosis.

44

Procedure

Step 1. The patient is placed in dorsal lithotomy position allowing simultaneous access to the abdomen and perineum. The patient is positioned awake to check for comfort in positioning with respect to the back, hips, and knees. Once the patient confirms that the positioning is comfortable, general anesthesia is induced. Care should be taken that there is no excessive pressure on the calves or the lateral aspect of the proximal leg after positioning to avoid peroneal nerve injury and compartmental syndrome postoperatively.

Step 2. The abdomen and perineum are widely prepped with a povidone/alcohol combination prep. The abdomen and perineum are draped to provide wide access to both areas.

Step 3. A lower midline incision is made, taking care to divide the midline fascia down to the pubic symphysis. A thorough abdominal exploration is undertaken to assess the extent of tumor involvement.

Step 4. The lateral and medial peritoneal reflections of the sigmoid colon are incised down past the rectosigmoid junction at the level of the sacral promontory. The left ureter is identified and displaced laterally along with the gonadal vessels.

Step 5. The superior hemorrhoidal vessels and distal sigmoidal vessels are ligated proximally at the base of the mesentery, taking care to identify and avoid the left ureter during this maneuver.

Northwestern Handbook of Surgical Procedures, 2nd Edition, edited by Nathaniel J. Soper and Dixon B. Kaufman. ©2011 Landes Bioscience.

Step 6. The descending colon and proximal rectum are mobilized as necessary to allow a tension-free anastomosis. This may require mobilization of the splenic flexure into the lesser sac. The mesocolon is divided up to the mesenteric aspect of the proximal margin. The bowel wall at the proximal margin of resection is cleared of pericolic fat for a short distance, and a pursestring suture placed on the proximal end as the bowel is transected. The anvil cap of an end-to-end stapling device is inserted into the open proximal end, and the pursestring suture is tied.

Step 7. The distal margin of resection is chosen, and the mesocolic fat is cleared for a short distance at this point. The bowel is divided with a stapling device in preparation for a double-stapling technique. The specimen is sent for pathologic analysis.

Step 8. The end-to-end stapler is placed transanally and carefully advanced to the stapled rectosigmoid junction. The trocar at the end of the stapler is advanced through the staple line just anterior or just posterior to its midpoint. The anvil cap and trocar end of the stapler are connected. The stapler is closed, fired, and slowly withdrawn through the anus.

Step 9. A proctoscope is inserted into the rectum and air is insufflated as the colon is occluded proximal to the anastomosis. The pelvis is filled with saline and the anastomosis is checked for an air leak. Minor air leaks are repaired. If no leak is seen, the rectum is deflated and the saline aspirated from the pelvis.

Step 10. The midline incision is closed in layers.

Postop

Intravenous antibiotics are continued for 24 hours postoperatively. Subcutaneous heparin is administered postoperatively. The patient ambulates on the evening of surgery. A urinary catheter is left in for 24-48 hours. Clear liquids are begun orally upon resumption of bowel activity.

Complications

Early postoperative complications include: atelectasis, intra-abdominal or wound infection, urinary tract infection, postoperative ileus, hemorrhage, venous thrombosis, anastomotic dehiscence, or ureteral injury. Late complications include adhesive small bowel obstruction and anastomotic stricture.

Follow-Up

Patients are checked in the office at 2-4 weeks postdischarge to assess wound healing and overall progress. If the resection has been performed for malignancy, the patient is seen at 3-month intervals for the first 2 years, at 4-month intervals for the 3rd year, and at 6-month intervals during the 4th and 5th years of follow-up.

Chapter 45

Sigmoid Colectomy: Laparoscopic

Alexander P. Nagle

Indications

Laparoscopic sigmoid resection is indicated for both benign and malignant diseases of the sigmoid colon. Indications for benign disease include diverticulitis, large polyps, localized Crohn's colitis, ischemic colitis, and as an adjunct to suture rectopexy for rectal prolapse.

Preop

In addition to standard history and physical, a patient's suitability for laparoscopic sigmoid colectomy is determined by a combination of patient factors and the surgeon's experience. Challenging patient factors include pulmonary or cardiac conditions that limit the patients' ability to tolerate pneumoperitoneum, morbid obesity, prior abdominal surgery, prior radiation, inflammatory process and abscess. Conversion rates are always higher under these circumstances and should be discussed with the patient during the informed consent process. Large bulky tumors should not be approached laparoscopically. The use of fiberoptic illuminating ureteral stents should also be considered. These stents facilitate rapid identification of the ureter and can be particularly helpful for cases of diverticulitis. A mechanical bowel prep is indicated. Preoperative antibiotics and DVT prophylaxis should be utilized.

Procedure

Step 1. The patient is positioned in the dorsal lithotomy position allowing simultaneous access to the abdomen and perineum. Care is taken to avoid excess pressure on the calves or the lateral aspect of the proximal leg to avoid peroneal nerve injury. The upper extremities are tucked to the patient's sides. The patient should be securely strapped to the table to allow tilting and manipulation of the operating table. A bean-bag on the operating table is also helpful to stabilize the patient.

Step 2. Trocar position is variable, depending on personal preference, but the majority of surgeons will use a trocar placed above the umbilicus for the camera port, a trocar in the left mid lateral quadrant, and two trocars on the right side to provide the surgeon with two hands for operating. If a hand-assisted approach is used, either a Pfannenstiel or vertical midline suprapubic incision can be created for the hand-access port.

Northwestern Handbook of Surgical Procedures, 2nd Edition, edited by Nathaniel J. Soper and Dixon B. Kaufman. ©2011 Landes Bioscience.

Step 3. The descending colon and sigmoid colon are mobilized by starting at the lateral peritoneal reflection (the white line of Toldt) and progressing medially to mobilize the colon off of the retroperitoenum. It is important to provide adequate retraction to visualize the appropriate avascular plane. It is also preferable not to grab to colon directly but rather grasp the peri-colic fat (epiploica). The descending colon should be fully mobilized in order to minimize tension when performing the colorectal anastomosis. In most cases of sigmoid resection, complete mobilization the splenic flexure is not necessary. An alternative approach of colon mobilization is the medial-to-lateral approach, which involves identification and control of the vascular pedicles prior to division of the lateral attachments.

Step 4. It is critically important to identify the ureter. It is often easiest to identify the ureter as it crosses anterior to the iliac artery and runs along the pelvic side wall.

Step 5. Identification and division of the mesenterie vessels. The superior hemorrhoidal vessels and distal sigmoidal vessels are ligated proximally at the base of the mesentery, taking care to identify and avoid the left ureter during this maneuver. In cases of cancer, it is important to divide the inferior mesenteric artery (IMA) close to its take-off from the aorta. The IMV is divided below the inferior border of the pancreas or above the left colic vein. The IMV is identified to the left of the IMA or in case of difficulty, higher, to the left of the ligament of Treitz (duodenojejunal flexure).

Step 6. Distal transection of the colon. The distal margin of resection is chosen, and the mesocolic fat is cleared for a short distance. The bowel is transected utilizing a laparoscopic roticulating GIA linear stapler. In cases of diverticulitis it important (to minimize recurrence) that the distal line of resection is on the rectum and that no sigmoid is left behind. For cancer, the distal line of resection should be least 5 cm below the tumor.

Step 7. Exteriorization of the specimen and proximal line of resection. A 4-5 cm extraction incision is made in the midline or low transverse position. A wound protector is recommended and provides excellent exposure. The sigmoid colon is delivered through the extraction site. The proximal line of resection is chosen and the colon is divided. The specimen is sent to pathology. In cases of diverticulitis, the bowel wall is palpated and the proximal line of transection is chosen where the bowel wall feels normal, i.e., no muscular hypertrophy or thickening. The left colon is sized with gradual dilators and the appropriate size EEA stapler is chosen. The EEA anvil is inserted into the left colon and secured with a pursestring suture. The left colon is returned to the peritoneal cavity, and the fascia of the extraction site is closed with permanent suture.

Step 8. Colorectal anastomosis. Pneumoperitonium is re-established. The EEA stapler is inserted transanal into the rectum and advanced under laparoscopic visualization to the rectal staple line. Once appropriately positioned the center spike of the stapler is extended and penetrates through the wall of the rectum. The anvil in the left colon is then connected to the spike of the stapler. Under laparoscopic visualization, the stapler is closed into firing range and the anastomosis is performed. The stapler is gently removed and the doughnuts are inspected. A completion rigid proctoscopy is performed and the rectum is insufflated to gently distend the anastomosis while it is submerged under saline irrigation.

Postop

Intravenous antibiotics are continued for 24 hours postoperatively. The patient ambulates on the day following surgery. A urinary catheter is left in for 24-48 hours. Clear liquids are begun orally upon resumption of bowel activity.

Complications

Early complications include: atelectasis, intra-abdominal or wound infection, urinary tract infection, postoperative ileus, hemorrhage, venous thrombosis, anastomotic dehiscence, or ureteral injury. Late complications include adhesive small bowel obstruction and anastomotic stricture.

Follow-Up

Patients are checked in the office at 2-4 weeks postdischarge to assess wound healing and overall progress. If the resection has been performed for malignancy, the patient is seen at 3-months intervals for the first 2 years, at 4-month intervals for the 3rd year, and at 6-month intervals during the 4th and 5th years of follow-up.

45

Chapter 46

Colostomy: Transverse Loop

Wilson Hartz

Indications

Transverse loop colostomy is created for fecal diversion and/or left colon decompression. It is indicated for protection of low anterior anastomoses or diversion from chronically, nonhealing anal conditions such as complex fistulas, pressure sores or chronic diverticular perforations in the presence of multiple co-morbid conditions contraindicating primary resection. It is also a remedy for left colon obstruction, especially when an obstructing lesion and a competent ileocecal valve create a closed loop.

Preop

In obstructed patients, fluid and electrolyte imbalances should be corrected to the extent possible in an expeditious manner prior to surgery. In general, it is desirable to preoperatively verify the presence and site of a mechanical obstruction using CT scanning w/rectal contrast, water soluble contrast enema, or colonoscopy.

A site for colostomy is chosen overlying the right rectus muscle, slightly above the level of the umbilicus, 2-3 fingers below the right costal margin. Prophylactic antibiotics should be given preoperatively.

Procedure

Step 1. The patient is positioned supine and general anesthesia is used to allow for adequate manipulation and exploration. Prep and drape the entire abdomen.

Step 2. A 10 cm transverse incision is made over the right rectus muscle just above the level of the umbilicus. It must be large enough to allow adequate access to the colon, especially if it is dilated.

Step 3. A transverse incision is made in the anterior rectus sheath and the rectus muscle is freed from the fascia above and below the incision, creating fascial flaps. The rectus muscle is spread vertically. 0-Proline sutures are placed at each end of the fascial incision, to be tied or removed during closure. The posterior sheath and peritoneum are incised transversely, and 0-Vicryl sutures are placed at the ends of that incision to be tied or removed during closure.

Step 4. The colon is found by following the surface of the omentum to the transverse colon. The *avascular* plane is incised to free the omentum and skeletonize the colon. The mesocolon near the colon wall is then opened in an *avascular* area, and a penrose drain is pulled through it to sling the colon.

Step 5. If the colon is massively dilated and thin, it can be decompressed with a large bore hypodermic needle puncture or by placing a pursestring in the anterior wall and decompressing with a pool suction. The closure will then be incorporated into the final stoma.

Northwestern Handbook of Surgical Procedures, 2nd Edition, edited by Nathaniel J. Soper and Dixon B. Kaufman. ©2011 Landes Bioscience.

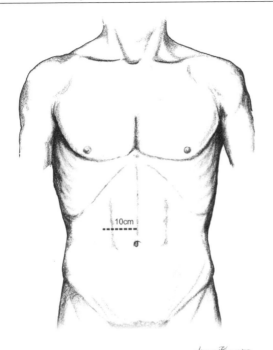

10cm

Figure 46.1. Transverse loop colostomy.

Step 6. Using the Penrose for traction, the transverse loop is brought up through the incision, being careful to maintain proper orientation. It is important to bring enough colon out of the abdominal cavity to eliminate tension once the bridge has been placed under the loop. Tension is reduced by lengthening the loop by mobilizing the hepatic flexure, dividing retroperitoneal attachments, or by making sure the segment utilized is far enough medial to the flexure to be more mobile.

Step 7. The sutures previously placed in the posterior sheath are tied to make the opening snug around the colon and a fingertip. Unnecessary sutures are removed. Similarly the anterior rectus sheath sutures are tied or removed creating a snug closure around the colon. 3-0 Vicryl suture can be used to tack the seromuscular colon to the facial defect to prevent herniation.

Step 8. Subcutaneous tissue is closed with interrupted 3-0 Vicryl. Skin is closed with 3-0 nylon vertical mattress sutures, easily allowing a finger to be passed through adjacent to the colon.

Step 9. The Penrose is withdrawn, pulling a bar placed into the drain, along with it. The bar is then secured to the skin with 3-0 nylon sutures.

Step 10. The colon is opened along the taenia coli incorporating any prior decompression site. The edges of the bowel are everted and approximated with full thickness colon bites to a small full thickness skin bite circumferentially.

Step 11. An opening is fashioned in the ostomy disc just large enough to fit around the stoma flush against the surface, and the disc is placed on the skin.

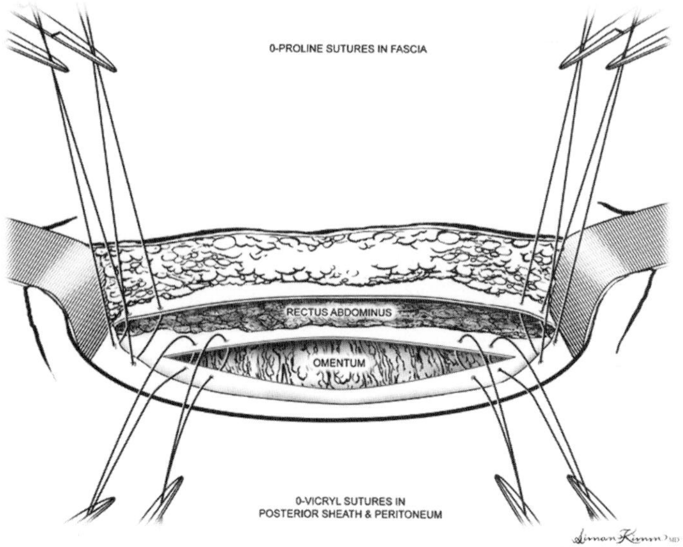

0-PROLINE SUTURES IN FASCIA

RECTUS ABDOMINUS

OMENTUM

0-VICRYL SUTURES IN
POSTERIOR SHEATH & PERITONEUM

Figure 46.2. Transverse loop colostomy.

46

Postop

The ostomy should be examined every day to assure viability and function. It will become edematous until the rod is removed at day 7-10. The wound is further checked for signs of infection which might necessitate removal of a suture or two to provide drainage.

Acknowledgment

The editors and author wish to acknowledge Richard S. Berk for contributing to the previous version of this chapter.

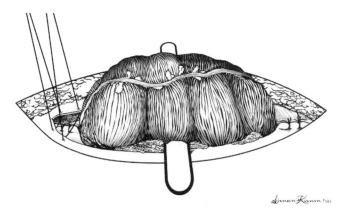

Figure 46.3. Transverse loop colostomy. Supporting rod.

Chapter 47

Colostomy: End Sigmoid with Hartmann's Procedure

Anne-Marie Boller

Indications

Indications for an end sigmoid colostomy include perforated diverticulitis, perforated sigmoid carcinoma, iatrogenic colon perforation following colonoscopy, obstructing unresectable rectal cancer, and rectal injury.

Preop

Most patients who require end sigmoid colostomy with Hartmann's procedure present acutely and therefore are unable to have any preoperative bowel preparation. Intravenous antibiotics should be given upon reaching the diagnosis if peritonitis is suspected or immediately preoperatively if it is not. Broad-spectrum coverage should include Gram-negative and anaerobic organisms. Deep venous thrombosis prophylaxis should include subcutaneous heparin and sequential compression devices. If possible, the patient should be evaluated by a stomal therapist preoperatively to mark the optimal stomal position. A urinary catheter should be placed. Ureteral stents should be considered preoperatively or even intraoperatively if the preoperative workup indicates an extensive inflammatory or neoplastic process.

47

Procedure

Step 1. The abdomen should be cleansed and draped to expose the area from the pubic symphysis to the nipples. A lower vertical midline incision is made. For obese patients, the incision may need to be extended cephalad. Explore the abdomen and evaluate the pathology. If purulent fluid is present, obtain aerobic and anaerobic cultures.

Step 2. Place a self-retaining retractor such as a Balfour or Omni retractor.

Step 3. Place moist laparotomy pads across the small intestine and pack it into the upper abdomen.

Step 4. Incise the peritoneal reflection of the sigmoid colon above and below the site of obstruction or perforation and mobilize the sigmoid to the extent possible. Attempt to identify the left ureter. If the inflammation is severe, maintain a close plane of dissection along the mesenteric border of the sigmoid colon to avoid injuring the ureter, particularly for benign disease. If inflammation or soiling is not severe, attempt to perform a lymphadenectomy for suspected malignancies. For benign disease, there should be no attempt to include the mesentery or lymph nodes with the resection.

Northwestern Handbook of Surgical Procedures, 2nd Edition, edited by Nathaniel J. Soper and Dixon B. Kaufman. ©2011 Landes Bioscience.

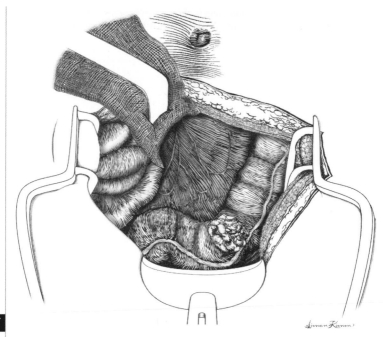

Figure 47.1. End sigmoid colostomy with Hartmann's procedure.

Step 5. Identify proximal and distal sites of the sigmoid for resection. The bowel should be as healthy as possible at the resection sites. For unusually severe inflammation or extensive tumor, it may be safest to divide the sigmoid colon proximal to the site of pathology and leave the pathology in place but diverted until a subsequent operation is performed.

Step 6. Divide the bowel in both locations using an automated stapling device.

Step 7. Clamp, ligate, and divide the mesenteric vessels with 2-0 silk suture.

Step 8. Remove the specimen. Irrigate the abdominal cavity with several liters of warm saline. Inspect for hemostasis.

Step 9. The distal sigmoid colon may be loosely tacked to the sacral promontory or psoas muscle to prevent retraction and facilitate subsequent colostomy reversal providing that adequate length remains after the resection. A long polypropylene suture attached to the Hartmann pouch will also facilitate future identification.

Step 10. Excise a circular piece of skin at the preoperatively marked stomal site. This should be approximately 2 cm in diameter. Make a cruciate incision in the anterior rectus sheath. Do not divide the muscle. Incise the peritoneum. These incisions should allow passage of two fingers.

Step 11. Bring the proximal sigmoid colon out through the stomal incision. The colon should lie freely at the surface of the skin without tension. If this is not possible, mobilize the descending colon prior to closing the abdominal fascia.

Step 12. Place a closed suction drain in the pelvis if an abscess was present. Close the abdominal fascia with a running #2 nylon suture with interrupted 0 polypropylene figure-of-eight sutures. The skin should be loosely approximated and packed with moist gauze in the presence of a perforation. Otherwise, the skin may be closed with staples. Dress the abdominal incision.

Step 13. Excise the staple line on the proximal colon. Sew the edges of the colostomy to the skin with 3-0 Vicryl suture, approximating the dermis to full-thickness purchases of the transected proximal sigmoid. Place a stomal appliance.

Postop

The patient should receive postoperative antibiotics in the setting of a perforation. Deep venous thrombosis prophylaxis should be continued until the patient is ambulating. Intravenous fluid should be continued until adequate oral intake has been established. The diet should be advanced according to the patient's appetite, lack of nausea and emesis, presence of bowel sounds, and bowel activity. The Foley catheter may be removed on postoperative day 3-5 depending upon the degree of pelvic dissection, the stability of the patient's vital signs, and urinary output.

Complications

Possible procedure-related complications include wound infection, pelvic abscess, abdominal wall dehiscence, prolonged postoperative ileus, and small bowel obstruction.

Follow-Up

Follow-up depends on the nature of the underlying condition. If the colostomy is to be closed, this is typically done 12 weeks after the initial procedure. A LGI should be performed prior to a takedown procedure and a colonoscopy performed through the stoma site to clear the remainder of the colon of any pathologic process.

47

Chapter 48

Colostomy Closure

Anne-Marie Boller

Indications

Colostomy closure is indicated when the underlying condition which required the colostomy allows it and there is adequate distal colon and rectum to safely re-establish gastrointestinal continuity. Colostomy closure is typically performed 12 weeks after creation of the stoma when the inflammation that may have been associated with colostomy placement has resolved.

Preop

The proximal and distal colon should be assessed by colonoscopy, proctoscopy, and/or contrast enema through both colostomy and rectum. Prior to surgery, both a mechanical and an antibiotic bowel prep may be performed. Preoperative enemas or a proctoscopic rectal wash out after induction should be considered. Prophylactic IV antibiotics should be given 30 minutes prior to surgery. A nasogastric tube is inserted. Bilateral pneumatic compression boots are placed and subcutaneous heparin is administered.

Procedure

48

Step 1. An adequate abdominal incision is made for exposure. For closure of a loop colostomy, this is an elliptical incision around the stoma. When the two ends of the bowel are separated, a midline incision or re-opening of the previous laparotomy incision is often required. When closing an end colostomy, the stoma is typically oversewn before prepping the abdomen to minimize spillage.

Step 2. The stoma is separated from surrounding skin, subcutaneous tissue, and fascia until it is possible to pass a finger into the peritoneal cavity circumferentially around the stoma.

Step 3. The proximal and distal colon are mobilized as much as necessary to allow the bowel to be reanastomosed without tension.

Step 4. The colon is anastomosed in a tension-free manner using sutures or stapling instruments. In a divided colostomy, the bowel ends should be resected back to healthy tissue prior to anastomosis. Either a one-layer or two-layer sutured anastomosis can be performed. If the anastomosis is stapled, the two ends of bowel are closed with a linear stapler and then a functional end-to-end anastomosis performed. When closing a loop colostomy, it may be possible to preserve the bridge of tissue connecting the two limbs and close the remainder of the circumference of the bowel. However, this must be done in a way that does not compromise the lumen, and it may be necessary to resect back to fresh ends to effect an adequate anastomosis.

Northwestern Handbook of Surgical Procedures, 2nd Edition, edited by Nathaniel J. Soper and Dixon B. Kaufman. ©2011 Landes Bioscience.

Step 5. The site of the stoma is typically left open and packed to heal by secondary intention. If a separate abdominal incision has been made, it can be closed in layers including skin if contamination has been minimal.

Postop

The nasogastric tube is removed and oral feeding resumed as bowel activity returns. Pain control is switched from parenteral to oral when bowel activity returns.

Complications

Complications of colostomy include wound infection and anastomotic dehiscence with fistula formation or peritonitis.

Follow-Up

Patients should be seen as appropriate in the weeks after colostomy closure to follow healing of the abdominal wound(s). Long-term follow-up is dictated by the underlying condition requiring colostomy.

48

Chapter 49

Laparoscopic Rectopexy

Amy L. Halverson

Indications

Full thickness rectal prolapse.

Preop

Full-thickness rectal prolapse may be identified by asking the patient to strain on a commode. If prolapse cannot be elicited on physical exam, prolapse may be demonstrated with a defecating proctogram. Individuals with rectal prolapse may also have anal incontinence or pelvic organ prolapse. Patients should be asked whether they have symptoms of anal incontinence, uterine prolapse, or urinary symptoms.

Procedure

Step 1. Abdominal insufflation.

Step 2. Insertion of trochars. This will include a supraumbilical camera port and two or three additional working ports.

Step 3. Incise right lateral peritoneal reflection and mobilize rectum anteriorly off of the sacrum.

Step 4. Incise left lateral peritoneum and anterior peritoneal reflection. (The anterior peritoneal reflection is identified by adipose tissue traversing the midline of the anterior rectum.)

Step 5. Continue anterior and posterior dissection down to the levator muscles. The lateral rectal stalks are left intact.

Step 6. After mobilization of the rectum, retract the rectum cephalad and secure the lateral mesorectum to the presacral fascia just below the sacral promontory. The rectum may be secured by placing bilateral sutures. A second option is to secure a piece of nonabsorbable mesh to the presacral fascia and then suture the mesh to the lateral mesorectum bilaterally. Some advocate concomitant sigmoid colon resection with rectopexy, especially in individuals with chronic constipation.

Postop

Initiate bowel management program to facilitate bowel movements. Educate patient to avoid straining with bowel movements.

Complications

Some patients will experience an increase in constipation following rectopexy. Recurrence rates of over 30% have been reported with long-term follow-up.

Northwestern Handbook of Surgical Procedures, 2nd Edition, edited by Nathaniel J. Soper and Dixon B. Kaufman. ©2011 Landes Bioscience.

Chapter 50

(Sub-) Total Colectomy with Ileorectal Anastomosis

Anne-Marie Boller

Indications

There are numerous indications for the procedure. As with all surgical interventions, it is most important to understand who would benefit from this procedure in the setting of their diagnosis, their quality of life and their functional preoperative status.

- Synchronous colon cancers or polyps
- HNPCC with colon cancer proximal to the rectum
- FAP with relative rectal sparing (<20 polyps)
- Slow transit constipation
- Crohn's disease with rectal sparing and without perianal disease
- Ulcerative colitis with rectal sparing

Preop

Preoperative colonoscopy to evaluate the entire colon is mandatory in order to evaluate, tattoo and biopsy all lesions. Rigid proctoscopy should be utilized in the case of lesions in the sigmoid colon or rectum to establish their exact distance from the anal verge, as this effects both surgical planning and possible neoadjuvant treatment decisions. Functional status in terms of the patients' continence and quality of life must be assessed preoperatively and utilized in deciding with the patient if this is an appropriate operation for them. Anal manometry should be utilized to quantify the functional status of the sphincter complex. Transanal ultrasound should be used to assess the integrity of the sphincter complex and transrectal ultrasound must be utilized in the case of a rectal neoplasm to document the level of tumor infiltration and the presence of any mesorectal lymph nodes. Dynamic proctography should be obtained if any pelvic floor dysfunction or evacuation disorder is suspected.

Mechanical bowel preparation is a matter of surgeon preference but is no longer an absolute mandate in colon surgery. However, if the subtotal colectomy is to be performed laparoscopically, mechanical bowel preparation is recommended as excessive stool in the colon makes laparoscopic handling of the colon difficult. Deep venous thrombosis prophylaxis should include subcutaneous heparin and sequential compression devices. If possible, the patient should be evaluated by a stomal therapist preoperatively to mark the optimal stomal positions. A urinary catheter should be placed. A naso- or oral-gastric tube should be placed after induction. Ureteral stents should be considered preoperatively or even intraoperatively if the preoperative workup indicates an extensive inflammatory or neoplastic process.

Northwestern Handbook of Surgical Procedures, 2nd Edition, edited by Nathaniel J. Soper and Dixon B. Kaufman. ©2011 Landes Bioscience.

Procedure

The extent of this procedure is different depending on the indication for the procedure. It may be performed through a laparotomy or as a laparoscopic procedure. We will discuss a standard laparotomy approach for a total abdominal colectomy with an end-to-end ileorectal anastomosis.

Step 1. The patient should be positioned in the lithotomy position with both hands tucked at the sides. A Foley catheter and a naso/orogastric tube are placed in all patients. Standard skin cleansing is performed.

Step 2. Periumbilical incision carried down to within 2 cm of the pubic bone. Extension above the umbilicus as needed.

Step 3. Placement of an appropriate retractor using laparotomy towels to hold away the small bowel as needed.

Step 4. Complete and meticulous exploration of the abdomen.

Step 5. Identification of the terminal ileum and evaluation of its mesentery. Transect bowel with linear cutting stapler.

Step 6. Mobilize the right colon from its retroperitoneal attachments moving cephalad toward the hepatic flexure. Dissect hepatic flexure away from duodenum and proceed medially. Be aware of the middle colic vessels.

Step 7. Enter lesser sac and complete dissection laterally.

Step 8. Take the middle colic vessels under direct visualization with an appropriate device or suture ligature. After these vessels are controlled, continue laterally toward the splenic flexure.

Step 9. Change your position and visualize the left colon. Dissect the sigmoid and descending colon off their retroperitoneal attachments, moving upwards along the white line of Toldt. Visualize the ureter and gonadal vessels. Retract the organ medially and complete your dissection up to the splenic flexure. Take down the flexure by carefully connecting your lateral proximal descending colon dissection and your lateral transverse colon resections. Do not put traction on the spleen.

Step 10. Once the entire abdominal colon is mobilized, the remaining mesentery may be addressed with an appropriate device or suture ligature.

Step 11. Appropriate mesorectal resection with careful attention to the ureters, controlled high ligation (if necessary) of the mesentery and complete mesorectal excision with intact fascia. Careful attention to the nerves will allow them to be swept away from the specimen unless they are involved in the disease process.

Step 12. Distal transaction of the specimen with a transverse stapling device, cutting or noncutting is acceptable depending on the surgeon's preference. Proctoscopic evaluation should be utilized if necessary to ensure adequate margins and complete resection.

Step 13. Removal of the specimen and evaluation for appropriate margins when necessary.

Step 14. Functional end-to-end ileorectal anastomosis with insertion of a circular stapler (EEA) through the rectum and placement of the anvil in the ileum after a prolene pursestring has been placed. Bring the EEA spike through the rectal stump under direct visualization and connect the anvil in the ileum to it in a tension free manner. Establish the anastomosis and assess the tissue donuts after careful, controlled removal of the EEA stapler.

Step 15. Proctoscopy with water-air testing of the anastomosis.

Step 16. Drain into pelvic if the peritoneal reflection was violated.

Step 17. Closure of laparotomy incision.

Step 18. Maturation of diverting loop ileostomy if indicated.

50

Postop

Deep venous thrombosis prophylaxis should be continued until the patient is ambulating. Intravenous fluid should be continued until adequate oral intake has been established. The diet should be advanced according to the patient's appetite, lack of nausea and emesis, presence of bowel sounds, and bowel activity. The Foley catheter may be removed on postoperative day 3-5 depending upon the degree of pelvic dissection, the stability of the patient's vital signs, and urinary output. The drain may be removed when output is less than 30 ml a day and decreasing. Steroid tapers should be implemented as needed. If a rod was used in creating the diverting loop ileostomy, remove it on the fifth day.

Complications

Possible procedure-related complications include bleeding, ureteral injury, wound infection, pelvic abscess, anastomotic dehiscence, abdominal wall dehiscence, prolonged postoperative ileus, and small bowel obstruction.

Follow-Up

For benign conditions, accepted screening measures should be followed. For malignancy, the patient should undergo proctoscopy one year following the operation and have appropriate surveillance thereafter. Pelvic floor physical therapy may be beneficial for patients postoperatively.

50

Chapter 51

Proctocolectomy with Ileal Pouch: Anal Anastomosis

Amy L. Halverson

Indications

Complete removal of the colon and rectum with ileal pouch reconstruction is indicated in individuals with ulcerative colitis who are not responding adequately to medical therapy or who have colonic dysplasia. It is also indicated in individuals with familial adenomatous polyposis (FAP).

Preop

Patients should have had a complete colonoscopy. An enterostomal therapist is helpful for marking an optimal site for the ileostomy and providing preoperative teaching regarding ileostomy care. To allow for a secure fit of the ostomy appliance, the site should be located over the rectus muscle and should be at the apex of any abdominal fold that occurs with the patient in the sitting position. The patient should be able to see the site while sitting and standing. Patients should receive mechanical bowel cleansing and intravenous antibiotics 30 minutes prior to operation. Deep vein thrombosis prophylaxis with subcutaneous heparin and/or sequential compression devices should be used based on the patient's risk factors.

Procedure

Step 1. The patient should be in the modified lithotomy position with the hips only slightly flexed. There should be no pressure on the upper or lateral calf. The leg should be supported primarily by the foot and lower calf. Incorrect positioning may result in injury to the peroneal nerve.

Step 2. The abdomen is entered though a midline incision and a complete exploration of the peritoneal cavity performed. The ascending colon is mobilized towards the midline by incising the right peritoneal reflection. Mobilization is continued around the hepatic flexure to the distal transverse colon. The omentum is dissected away from the transverse colon across its entire length and the omentum preserved unless malignancy is suspected. The left colon and splenic flexure are then mobilized.

Step 3. The ileum is divided just proximal to the ileocecal valve. The mesocolon is then divided starting at the point of terminal ileal division and extending to the proximal rectum. If the mesocolon is thickened, it may be prudent to employ suture ligatures on the base of the mesocolon.

Northwestern Handbook of Surgical Procedures, 2nd Edition, edited by Nathaniel J. Soper and Dixon B. Kaufman. ©2011 Landes Bioscience.

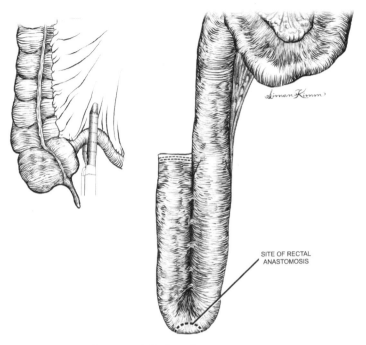

Figure 51.1. Proctocolectomy with ileal pouch.

Step 4. The upper and mid-rectum is mobilized by dividing the peritoneum on either side of the rectum as it descends into the pelvis. The avascular plane between the rectum and the sacrum is identified and the rectum is mobilized anteriorly down to the level of the coccyx. The lateral rectal attachments are divided with electrocautery. The anterior peritoneal reflection is identified where the perirectal fat extends across the midline. The lateral incisions are continued anteriorly in this plane. Dissecting too far anteriorly may result in injury to the vagina in women or bleeding from the seminal vesicles and prostate in men. Circumferential mobilization of the rectum is continued down to the level of the levator muscles. With upward traction on the rectum, the distal rectum is stapled just above the proximal extent of the anal canal.

Step 5. To begin creation of the ileal pouch, the terminal 30 cm of ileum is irrigated with saline.

Step 6. The ileum is then folded into a J-configuration and the two antimesenteric borders of the limbs placed in apposition. The limbs are then stapled along the antimesenteric borders, incorporating both walls with repeated firings of an 8 cm linear stapler. The length of the pouch created is approximately 15 cm.

Step 7. The open end of the J limb is closed with a double-row stapler.

Step 8. An enterotomy is created in the bend at the apex of the pouch, and a pursestring suture is placed to hold the anvil of the circular stapler. Prior to inserting the anvil of the stapler, the anterior staple line of the pouch should be checked for bleeding.

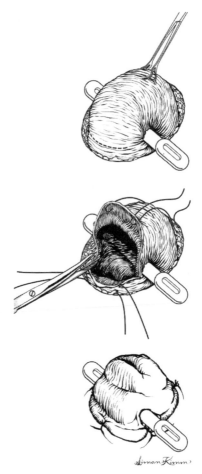

Figure 51.2. Proctocolectomy with ileal pouch. Temporary diverting loop ileostomy.

Step 9. The end-to-end circular stapler is inserted through the anus. Take care to not disrupt the closed end of the rectal stump with the stapling device. Using the sharp pin, the rectal stump is penetrated from below. The anvil protruding from the J-pouch is attached to the stapler and the stapler is fired.

Step 10. A temporary diverting loop ileostomy may be created by excising a 1.5 cm diameter disc of skin at a preoperatively marked stoma site. The subcutaneous tissue and the anterior rectus sheath are divided using electrocautery. The fibers of the rectus muscle are separated using a blunt curved clamp and the abdomen is entered by incising the posterior rectus sheath and peritoneum.

Step 11. A loop of terminal ileum is brought through the opening in the abdominal wall and a plastic rod for support is placed under the mesenteric edge of the bowel through a small hole in the mesentery immediately adjacent to the bowel wall.

Step 12. A closed suction drain is placed in the pelvis and brought out through a separate stab wound. The abdominal incision is closed in layers.

Step 13. A transverse enterotomy is made just above the skin on the distal limb and the bowel everted over the proximal limb. The edges of the opened ileum are sutured to the dermis circumferentially with interrupted 4-0 absorbable sutures.

Postop

The rod under the loop ileostomy may be removed on postoperative day 3. The patient should be examined after 6 weeks to dilate any anastomotic stricture. This can usually be done with gentle digital pressure. The loop ileostomy is closed after approximately 12 weeks. A gastrograffin enema should be performed to rule out a leak prior to loop ileostomy closure.

Complications

Complications of proctocolectomy with ileal pouch-anal anastomosis include anastomotic leak , bleeding from the pouch, pelvic abscess, anal stricture, and wound infection.

Follow-Up

Endoscopic evaluation of the pouch with biopsy of the anal transition zone is performed on an annual or biannual basis to evaluate for dysplasia. Long-term problems following ileal pouch-anal anastomosis include fistula formation, which may occur in patients with Crohn's disease who were initially thought to have ulcerative colitis or indeterminate colitis. Anal strictures may require dilation. Approximately 50% of individuals will experience at least one episode of pouchitis.

51

Chapter 52

Proctocolectomy: Total with Ileostomy

Amy L. Halverson

Indications

Complete removal of the colon and rectum is indicated in patients with Crohn's colitis involving the colon and the rectum. It may also be appropriate for selected individuals with ulcerative colitis or familial adenomatous polyposis (FAP) who are not candidates for reconstruction with an ileal pouch-anal anastomosis. Indications for proctocolectomy in patients with Crohn's disease or ulcerative colitis include symptoms not responding to medical therapy or dysplasia identified on surveillance biopsy.

Preop

A small bowel contrast study is indicated to look for additional disease in patients with Crohn's disease or to identify small bowel polyps in patients with FAP. An enterostomal therapist is helpful for marking an optimal site for the ileostomy and providing preoperative teaching regarding ileostomy care. To allow for a secure fit of the ostomy appliance, the site should be located over the rectus muscle and should be at the apex of any abdominal fold that occurs with the patient in the sitting position. The patient should be able to see the stoma site while sitting or standing. Patients should receive mechanical bowel cleansing and intravenous antibiotics 30 minutes prior to operation. Deep vein thrombosis prophylaxis with subcutaneous heparin and/or sequential compression devices should be used based on patient's risk factors.

Procedure

52

Step 1. The patient should be placed in the modified lithotomy position with hips only slightly flexed. There should be no pressure on the upper or lateral calf. The leg should be supported primarily by the foot and lower calf. Incorrect positioning may result in damage to the peroneal nerve.

Step 2. The abdomen is entered through a midline incision and a complete exploration of the peritoneal cavity performed. The ascending colon is mobilized towards the midline by incising the right peritoneal reflection. Mobilization is continued around the hepatic flexure to the distal transverse colon. The omentum is dissected away from the transverse colon across its entire length and the omentum preserved unless malignancy is suspected. The left colon and splenic flexure are then mobilized. The ileum is divided just proximal to the ileocecal valve. The mesocolon is then divided starting at the point of terminal ileal division and extending to the proximal rectum. If the mesocolon is thickened, it may be prudent to employ suture ligatures on the base of the mesocolon.

Northwestern Handbook of Surgical Procedures, 2nd Edition, edited by Nathaniel J. Soper and Dixon B. Kaufman. ©2011 Landes Bioscience.

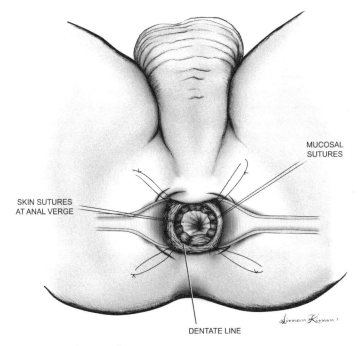

MUCOSAL
SUTURES

SKIN SUTURES
AT ANAL VERGE

DENTATE LINE

Figure 52.1. Total proctocolectomy.

Step 3. The rectum is mobilized by dividing the peritoneum on either side of the rectum as it descends into the pelvis. The avascular plane between the rectum and the sacrum is identified, and the rectum is mobilized anteriorly down to the level of the coccyx. The lateral rectal attachments are divided with electrocautery. The anterior peritoneal reflection is identified where the perirectal fat extends across the midline. The lateral incisions are continued anteriorly in this plane. Dissecting too far anteriorly may result in injury to the vagina in women or bleeding from the seminal vesicles and prostate in men. Circumferential mobilization of the rectum is continued down to the level of the levator muscles.

Step 4. Moving to the perineum, exposure is obtained by effacing the anal canal with circumferential sutures tied to the skin of the buttocks or a Lone Star Retractor System™.

Step 5. Submucosal injection of saline with epinephrine facilitates dissecting the mucosa off of the underlying internal sphincter muscle. Transanal mucosectomy is begun with a circumferential incision at the bottom of the mucosa and working proximally. No mucosa should be left behind.

Step 6. After completely removing the mucosa and obtaining hemostasis, the effacing sutures are removed and the anal canal is closed with 2 to 3 rows of absorbable suture. The skin is closed with 3-0 absorbable interrupted vertical mattress sutures.

Step 7. At the site chosen for the ileostomy, a 1.5 cm diameter disc of skin is excised. The subcutaneous tissue and the anterior rectus sheath are divided using electrocautery. The fibers of the rectus muscle are separated using a blunt curved clamp and the posterior rectus sheath and peritoneum opened.

52

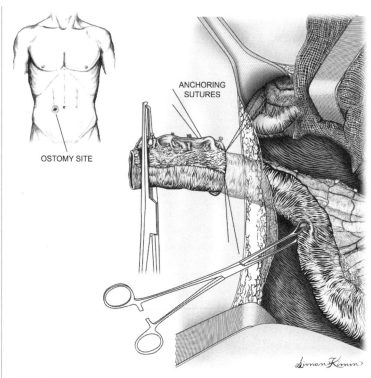

Figure 52.2. Total proctocolectomy. Ileostomy.

Step 8. The cut end of the ileum is then brought through the defect in the abdominal wall. The ileal mesentery is secured to the subcutaneous tissue with interrupted sutures at each border between serosa and mesentery and at the cut edge of the mesentery.

Step 9. The cut edge of mesentery is tacked to the anterior abdominal wall. A closed suction drain is placed in the pelvis.

Step 10. After closure of the abdominal wound, the end of the ileum is everted and secured to the dermis with circumferential interrupted absorbable sutures.

Postop
Once the ileostomy functions, the diet is advanced as tolerated. Ileostomy care teaching should be continued in the postoperative period.

Complications
Bleeding may occur internally or from the stoma. Pelvic infection, perineal wound dehiscence, or stomal necrosis/retraction are other potential complications.

Follow-Up
Patients with Crohn's disease should be followed indefinitely, since they may be at risk for developing small bowel Crohn's disease. Patients with FAP may develop polyps elsewhere in the gastrointestinal tract.

Chapter 53

Internal Hemorrhoids: Band Ligation

Steven J. Stryker

Indications

Rubber band ligation of internal hemorrhoids is indicated for bleeding or prolapse. It is most effective for Grade 1 and 2 internal hemorrhoids and less effective for Grade 3 and 4. Band ligation is contraindicated for the treatment of external hemorrhoids.

Preop

A detailed history is taken to confirm that the internal hemorrhoids are symptomatic. In the case of bleeding, examination of the colon and rectum is performed as appropriate to exclude other pathology. Immediately prior to surgery, a disposable enema is administered to evacuate the rectum.

Step 1. Sedation and local anesthesia are not necessary. The patient is positioned in the prone-jackknife position. Two bands are loaded on the ligating instrument using the cone-shaped loader.

Step 2. An anoscope is inserted and the internal hemorrhoidal quadrants are identified.

Step 3. In general, it is best to perform band ligation of only one quadrant at a session, typically the largest hemorrhoid. Less often, ligation of two or even three quadrants can be accomplished at a single session.

Step 4. The grasper is passed through the ligating instrument, and the columnar mucosa just proximal to the targeted hemorrhoid is grasped. This mucosa is drawn well into the ligating instrument, and the instrument is deployed, firing the band across this tissue. Great care must be taken to avoid inadvertently capturing any anoderm within the band to minimize postprocedural discomfort.

Step 5. The strangulated tissue is inspected and should be seen just proximal to the hemorrhoid. The blood supply to the targeted hemorrhoid is theoretically interrupted, and the hemorrhoid should regress over the ensuing days and weeks.

Postop

The patient is asked to limit physical activity the day of the procedure, but can resume normal activities the following day. Bulk-forming laxatives and nonnarcotic analgesics are recommended.

Northwestern Handbook of Surgical Procedures, 2nd Edition, edited by Nathaniel J. Soper and Dixon B. Kaufman. ©2011 Landes Bioscience.

Complications

Immediate, severe pain can occur and is a sign that the band has been placed too distal in the anal canal. This usually requires the removal of the band by carefully incising it. Postoperative sepsis is a rare, but life-threatening complication, and use of band ligation in immunocompromised individuals is strongly discouraged. Signs of postbanding sepsis include delayed anorectal pain, urinary retention, and fever. Finally, delayed hemorrhage can occur in the first 7-10 days.

Follow-Up

The patient is seen 4-8 weeks after band ligation and the adequacy of the initial treatment determined. Additional band ligation can be performed at this time, if necessary.

Chapter 54

Anal Fissure: Lateral Internal Sphincterotomy

Anne-Marie Boller

Indications

Lateral internal sphincterotomy is indicated for the treatment of symptomatic chronic anal fissure. There are open and closed techniques; we will discuss the open technique.

Preop

Nonsurgical management prior to surgery is prudent, including stool management, lifestyle changes, topical medications and Botulinum toxin A injection. Preoperative functional evaluation should be completed prior to the procedure. Transanal ultrasound should be utilitzed to establish the length of the anal sphincter complex and the health and integrity of both internal and external sphincters. The patient should take a disposable enema 2 hours before surgery. Anesthetic choices are local with intravenous sedation (preferred), regional, or general. Antibiotics should be administered within thirty minutes prior to surgery.

Procedure

Step 1. The patient is placed in the prone jackknife position, buttocks separated using tape. The perianal area is prepped with povidone-iodine.

Step 2. An anoscopic evaluation of the anal canal is performed.

Step 3. The intersphincteric groove is identified by palpation on the right side of the anal canal.

Step 4. A longitudinal incision is made in the mucosa overlying the intersphincteric groove.

Step 5. The intersphincteric plane is developed bluntly with a hemostat.

Step 6. The internal sphincter is divided. The incision is started distally and carried proximally to the level of the dentate line, or less than 30% of the entire internal anal sphincter.

Step 7. The wound is irrigated with saline and hemostasis assessed.

Step 8. The mucosal defect is closed with a 3-0 chromic suture.

54

Postop

This is an outpatient procedure. The patient should be maintained on stool softeners, and sitz baths should be started the following day.

Complications

Complications include bleeding, perianal abscess or fistula, urinary retention, persistent fissure and fecal incontinence.

Northwestern Handbook of Surgical Procedures, 2nd Edition, edited by Nathaniel J. Soper and Dixon B. Kaufman. ©2011 Landes Bioscience.

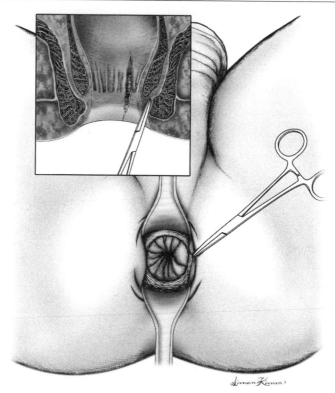

Figure 54.1. Anal fissure. Lateral internal sphincterotomy.

Follow-Up

Patients should be seen in one week and then regularly until healing occurs. Fecal continence should be confirmed by history in the postoperative period.

54

Chapter 55

Anorectal Abscess: Drainage Procedure

Amy L. Halverson

Indications

Acute perirectal abscess.

Preop

This procedure may be performed using local anesthesia with or without intravenous sedation, although some patients may require general anesthesia. A digital rectal examination may be too painful and is not necessary if the site of the abscess is readily apparent.

Procedure

Step 1. Identify area of maximal erythema, induration, and/or fluctuance.

Step 2. Prep skin and infiltrate with local anesthetic.

Step 3. Incise the area identified in Step 1 with a 1-2 cm incision placed over the abscess at the point closest to the anal verge.

Step 4. Allow drainage of purulent fluid. This may be facilitated with gentle insertion of a curved clamp.

Step 5. Insert a 10-14 F mushroom catheter into the abscess cavity and trim drain.

Postop

Consider systemic antibiotics for patients with fever, cellulitis, diabetes, or those who are otherwise immunocompromised.

Complications

Incomplete drainage may result in progression of infection. Inadvertent division of anal sphincter muscles may cause anal incontinence. If symptoms do not resolve after drainage, consider examination under anesthesia or imaging (CT or MRI) to identify an undrained collection.

Follow-Up

Examine the patient 2-4 weeks after procedure. Remove drain and allow wound to heal secondarily. Persistent drainage may require examination under anesthesia to identify a fistula tract.

Northwestern Handbook of Surgical Procedures, 2nd Edition, edited by Nathaniel J. Soper and Dixon B. Kaufman. ©2011 Landes Bioscience.

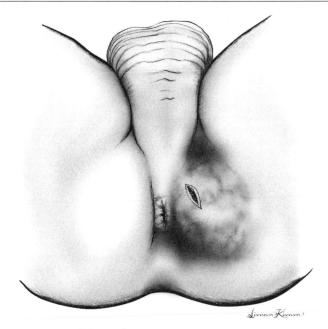

Figure 55.1. Anorectal abscess drainage.

Chapter 56

Anal Fistulotomy

Anne-Marie Boller

Indications
Anal fistula

Preop
Routine preoperative evaluation is required. Digital examination and anoscopy should be used in the preoperative clinical setting to establish the location of the internal opening. If this cannot be established by these means, transanal ultrasound should be used to locate the internal opening and quantify the amount of sphincter involved. A 10 ml syringe with an 18-gauge angiocatheter should be filled with dilute hydrogen peroxide. This is then injected into the external opening with the aid of suction so that the peroxide bubbles are only imaged through the tract and not the skin. Establishing the location of the internal opening, mapping the fistula tract, and quantifying the amount of sphincter involved in the fistula are critical to the surgeon's decision-making process.

Regional anesthesia is preferred, but general anesthesia and monitored local anesthesia with sedation are also acceptable choices.

Procedure
Step 1. The patient is placed in the prone jackknife position, with the buttocks separated using tape. The perianal area is prepped with povidone-iodine.

Step 2. The external opening of the fistula is identified on the perianal skin and a probe passed to assess direction of the tract into the rectum.

Step 3. An anoscopic evaluation of the anal canal and lower rectum is performed, and the internal opening of the fistula is searched for. If the opening is not obvious, dilute hydrogen peroxide is injected through the external opening using a 10 ml syringe and an 18-gauge angiocath while the anoscope is in place. Bubbles will mark the site of the internal opening.

Step 4. The extent of incorporated sphincter muscle is assessed by palpation.

Step 5. If less than 30% of sphincter muscle will be incorporated by a fistulotomy, the metal probe is then passed through the length of the fistula tract and the overlying tissues divided down to the probe. If greater than 30% of sphincter muscle is at risk of division, the distal (epidermal) end of the fistulous tract is opened and a vessel loop placed around the sphincter as a seton, and the muscle is not divided at this time.

Step 6. The opened tract is curetted to remove granulation tissue. Hemostasis is obtained using electrocautery.

56

Northwestern Handbook of Surgical Procedures, 2nd Edition, edited by Nathaniel J. Soper and Dixon B. Kaufman. ©2011 Landes Bioscience.

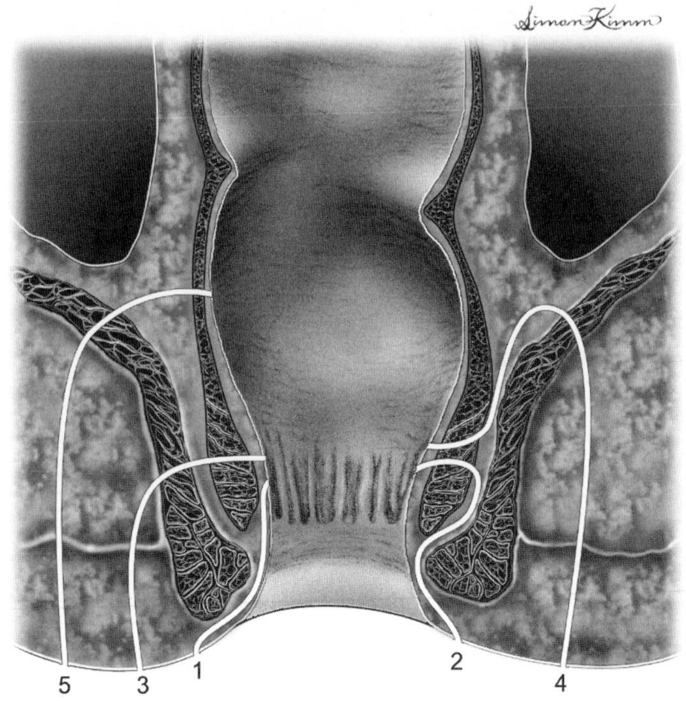

1 SUBCUTANEOUS
2 INTERSPHINCTERIC
3 TRANSSPHINCTERIC
4 SUPRALEVATOR
5 EXTRASPHINCTERIC

Figure 56.1. Anal fistulotomy. Common locations of anal fistulae.

Postop

Anal fistulotomy is an outpatient procedure. The patient should be maintained on stool softeners and given pain medication. Sitz baths are started the following day.

Complications

Complications include bleeding, perianal abscess, recurrent fistula, and fecal incontinence.

Follow-Up

Patients should be seen approximately one week after the procedure and then followed as appropriate until healing occurs. Fecal continence should be confirmed by history in the post-operative period.

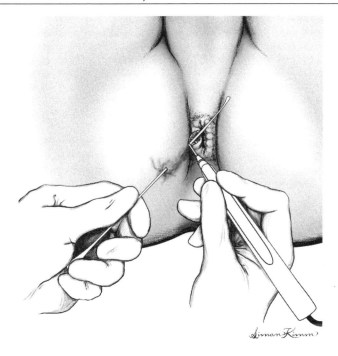

Figure 56.2. Anal fistulotomy.

56

Chapter 57

Breast Duct Excision

Seema A. Khan

Indications

Spontaneous, single duct, serous or bloody (or guiac positive) nipple discharge, also described as pathologic nipple discharge. Such discharge is more likely to be related to intraductal neoplasia, either benign (papilloma) or malignant (duct carcinoma in situ), particularly when it occurs in older women. In younger women, green, blue or yellow discharge is often nonspontaneous, expressible from several ducts, is caused by ductectasia and is rarely associated with significant pathology.

Preop

If mammographic calcifications are present, particularly in the retroareolar region, histological sampling with sterotactic core biopsy may obviate the need for duct excision. Papillomata may be visualized ultrasonographically as a mass in a fluid-filled tubular structure and can be sampled by core needle biopsy. Yet these will still need excision particularly if large or if cytological atypia is present. However, about two-thirds of women with pathologic nipple discharge have no imaging abnormalities. A preoperative ductogram can be performed, although the sensitivity of ductography is only in the 70% range. If a filling defect is demonstrated and is peripherally located (i.e., more than 4-5 cm from the nipple duct orifice), the study can be repeated on the morning of surgery and the lesion wire localized. This is not necessary for centrally located lesions. The procedure is usually performed under local anesthesia, with sedation. Prophylactic antibiotics are not indicated. Do not try to elicit discharge in the preoperative holding area on the morning of surgery.

Procedure

Step 1. Prone position, ipsilateral arm extended on arm board, at right angles to operating table. Skin preparation of entire breast.

Step 2. Place a nipple block by infiltration of 1% xylocaine circumferentially at the areolar margin. Locate the discharging orifice by expressing the discharge. Cannulate the discharging duct with a 00 lachrymal probe, being careful not to create a false channel. Palpate the direction of the lachrymal probe under the areola, and mark your incision at the areolar margin, centered over the course of the probe.

Step 3. Make a skin incision through the dermis with a 15 blade. Elevate an areolar skin flap using skin hooks for traction; flap elevation can be performed with a 15 blade, iris scissors, or the cutting current of the electrocautery (to avoid charring and for better visualization of tissue planes). Take care to keep the field dry, using electrocautery for hemostasis as needed. Palpate the lachrymal probe intermittently. A dilated duct is often visualized as one approaches the base of the nipple papilla.

Northwestern Handbook of Surgical Procedures, 2nd Edition, edited by Nathaniel J. Soper and Dixon B. Kaufman. ©2011 Landes Bioscience.

An alternative or additional approach. Once the duct orifice has been dilated with graded lachrymal probes, insert a 24 gauge angiocath into the duct, and inject lymphazurin into the ductal tree. This will help in the identification of the affected duct and help trace its branches, and it is particularly useful when the lesion is not within the first 2 cm of the nipple duct.

Step 4. Use a mosquito hemostat to dissect the dilated duct, containing the probe, from surrounding ducts and areolar tissue. Trace the duct into the nipple papilla and transect it just below its orifice. A ligature can be placed on the duct before the transection and left long. This aids in the subsequent mobilization of the duct and can be used for gentle traction.

Step 5. When the terminal duct starts to branch into the surrounding breast parenchyma, usually at the level of the areolar margin, mobilize a narrow cone-shaped segment of breast parenchyma extending for about 2 cm peripherally. Deliver the specimen, consisting of the terminal duct and the attached 2 cm cone of breast tissue, and orient the specimen for the pathologist.

Step 6. Inspect the excisional cavity to see if an addition dilated or blue-dyed ductal orifice can be identified in the wall of the cavity. If one is seen, cannulate it with a probe and resect the additional dilated length of the duct.

Step 7. Complete hemostasis in the excisional cavity. Additional infiltration of 0.5% bupivicaine is optional at this point.

Step 8. Close the skin using a subcuticular closure. Subcutaneous sutures may be used if there is a sufficient layer of subcutaneous fat, but are not essential. Place steri-strips and apply dressing.

Postop

Most patients are discharged on the day of surgery; the dressing can be removed and the patient allowed to shower within 1 or 2 days of surgery. Analgesic prescriptions are required.

Complications

Hematoma: significant hematomata will usually present within the first 24 hours, often before the patient is discharged, and are best evacuated. Smaller ones can be allowed to resolve spontaneously.

Infection (extremely rare).

Follow-Up

Additional therapy is dictated by final pathologic diagnosis. A papilloma without atypia needs no further therapy; if atypia is present, the use of chemopreventive agents such as tamoxifen or raloxifene should be discussed. If duct carcinoma is diagnosed, it will require re-excision if margins are involved, and radiotherapy with or without tamoxifen.

57

Chapter 58

Pilonidal Cystectomy

Anne-Marie Boller

Indications

Pilonidal cystectomy is indicated in the setting of acute pilonidal disease, chronic pilonidal sinus/fistula and in the absence of an acute infection.

Preop

Two disposable enemas should be given on the morning prior to surgery. Sequential compression devices are applied to all patients and subcutaneous heparin administered. Prophylactic antibiotics should be given 30 minutes prior to the incision. Anesthesia choices are local with intravenous sedation or general anesthesia.

Procedure

Step 1. The patient is placed in the prone jackknife position, buttocks separated using tape. The perianal area is prepped with povidone-iodine.

Step 2. The procedure begins with an anoscopic evaluation of the anal canal. A lacrimal probe is placed in the pilonidal pits to establish the extent of the cyst.

Step 3. Marking of an elliptical incision is performed to include all sinus tracts and the entirety of the pilonidal cyst.

Step 4. Incise the skin respecting the markings previously placed.

Step 5. Carry the incision down to include the entirety of the cyst. Remove the whole specimen. Ensure that you have all tracts and sinuses in the specimen. Pass the specimen off the field and achieve hemostasis.

Step 6. The edges of the excised tract should be trimmed and cleaned. A 2-0 or 3-0 Vicryl suture should be utilized in a running, locking fashion to marsupialize the edges of the wound down to the wound base. Depending on the wound size, multiple running sutures may be used. A long tail should be left on each running suture so that each subsequent running suture can be tied to the previous suture.

Step 7. Achieve hemostasis.

Step 8. Pack wound with damp 4x4, depending on wound size.

Postop

This is an outpatient procedure. Daily sitz baths, stool softners and pain medications should be advised. Physical activity should be limited for approximately one month.

Complications

Complications include bleeding, infection, dehiscence of the wound, recurrent fistula or sinus tract and delayed wound healing.

Northwestern Handbook of Surgical Procedures, 2nd Edition, edited by Nathaniel J. Soper and Dixon B. Kaufman. ©2011 Landes Bioscience.

Follow-Up

Patients should be seen in clinic within two weeks to check on healing and then regularly to follow the wound. These wounds are prone to dehiscence and daily wound care may be needed.

58

Chapter 59

Major Hepatic Laceration: Open Repair

Michael B. Shapiro

Indications

A large proportion of liver injuries can be managed nonoperatively. The presence of a liver injury is not, by itself, an indication for operative exploration or repair. Operative control of liver hemorrhage is indicated with evidence of ongoing, hemodynamically significant bleeding.

Preop

Prior to performing exploratory laparotomy in a hemodynamically unstable patient, appropriate large-bore venous access must be present and resuscitation initiated. Blood for type and cross-match should be sent. A Foley catheter and nasogastric tube are placed prior to abdominal exploration. When performing exploratory laparotomy for trauma the surgeon should have adequate operative suction (two suctions), lighting, and the patient positioned so that the chest and/or mediastinum can be accessed intraoperatively. Antibiotic prophylaxis should be given prior to making the incision. A second-generation cephalosporin or other agents that cover aerobic and anaerobic enteric pathogens are generally used.

Step 1. Exploratory laparotomy is performed.

Step 2. A retractor is inserted below the right costal margin to permit visual examination of the liver. Initially the liver is examined in situ, looking for evidence of deep lacerations and active bleeding. Gentle upward (anterior and cephalad) retraction may permit visualization of the undersurface of the liver. The condition of the gallbladder should also be carefully noted. It is not necessary to mobilize the liver in the absence of an actively bleeding injury because mobilization itself may precipitate bleeding. Clots should not be removed if there is no evidence of ongoing bleeding. If there is hepatic injury and ongoing hemorrhage, operative control may be required.

Step 3. Temporary control of active liver bleeding can often be accomplished using a combination of direct manual compression and packing. Laparotomy pads are packed between the liver and the diaphragm and between the liver and the retroperitoneum. Additionally, direct manual compression can be applied using two hands to compress the liver. If temporary control of bleeding is attained, direct pressure should be held for 10 minutes. Packing may be facilitated by partial mobilization of the liver (see below). After an appropriate interval, release pressure to determine if bleeding resumes. If not, cautiously remove the laparotomy packs. If bleeding resumes, replace the packs and restore hemostasis. Consider leaving the packs in place and returning for a "second look" laparotomy in 24-72 hours.

Northwestern Handbook of Surgical Procedures, 2nd Edition, edited by Nathaniel J. Soper and Dixon B. Kaufman. ©2011 Landes Bioscience.

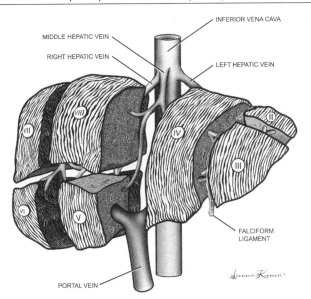

Figure 59.1. Liver segmental anatomy.

Step 4. If direct pressure and packing does not achieve hemostasis, a Pringle maneuver should be performed. The Pringle maneuver involves using digital pressure or an atraumatic clamp across the portal triad to occlude the portal vein and hepatic artery. If this maneuver significantly slows hemorrhage it suggests that the injury within the parenchyma involves branches of these vessels. Several methods to identify and control such injuries are described below. More importantly, if the Pringle maneuver does not halt or significantly slow bleeding from the liver it suggests that hemorrhage is arising from the hepatic veins or retrohepatic vena cava.

Step 5. If bleeding continues after the Pringle maneuver, try repacking the patient. If bleeding is controlled or significantly slowed, consider closing the patient with the packs in place and returning in 24-72 hours to reassess and remove the packs. If repacking does not control the hemorrhage, an attempt at operative control is indicated. Operative measures to control hemorrhage frequently require complete mobilization of the liver. The surgeon should have long vascular instruments and a needle holder preloaded with a 5-0 vascular suture available when liver mobilization for bleeding is initiated.

Step 6. Hepatic mobilization is initiated by dividing the ligamentum teres and falciform ligament. This permits complete visualization of the anterior surface of the liver. Next, the lateral attachments are released by incising the triangular ligaments. Usually the left triangular ligament is divided first. The right triangular ligament is incised next. In many instances it is easiest to begin incising the right triangular ligament at the inferior edge. When fully mobilized the liver can be rotated (gently) to assess the posterior aspect of the right and left lobes. Mobilization is required to access the hepatic veins and retrohepatic vena cava. If the surgeon strongly suspects a major retrohepatic injury, consideration should be given to performing an atrial-caval shunt (see Step 12) prior to complete mobilization to avoid exsanguinating hemorrhage.

59

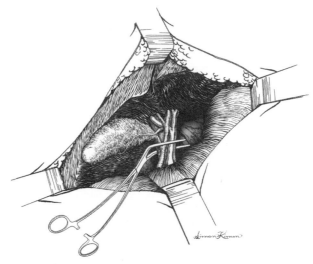

Figure 59.2. Pringle maneuver.

Step 7. Intrahepatic hemostasis can often be accomplished using the "finger-fracture" technique. This involves blunt division of the hepatic parenchyma in the direction of the laceration to expose bleeding intrahepatic vessels. Once identified, sites of bleeding can be controlled with clamps, ligatures, or surgical clips. Deep liver sutures may also be useful to control hemorrhage, but can produce significant ischemic damage to adjacent healthy parenchyma. Deep liver sutures (typically a 0-chromic suture on a round-tip needle) are placed in a horizontal mattress fashion on either side of the laceration. (Demonstration that manual compression controls hemorrhage should precede suture placement). Sutures are placed so as to achieve gentle compression of the interposed tissue. In some instances an absorbable mesh (e.g., polyglycolic acid) can be used to buttress the sutures.

Step 8. Electrocautery or the argon beam coagulator are frequently very useful for control of surface bleeding or small vessel bleeding from raw surfaces. When using conventional electrocautery, improved hemostasis is often achieved by turning the coagulation level to a very high level and positioning the cautery tip 2-3 mm above the bleeding surface. The resulting arc coagulates a wider superficial area. Care must be taken to remember to turn the cautery level back down before using the cautery on other structures. The gas jet of the argon beam coagulator permits the surgeon to more easily visualize the bleeding surgical field.

Step 9. In instances where multiple or deeper lacerations are found, bleeding can sometimes be controlled by wrapping the injured portion of the liver with polyglycolic acid mesh. The mesh should be cut, folded, and the edges sutured with a running absorbable suture so as to envelop a lobe and to achieve a slight degree of compression. However, the surgeon must be careful to avoid excess compression that may lead to infarction.

Step 10. An important adjunct to the hemostatic techniques described above is placement of a pedicle of omentum into the laceration. The omentum stimulates fibrin polymerization, provides a tissue seal for severed biliary radicles, minimizes dead space within the injured liver, and may hasten healing.

59

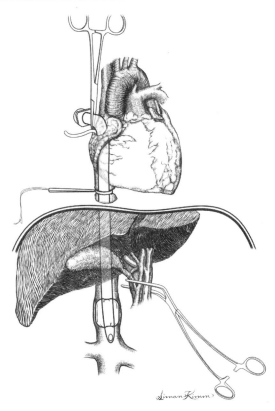

Figure 59.3. Atrial-caval shunt.

Step 11. If hemostasis has been achieved and if the patient is hemodynamically stable, nonviable liver tissue should be debrided to minimize late infectious complications. Debridement is not a priority and should not be performed during the initial "damage control" operation in hemodynamically unstable patients.

Step 12. Severe retrohepatic injuries have an extraordinarily high mortality irrespective of the surgical management employed. Prompt recognition of this lethal injury (continued venous bleeding with Pringle maneuver) is essential to achieve survival. Atrial-caval shunt permits vascular isolation of the liver and may permit the surgeon to visualize and repair the injury. A median sternotomy is performed and a 7-mm endotracheal (ET) tube (or a 36 F chest tube) is inserted into the inferior vena cava via the right atrial appendage. A hole must be carefully cut in the side of the tube (avoiding the balloon lumen of the ET tube) that will lie in the right atrium. An encircling tourniquet is passed around the vena cava cephalad to the injury (this is most easily achieved within the pericardial sac). A Pringle maneuver is performed along with inflation of the ET balloon in the infrahepatic cava (an additional encircling tourniquet is required below the liver if a chest tube is used), and clamping of the open end of the tube is performed to achieve vascular isolation. The injured vein is then repaired expeditiously to minimize warm liver ischemia.

Step 13. Hepatic resection should be reserved for the rare instances where the injury has already performed most of the resection or as a last resort to control uncontrollable exsanguinating hemorrhage.

Step 14. The role of liver drainage remains controversial. Drainage is not needed for most Grade I/II liver injuries. When drainage appears appropriate for Grade III-V injuries, one or more closed suction drains should be used. Open drainage techniques (e.g., Penrose drains) have a high incidence of septic complications and should be avoided.

Operative Principles

Complete abdominal exploration should be performed prior to attempting to control bleeding from liver injuries. Most liver injuries result in bleeding from relatively low-pressure vessels (portal or hepatic veins). Therefore, packing and correction of coagulation abnormalities is frequently crucial to successful control of liver hemorrhage. Damage control techniques may be life-saving with severe hepatic injuries. A good understanding of hepatic vascular anatomy is important prior to attempting open repair of liver injury. Coagulopathy from hypothermia, hemodilution, massive blood transfusion, and coagulation factor depletion may contribute to ongoing hemorrhage. Angiographic techniques may be useful in selected unstable patients with extensive liver injuries or when operative techniques have failed.

Postop

Careful postoperative hemodynamic management and fluid management are indicated following repair of major hepatic injuries. Hypertension may aggravate bleeding and should be avoided. Hypothermia should be assiduously avoided. Coagulation parameters should be aggressively monitored and abnormalities corrected. Alterations in renal function are frequently encountered after severe hemorrhagic shock. Unstable patients should be monitored in an ICU, and strong consideration should be given to utilizing a pulmonary artery catheter to guide resuscitation, even in young, otherwise healthy patients. Marked elevations of transaminase levels suggest significant ischemic injury and may signal evidence of hepatic insufficiency.

Complications

Complications include rebleeding, hemobilia, and intrahepatic arteriovenous fistulae. Postoperative bile leaks are common and can often (but not always) be controlled with biliary stents or percutaneous drains. Infectious complications, such as subphrenic abscess or intrahepatic abscesses, are also frequent. Hepatic insufficiency, acute renal failure, ARDS, multiple organ dysfunction syndrome, and death may occur.

59 Follow-Up

The patient should be followed until all wounds have healed. Long-term follow-up depends on the nature of the underlying disease/injury.

Acknowledgment

The editors and author wish to acknowledge Michael A. West for contributing to the previous version of this chapter.

SECTION 2: ENDOCRINE

Section Editor: Cord Sturgeon

Chapter 60

Adrenalectomy: Laparoscopic

Dina Elaraj and Cord Sturgeon

Indications

The differential diagnosis of an adrenal mass includes benign and malignant functional or nonfunctional primary adrenal tumors, cysts, angiomyolipomas, or metastases. Adrenalectomy is indicated for the removal of functional adrenal tumors, nonfunctional tumors that are larger than 3-4 cm or increasing in size, adrenal cancers, or isolated metastases in the setting of adequate treatment of the primary site. If the tumor is large or if there is preoperative suspicion for malignancy or evidence of invasion into surrounding structures, adrenalectomy should be performed via an open approach.

Preop

The preoperative workup of an adrenal mass includes high-resolution cross-sectional imaging and biochemical testing to evaluate for functionality. The most common functional tumors and their common screening tests are as follows:

1. Pheochromocytoma—Plasma-free metanephrines or 24 hour urine collection for vanillylmandelic acid, catecholamines, and metanephrines.
2. Aldosteronoma (Conn's tumor)—Measurement of serum potassium, serum aldosterone, and plasma renin activity.
3. Cushing's syndrome—Low-dose dexamethasone suppression test, midnight salivary cortisol test, or 24 hour urine collection for cortisol.
4. Virilizing and feminizing tumors—usually screened clinically.

In addition, all patients with nonfunctional adrenal masses should undergo age, gender, and risk factor-appropriate screening tests such as mammography, colonoscopy, and chest X-ray. If the patient does have a pheochromocytoma, preoperative alpha-adrenergic blockade for a period of 1-3 weeks and rehydration are mandatory prior to surgical resection. Beta-adrenergic blockade should be added if there is persistent tachycardia or dysrhythmia after alpha blockade. An arterial line and central line are mandatory when performing adrenalectomy for pheochromocytoma. If an aldosterone-secreting tumor is diagnosed, potassium should be normalized preoperatively. Only patients with Cushing's syndrome require a preincision dose of prophylactic antibiotics. All patients should receive perioperative deep venous thrombosis prophylaxis.

Procedure

There are multiple approaches for laparoscopic adrenalectomy. The lateral transabdominal approach is the most common for benign adrenal tumors and is described herein.

Northwestern Handbook of Surgical Procedures, 2nd Edition, edited by Nathaniel J. Soper and Dixon B. Kaufman. ©2011 Landes Bioscience.

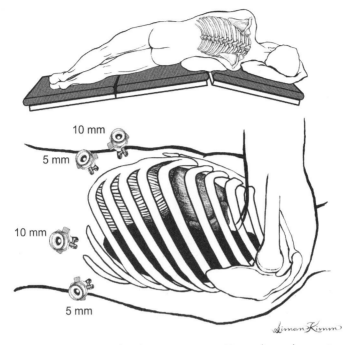

Figure 60.1. Laparoscopic adrenalectomy. Patient position and port placement.

Step 1. The operating room is set up with the monitors at the head of the bed. The patient is placed in the lateral decubitus position on a beanbag with the side of the tumor up and the break in the table located between the costal margin and the anterior superior iliac spine (ASIS). The table is flexed and the kidney rest elevated, thus opening and maximizing the space between the costal margin and the ASIS. The surgeon stands facing the patient's abdomen.

Step 2. The entire flank is prepped and draped from the spine to the umbilicus, and from the nipple to the iliac crest. The lower chest and entire abdomen are draped into the field to allow maximal access.

Step 3. The positions for port sites are marked approximately 1-2 fingerbreadths below the costal margin extending from the mid-axillary line to the lateral border of the rectus muscle. Pneumoperitoneum is then established.

Step 4. A 30° laparoscope is then inserted, and three additional ports are placed along the subcostal margin. It may be necessary to take down the lateral attachments of the left colon to place the lateral port on the left side. It is very rare to need to take down the hepatic flexure to get further exposure.

Step 5. For left adrenalectomy, the lateral attachments of the spleen are divided, extending the dissection cephalad to the level of the diaphragm until the stomach is visualized. This allows the spleen and tail of pancreas to fall anteromedially and opens up the retroperitoneal space. On the right side, the lateral attachments of the liver and right triangular ligament are divided in order to open the retroperitoneal space. Gerota's fascia is opened over the adrenal and superior pole of the kidney.

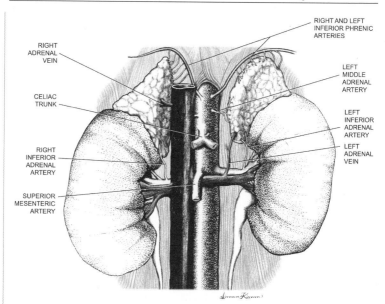

RIGHT AND LEFT
INFERIOR PHRENIC
ARTERIES

RIGHT
ADRENAL
VEIN

LEFT
MIDDLE
ADRENAL
ARTERY

CELIAC
TRUNK

LEFT
INFERIOR
ADRENAL
ARTERY

LEFT
ADRENAL
VEIN

RIGHT
INFERIOR
ADRENAL
ARTERY

SUPERIOR
MESENTERIC
ARTERY

Figure 60.2. Laparoscopic adrenalectomy. Adrenal anatomy.

Step 6. The superomedial aspect of the adrenal gland and surrounding periadrenal fat are dissected in a "top-down" (cephalad to caudad) direction until the diaphragm and posterior abdominal wall muscles are seen. On the left side, the dissection is done along the medial aspect of the adrenal gland until the adrenal vein is encountered. The inferior phrenic vein may be encountered in this plane and traced inferiorly to the adrenal vein. If there is difficulty identifying the adrenal gland, a laparoscopic ultrasound can be used to identify an adrenal mass in the retroperitoneal fat. On the right side, the dissection is also done in a cephalad to caudad direction along the medial aspect of the adrenal gland and lateral aspect of the inferior vena cava (IVC). Note that the right adrenal vein is short and inserts into the posterolateral aspect of the IVC. It is not uncommon to encounter an accessory right adrenal vein draining into the right hepatic vein.

Step 7. For pheochromocytoma, the adrenal vein should be controlled as soon as is safely possible, keeping in mind that once this vein is divided the blood pressure may fall dramatically, and due to impaired venous outflow the adrenal tumor will become engorged and the dissection more bloody.

Step 8. The adrenal arteries are often quite small and are usually controlled by hemostatically dividing the periadrenal tissues. The plane posterior to the adrenal is essentially avascular. The lateral attachments of the adrenal are divided last, thus completing the dissection.

Step 9. Once the gland with the periadrenal fat is completely resected en bloc, the specimen is placed into an impermeable nylon bag and then removed through one of the port sites.

Step 10. This resection bed is inspected, irrigated, and drained of fluid to confirm adequate hemostasis. Drains are not necessary.

60

Postop

Pheochromocytoma patients are admitted to the ICU for hemodynamic monitoring, adequate fluid resuscitation, and vasopressor therapy if necessary. Patients with Cushing's syndrome will require a prolonged (up to 1 year in some cases) steroid taper due to suppression of the contralateral adrenal gland. Furthermore, glucose homeostasis is usually improved and diabetic regimens often require modulation. Aldosteronoma patients should be taken off potassium and evaluated for hyperkalemia. To avoid hypotension, pheochromocytoma and aldosteronoma patients are taken off their antihypertensives after surgery and blood pressure is closely monitored. Most patients can be safely discharged 1-2 days after a laparoscopic adrenalectomy.

Complications

Patients should be closely followed for any signs of hemorrhage or peritonitis due to inadvertent injury of any of the organs in proximity to the adrenal gland, such as the colon, stomach, pancreas, spleen, or liver. Adrenal insufficiency is a rare late complication after unilateral adrenalectomy.

Follow-Up

Patients with functional tumors should be followed to ascertain resolution of symptoms and signs of excess hormone secretion. Patients with pheochromocytoma should have annual screening to monitor for recurrence, which may represent malignancy.

Acknowledgment

The editors and authors wish to acknowledge Peter Angelos for contributing to the previous version of this chapter.

60

Chapter 61

Enucleation of Insulinoma

Dina Elaraj and Cord Sturgeon

Indications

Insulinoma is the most common neoplasm of the endocrine pancreas. They are found with an equal distribution within the head, body, and tail of the gland. Since 90% of insulinomas are solitary, benign, and measure <2 cm, enucleation is an appropriate treatment strategy as long as there is no evidence of malignancy such as invasion into surrounding structures or metastatic disease. Moreover, those tumors that are located deep within the gland or intimately related to the main pancreatic duct on imaging are not amenable to enucleation.

Preop

The diagnosis of insulinoma is a biochemical one and is made on the basis of an inappropriately high circulating insulin level at the time of hypoglycemia, usually measured during a 72 hour fast. Urinary sulfonylureas and C-peptide levels are also measured to exclude factitious hyperinsulinism.

Preoperative localization initially consists of cross-sectional imaging via pancreatic protocol triple phase computed tomographic (CT) scan or magnetic resonance imaging (MRI). Endoscopic ultrasound is excellent for evaluating the head of the pancreas and imaging the pancreatic duct. Sometimes more invasive procedures such as percutaneous transhepatic portal vein sampling for insulin or selective arterial injection of calcium with measurement of insulin in the hepatic veins is helpful to localize small tumors.

It is important to ensure that NPO status does not cause severe hypoglycemia. Dextrose-containing intravenous fluid should be given preoperatively and blood sugars maintained in the 60-80 mg/dL range. Deep vein thrombosis prophylaxis with sequential compression devices or subcutaneous heparin should be given.

Procedure

Enucleation may be done either open or laparoscopically, depending on the location of the tumor and the surgeon's experience with advanced laparoscopy.

Step 1. For an open approach, the abdomen is prepped and draped for an upper midline or bilateral subcostal incision. For a laparoscopic approach, the patient's legs are placed in stirrups to allow the surgeon to stand between the legs. A 30° laparoscope is placed at the umbilicus, a second port close to the xiphoid process to retract the stomach, and two additional ports are placed in the right and left upper quadrants to form a diamond configuration.

Northwestern Handbook of Surgical Procedures, 2nd Edition, edited by Nathaniel J. Soper and Dixon B. Kaufman. ©2011 Landes Bioscience.

Step 2. The abdomen is fully explored. Metastases to the liver and regional lymph nodes must be excluded, as their presence is likely to change the planned operation. Although distant metastatic disease usually precludes cure, enucleation or formal pancreatic resection with or without resection of metastatic deposits may be indicated for symptom control, taking into account the extent of disease and the patient's functional status and operative risk.

Step 3. The first step of the dissection involves exposure of the body and tail of the pancreas by entering the lesser sac. The stomach is retracted cephalad, the transverse colon retracted caudad, and the gastrocolic omentum incised. It is sometimes necessary to take down the splenic flexure of the colon in order to completely expose of the tail of the pancreas. Attachments between the pancreas and posterior stomach are then taken down to expose the entire anterior surface of the pancreas.

Step 4. The next step involves exposure of the posterior head of the pancreas. The hepatic flexure of the colon is taken down and a wide Kocher maneuver is performed by dividing the hepatoduodenal ligament in its avascular plane from the second portion of the duodenum superiorly to the inferior vena cava and aorta inferiorly.

Step 5. Intraoperative ultrasound, either open or laparoscopic, is then used to confirm the location and focality of the tumor and assess its proximity to the main pancreatic duct. If the tumor is located deep within the pancreatic parenchyma or too close to the main pancreatic duct, then pancreatectomy rather than enucleation should be done.

Step 6. If no tumor is seen, a laparoscopic procedure should be converted to an open procedure in order to explore the pancreas by bimanual palpation. The body and tail of the pancreas are further mobilized by incising the peritoneum along the inferior border of the pancreas and bluntly dissecting the retropancreatic space. The inferior mesenteric vein may be divided, if necessary. Further exposure of the posterior surface of the pancreas may be achieved by taking down the lateral attachments of the spleen and rotating the spleen and tail of the pancreas anteromedially.

Step 7. The head, uncinate process, body, and tail of the pancreas are then carefully evaluated by bimanual palpation. If neither ultrasound nor bimanual palpation is successful in finding the tumor, the operation should be terminated and the patient referred for further testing. Blind distal pancreatectomy should not be done.

Step 8. Once the tumor has been identified, enucleation is performed using blunt dissection immediately on the tumor capsule. If the edges of the tumor are not apparent or the tumor appears to be irregular or infiltrating, enucleation should be abandoned and a formal resection performed.

Step 9. The tumor bed is then inspected for hemostasis and for any evidence of a major pancreatic duct injury. Any suspected ductal injury should be repaired over a stent if possible, passing the tip of the stent into the duodenum for later retrieval. If a major duct injury is present and the surgeon is unable to repair it without difficulty, it is best to proceed with resection of the involved area.

Step 10. A closed-suction drain should be placed near the enucleation site and brought out through a separate stab incision.

61

Postop

Dextrose-free solutions should be used for intravenous fluid replacement. Blood sugars should be monitored frequently because they typically rise quickly (even while still in the operating room). Overnight, blood sugar elevations may reach the mid-200s and require insulin. Blood sugar should be checked three times per day until stable. Patients are requested to check a fasting blood sugar daily until their follow-up clinic visit.

Patients may be fed as soon as there is return of bowel function. The drain is kept in place until the patient is tolerating food and there is no amylase-rich drainage. If there is a pancreatic leak, the drain is kept in place until the fistula resolves, usually when the output is less than 30 ml/day.

Complications

Complications of enucleation are relatively frequent and include pancreatic duct injury with pancreatic fistula and/or pseudocyst formation, peripancreatic abscess, and pancreatitis.

Follow-Up

Patients with sporadic, benign insulinomas are not likely to recur. Patients with multiple endocrine neoplasia Type 1 (MEN 1) often have multiple functional and nonfunctional islet cell tumors and are usually treated with generous distal pancreatectomy along with enucleation of tumors from the head of the pancreas. These patients must be followed for endocrine and exocrine insufficiency. Malignant tumors require long-term follow-up for disease recurrence.

Acknowledgment

The editors and authors wish to acknowledge Daphne W. Denham for contributing to the previous version of this chapter.

Chapter 62

Parathyroidectomy: Four Gland Exploration

Cord Sturgeon and Dina Elaraj

Indications

Parathyroidectomy is the only curative therapy for primary hyperparathyroidism (PHPT). There is no debate in the surgical or medical community that parathyroidectomy is indicated for patients with symptomatic PHPT. Many specialists also believe that parathyroidectomy is indicated for patients who clearly have a biochemical diagnosis of PHPT, but are asymptomatic. The success of parathyroidectomy for PHPT, with or without preoperative localization, is higher than 95% in the hands of experienced surgeons.

Preop

Approximately 80-85% of cases of PHPT are caused by a single parathyroid adenoma. Hyperplasia accounts for about 12%, double or triple adenomas for 2-3%, and parathyroid cancer for less than 1% of cases. Since the majority of patients have single gland disease, it is appropriate to obtain imaging studies prior to surgery to attempt to localize the abnormal gland and facilitate a focused operation. The authors routinely obtain preoperative sestamibi scanning and neck ultrasonography. For those patients in whom the imaging studies are nonlocalizing, a four-gland exploration is recommended. For patients found to have multigland disease, a four-gland exploration is likewise mandatory.

Knowledge of the embryologic development and descent of the parathyroid glands is essential for successful parathyroidectomy. The lower parathyroid gland migrates with the thymus (3rd branchial pouch) and has the greatest potential for ectopic location; it may be located anywhere from the angle of the mandible to the arch of the aorta. The upper parathyroid gland migrates with the thyroid gland (4th branchial pouch) along a relatively short path and is less variable in location.

Procedure

Step 1. The patient is placed supine, with a shoulder roll horizontally under both scapulae and the neck extended.

Step 2. A transverse cervical incision is made 1 cm inferior to the cricoid cartilage in a Langer's line if possible.

Step 3. The subcutaneous tissue and platysma are opened and subplatysmal flaps are raised.

Step 4. The strap muscles (sternohyoid and sternothyroid) are opened in the midline raphe and elevated from the surface of the thyroid.

Step 5. The loose areolar tissues along the lateral aspect of the thyroid are gently swept laterally and the middle thyroid vein is ligated and divided.

62

Northwestern Handbook of Surgical Procedures, 2nd Edition, edited by Nathaniel J. Soper and Dixon B. Kaufman. ©2011 Landes Bioscience.

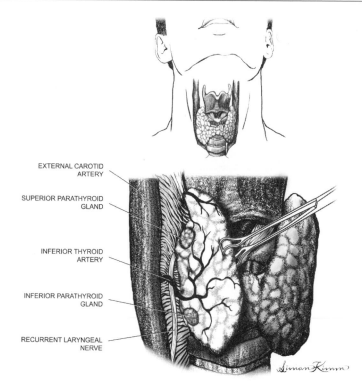

Figure 62.1. Parathyroid anatomy.

Step 6. The thyroid lobe is mobilized anteromedially. A bloodless field is essential to allow good visualization of the surgical landmarks.

Step 7. The inferior thyroid artery is identified. The recurrent laryngeal nerve (RLN) is identified. In 80% of cases each parathyroid gland is supplied by a single branch off the inferior thyroid artery. The blood supply to the upper gland may also come from the superior thyroid artery. It is important to not devascularize normal glands during a four-gland exploration.

Step 8. Both parathyroid glands can usually be found within 1 cm of the crossing of the inferior thyroid artery and the RLN. The upper parathyroid is usually located posterior to the RLN, and the lower parathyroid is located anterior to the RLN. Normal parathyroid glands are up to 6 or 7 mm in length and weigh 40-60 mg.

Step 9. Most parathyroid glands are found in the normal anatomic location. Occasionally parathyroid glands are encountered in ectopic sites such as the thyrothymic ligament, thymus, mediastinum, retroesophageal space, thyroid gland, carotid sheath, or undescended at the skull base. Ectopic glands can usually be identified by preoperative localization studies but may require careful systematic dissection in the areas listed above. Supernumerary parathyroid glands may be encountered in up to 15% of patients.

Step 10. Begin the search for an upper gland on the posterior surface of the thyroid, cephalad to the inferior thyroid artery and posterior to the RLN. In order to identify the upper glands, it is sometimes necessary to mobilize the superior pole of the thyroid by ligating and dividing the superior pole vessels. If no gland is found in this area, then explore the tracheoesophageal groove and para- and retro-esophageal spaces.

Step 11. Begin the search for a lower gland near the lower pole of the thyroid lobe. Look in the fat between the inferior thyroid veins, and in the thyrothymic ligament. If no gland is found in this area, perform a transcervical thymectomy. If no gland is found in the thymus, explore the carotid sheath from the level of the carotid bifurcation to the base of the neck. If this exploration is negative, consider that the missing lower gland may be intrathyroidal.

Step 12. Identify all four glands prior to removal of any parathyroid tissue. Aberrant glands may sometimes be identified by looking for a large vascular pedicle originating from the inferior thyroid artery. Furthermore, an enlarged parathyroid will often "float" freely in the perithyroidal connective tissue, whereas lymph nodes and thyroid nodules are usually more fixed to the surrounding structures and do not have a "float sign". There is no need to routinely biopsy normal parathyroid glands.

Step 13. Intraoperative adjuncts such as ultrasound, gamma probe, and rapid PTH determination may be employed to facilitate the operation.

Step 14. In the case of a single adenoma, gently dissect along the surface of the tumor until it is completely exposed, leaving the vascular pedicle for last. The tumor should not be grasped because of the potential for rupture and subsequent parathyromatosis. A firm tumor that is difficult to dissect due to adherence should raise consideration of parathyroid carcinoma, which requires en bloc resection of the parathyroid and ipsilateral thyroid lobe.

Step 15. In the case of multigland disease, the diseased glands are removed, but the equivalent of one normal parathyroid gland must stay in the patient to avoid permanent hypoparathyroidism. In this situation, dissect all four glands, leaving them on intact vascular pedicles. Place large hemoclips across the middle of the two most normal-appearing and well-vascularized parathyroid glands. Subtotally resect these two glands and entirely remove the other two glands. Send biopsies from each gland for histopathology. If possible, the remainder of the resected parathyroid tissue should be cryopreserved. Observe the two remaining glands for viability and retain only the subtotally resected gland that appears to have the best blood supply.

Step 16. The strap muscles, platysma and skin are reapproximated.

Postop

Most patients can be discharged within 23 hours. Patients are started on oral calcium supplementation after surgery. Patients are taught to recognize the symptoms of hypocalcemia. Serum calcium levels are drawn in the postoperative period based on institutional protocol or in response to symptoms of hypocalcemia.

Complications

62

Complications of parathyroidectomy are similar to those of thyroidectomy and include temporary or permanent hypocalcemia, temporary or permanent recurrent or superior laryngeal nerve injury, neck hematoma, infection, and persistent or recurrent hyperparathyroidism.

Follow-Up

Most clinicians check calcium once or twice in the first month, unless the patient suffers profound hypocalcemia which would necessitate more frequent testing. Normocalcemia at six months after surgery is the benchmark for cure; however, yearly calcium measurements are taken indefinitely to evaluate for recurrent hypercalcemia. Patients are encouraged to consume 1500 mg of calcium daily, get adequate vitamin D, and engage in weight-bearing exercise. Bone density scanning should be considered one year after surgery to establish the new bone density baseline.

Acknowledgment

The editors and authors wish to acknowledge Daphne W. Denham for contributing to the previous version of this chapter.

Chapter 63

Focused Parathyroidectomy for Primary Hyperparathyroidism

Dina Elaraj and Cord Sturgeon

Indications

As discussed in the previous chapter, 80-85% of cases of primary hyperparathyroidism (PHPT) are due to a solitary parathyroid adenoma. With the development of accurate preoperative and intraoperative localization techniques, a focused, or minimally-invasive, dissection has become feasible for the majority of these patients. Focused approach parathyroidectomy is indicated for patients with sporadic PHPT who have a single adenoma localized by preoperative imaging studies. Patients with familial hyperparathyroidism and multiple endocrine neoplasia (MEN) have multi-gland disease and thus are not appropriate candidates for the focused approach.

Preop

Our preference is to routinely obtain preoperative sestamibi scanning and neck ultrasonography. Sestamibi scanning provides both functional and anatomic information. A positive scan is defined as one showing a focus of persistent activity. Ultrasound provides anatomic information such as size, location, and depth of the parathyroid adenoma. These tests are considered concordant when they each identify an abnormality in the same location.

Procedure

The initial steps of a focused parathyroidectomy via an anterior approach are identical to that of a bilateral neck exploration, except that more limited mobilization of the thyroid gland and surrounding perithyroidal tissues can be done. Some surgeons offer the procedure under local or regional anesthesia.

Step 1. The patient is placed supine, with a shoulder roll placed horizontally under both scapulae and the neck extended.

Step 2. Pre-incision, intraoperative ultrasound is useful to help guide the location of the incision. A transverse cervical incision is made near the location of the parathyroid adenoma, in a Langer's line if possible.

Step 3. The subcutaneous tissue and platysma are opened and limited subplatysmal flaps are raised.

Step 4. The strap muscles (sternohyoid and sternothyroid) are opened through the midline raphe and elevated from the surface of the thyroid.

Step 5. If an upper parathyroid adenoma in a normal location is sought, then the thyroid lobe must be mobilized. The loose areolar tissues along the lateral aspect of the thyroid are gently swept laterally and the middle thyroid vein is ligated and divided.

63

Northwestern Handbook of Surgical Procedures, 2nd Edition, edited by Nathaniel J. Soper and Dixon B. Kaufman. ©2011 Landes Bioscience.

Step 6. Begin the search for the upper gland on the posterior surface of the thyroid, cephalad to the inferior thyroid artery and posterior to the RLN. In order to identify an upper gland it is sometimes necessary to mobilize the superior pole of the thyroid by ligating and dividing the superior pole vessels.

Step 7. If a lower parathyroid adenoma in a normal location is sought, the thyroid lobe does not need to be mobilized so extensively. Begin the search for the lower gland near the lower pole of the thyroid lobe, anterior to the recurrent laryngeal nerve. Look in the fat between the inferior thyroid veins, and in the thyrothymic ligament.

Step 8. After identification of the parathyroid adenoma, it is gently dissected from the surrounding tissues, the vascular pedicle ligated, and the adenoma removed.

Step 9. Intraoperative adjuncts such as ultrasound, gamma probe, and rapid PTH determination may be employed to facilitate the operation.

Step 10. The strap muscles, platysma and skin are reapproximated.

Postop

Most patients can be discharged within 23 hours. Patients are started on oral calcium supplementation and are taught to recognize the symptoms of hypocalcemia. Serum calcium levels are drawn in the postoperative period based on institutional protocol or in response to symptoms of hypocalcemia.

Complications

The risks of parathyroidectomy include temporary or permanent hypocalcemia, temporary or permanent recurrent or superior laryngeal nerve injury, neck hematoma, infection, and persistent or recurrent hyperparathyroidism.

Follow-Up

Normocalcemia at six months after surgery is the benchmark for cure. Yearly calcium measurements are taken indefinitely to evaluate for recurrent hypercalcemia. Patients are encouraged to get 1500 mg of calcium intake daily, get adequate vitamin D, and engage in weight-bearing exercise. Bone density scanning should be considered one year after surgery to establish the new bone density baseline.

Acknowledgment

The editors and authors wish to acknowledge Daphne W. Denham for contributing to the previous version of this chapter.

Chapter 64

Thyroidectomy

Cord Sturgeon and Dina Elaraj

Indications

Thyroid operations are classified by the amount of thyroid tissue removed. The removal of one thyroid lobe with or without the isthmus is described as a hemithyroidectomy or thyroid lobectomy. Subtotal thyroidectomy means the bilateral removal of greater than 50% of each thyroid lobe as well as removal of the isthmus. The total removal of one lobe and the subtotal removal of the other is a specific type of subtotal thyroidectomy also known as the Hartley-Dunhill procedure. The total removal of both lobes is properly termed a total thyroidectomy. If a small amount of thyroid tissue (<1 gram) is left in the region of the ligament of Berry on one or both sides the operation should be described as a near-total thyroidectomy.

A diagnostic hemithyroidectomy is usually performed when fine needle aspiration (FNA) has been unable to exclude malignancy within a thyroid mass. Hemithyroidectomy is also an appropriate treatment for uninodular goiter. A total or near-total thyroidectomy is indicated for the treatment of thyroid cancer, symptomatic multinodular goiter, or bilateral nodular disease when cancer is a concern. Subtotal or near-total thyroidectomy is a treatment option for most patients with Graves' disease.

Preop

The standard preoperative evaluation for thyroidectomy includes physical exam of the head, neck and chest. Particular attention should be paid to the size of the thyroid, location and size of thyroid nodules, presence of cervical lymphadenopathy, tracheal deviation, and possible extension of the thyroid into the chest. FNA biopsy is the best test to evaluate for the possibility of malignancy within thyroid nodules or cervical lymph nodes. Routine laboratory examination includes thyroid function studies and serum calcium. Ultrasonography of the thyroid should always be performed. If substernal extension is suspected by findings on physical exam or ultrasound, a noncontrast CT of the chest should be performed. If the patient has ever had neck surgery, it is essential to perform preoperative laryngoscopy to rule out a subclinical unilateral recurrent laryngeal nerve injury.

Procedure

Step 1. After the induction of general endotracheal anesthesia, the patient's neck is extended and the head supported in nearly full extension. The patient is placed in semi-Fowler's position to decompress the neck veins.

Step 2. The entire neck up to the chin, laterally to the shoulders, and down onto the upper chest is prepped.

64

Northwestern Handbook of Surgical Procedures, 2nd Edition, edited by Nathaniel J. Soper and Dixon B. Kaufman. ©2011 Landes Bioscience.

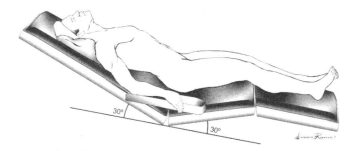

Figure 64.1. Thyroid lobectomy and total thyroidectomy. Patient position.

Step 3. A transverse collar incision is made 1-2 fingerbreadths above the sternal notch. The incision should be centered in the midline, approximately 1 cm inferior to the cricoid cartilage, over the isthmus, in a skin crease if possible. Most thyroid resections can be done safely through 5 cm or smaller incisions.

Step 4. Subplatysmal flaps are created sharply or with electrocautery. The limits of the subplatysmal flaps are the notch of the thyroid cartilage superiorly, the sternal notch inferiorly, and the medial borders of the sternocleidomastoid muscles bilaterally.

Step 5. The strap muscles (sternohyoid and sternothyroid) are separated in the midline with electrocautery and retracted laterally. This allows exposure of the thyroid gland. The strap muscles are then elevated off the thyroid lobe. It is usually not necessary to divide the strap muscles to provide exposure. Dissection should begin on the side with the tumor. The lateral edge of the thyroid gland is mobilized by dividing the middle thyroid vein (or veins) and gently sweeping away the loose areolar tissue along the lateral aspect of the gland to the top of the superior pole. This allows the thyroid gland to be rotated medially.

Step 6. Attention is then directed toward the superior pole of the thyroid lobe. The avascular plane between the superior pole and the cricothyroid muscle is entered to allow the superior pole of the thyroid to be retracted in a caudal direction. This allows careful exposure of the superior pole vessels as they enter the thyroid capsule. The vessels are divided and ligated individually along the surface of the gland in order to prevent injury to the external branch of the superior laryngeal nerve.

Step 7. With the superior vessels divided, the thyroid lobe can be rotated anteromedially. Dissection continues along the capsule of the thyroid. The parathyroid glands are then identified and carefully dissected off the capsule of the thyroid gland leaving their vascular pedicles intact. If the parathyroid glands are not immediately apparent it is helpful to identify the recurrent nerve and the inferior thyroid artery. Both parathyroid glands are usually found within a 1 cm radius of the crossing of these two structures.

Step 8. Once the parathyroid tissue is safely swept off the thyroid, gentle dissection in the tracheoesophageal groove will expose the recurrent laryngeal nerve. Once the nerve is identified and traced to its insertion into the cricothyroid muscle, branches of the inferior thyroid artery are individually divided as close to the thyroid as possible taking care to prevent devascularization of the parathyroid glands.

Step 9. Once the course of the recurrent laryngeal nerve is clearly visualized, the inferior pole vessels can be ligated. Next, the ligament of Berry is carefully divided making sure that the dissection plane is right on the surface of the trachea. The thyroid lobe can then be dissected from the surface of the trachea.

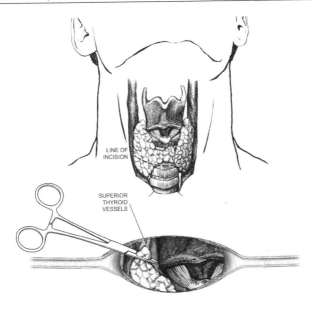

Figure 64.2. Thyroid lobectomy and total thyroidectomy.

Step 10. A thyroid lobectomy is completed by dividing the isthmus. For total thyroidectomy, Steps 5-9 are repeated for the opposite lobe. Many surgeons prefer to remove the thyroid en bloc without dividing the isthmus, especially when a total thyroidectomy is performed for cancer.

Step 11. The thyroid bed is irrigated and carefully inspected for hemostasis. Parathyroid glands which appear to be devascularized, should be removed from the patient, minced into multiple submillimeter pieces, and autotransplanted into multiple pockets in the sternocleidomastoid muscle. Each pocket should be oversewn in a figure-of-eight fashion and marked with a titanium clip.

Step 12. In the case of thyroid cancer, lymph nodes in the central compartment should be carefully evaluated by inspection and palpation. Any abnormal lymph nodes should be removed and sent for frozen section. If metastatic cancer is identified, a formal compartmental nodal dissection is warranted.

Step 13. The strap muscles and platysma are reapproximated. Skin closure can be performed with monofilament subcuticular sutures, clips, or surgical adhesive. The authors favor minimal neck dressings to allow for rapid identification of a hematoma should one develop. Surgical drains are almost never necessary, but may be considered for very large goiters, extensive nodal dissection, or if there is suspected injury to the aerodigestive tract.

Postop

After thyroidectomy the authors routinely admit patients overnight. If the patient has undergone a total or completion thyroidectomy, we obtain calcium measurements on postoperative day #1 and routinely discharge patients on calcium supplementation.

64

Complications

The major complications of thyroidectomy are neck hematoma, temporary or permanent hypocalcemia, and unilateral or bilateral recurrent laryngeal nerve injury. If a cervical hematoma, respiratory distress, or stridor develops, the incision and all layers of the neck should be opened without delay at the bedside, and the airway controlled. Hypocalcemia is initially treated with oral calcium supplementation. More profound hypocalcemia is treated with intravenous calcium or oral calcitriol. Recurrent laryngeal nerve injury is suspected in cases of dysphonia, aspiration, or dyspnea, and is best assessed by laryngoscopy. Aspiration precautions and airway protection (tracheostomy) may be necessary in cases of unilateral or bilateral recurrent nerve injury.

Follow-Up

After uncomplicated thyroid surgery, patients are seen for an initial follow-up visit at approximately 2-3 weeks postoperatively. If a thyroid lobectomy was performed, approximately 40% of patients will require thyroid hormone replacement, which can be determined by monitoring the TSH level 4-6 weeks after surgery. Patients with postoperative hypocalcemia are closely monitored with serial calcium measurements and titration of calcium and vitamin D doses.

Acknowledgment

The editors and authors wish to acknowledge Peter Angelos for contributing to the previous version of this chapter.

Functional Neck Dissection for Thyroid Cancer

Cord Sturgeon and Dina Elaraj

65

Indications

The nomenclature most frequently used to describe the neck nodal basins was developed by the Memorial Sloan-Kettering Head and Neck Surgery Service. The neck is divided into seven node-bearing regions, and each is referred to by a Roman numeral.

I.   Submental and submandibular triangles.

II. Upper jugular nodes located between the digastric muscle superiorly and the hyoid bone (clinical landmark) or carotid bifurcation (surgical landmark) inferiorly. Level II nodes are divided into IIa which includes nodes anterior to the spinal accessory nerve, and IIb posterior to the nerve.

III. Middle jugular nodes found between the carotid bifurcation and the cricothyroid (clinical landmark) or omohyoid muscle (surgical landmark).

IV. Lower jugular nodes found between the omohyoid muscle superiorly and the clavicle.

V.   Posterior triangle nodes found between the anterior border of the trapezius muscle, the posterior border of the sternocleidomastoid muscle (SCM), and the clavicle. This group is also subdivided by the omohyoid muscle into a supraclavicular (Va) and spinal accessory (Vb) group.

VI. Central neck lymph nodes located between the carotid sheaths extending from the level of the hyoid bone superiorly to the suprasternal notch inferiorly.

VII. Superior mediastinal nodes from manubrium to innominate vein.

In the classic "radical neck dissection" described by George Crile in 1906, all the node-bearing tissue of the lateral neck (regions I-V) is removed along with the SCM, spinal accessory nerve, and internal jugular vein. This operation is not indicated for thyroid cancer metastases. Modifications of this approach, sparing these important structures and/or limiting the dissection to specific nodal regions are known as "modified radical neck dissections" (MRND). The most commonly employed MRND for removal of nodal metastases from thyroid cancer is the "functional" or "lateral" neck dissection. This modification limits the operative field to Levels II, III, and IV and spares the SCM, spinal accessory nerve, and internal jugular vein. The authors typically perform formal nodal clearance of Levels IIa, III, IV, and Va for thyroid cancer metastases in the lateral neck. This chapter will describe this technique as it applies to formal clearance of nodal metastases from thyroid cancer. Selective excision of individual nodes (berry picking) is not considered appropriate in this context.

Northwestern Handbook of Surgical Procedures, 2nd Edition, edited by Nathaniel J. Soper and Dixon B. Kaufman. ©2011 Landes Bioscience.

Preop

If nodal metastases from thyroid cancer are suspected based on physical exam, ultrasound, or other imaging modality, the most appropriate first step is fine needle aspiration biopsy of the suspicious node. For patients with biopsy-proven nodal metastases, the authors recommend formal compartmental clearance of the lateral neck. As above, Levels IIa, III, IV, and Va are usually included in the dissection. A detailed discussion of the risks of surgery is mandatory (see below). Formal imaging of the lateral neck with ultrasound is very helpful to plan the operation. Chest radiography is helpful to establish that both phrenic nerves are functioning. If the patient has had prior neck surgery, laryngoscopy should be performed to rule out a subclinical vocal cord paresis.

Procedure

Step 1. The operating room setup is similar to thyroidectomy. General anesthesia is used without paralytic agents. The head is turned to extend and expose the lateral aspect of the neck. The corner of the mouth should be visible through clear draping.

Step 2. The incision is carefully planned based on extent of nodal dissection. Many incisions have been described. The authors favor either a Kocher incision with or without a cranial parallel counter-incision, or a hemi-apron (hockey stick) incision along the posterior border of the sternocleidomastoid muscle extending anteriorly into a Kocher incision. For simplicity, the approach via a hockey stick incision will be described. It is helpful to infiltrate the skin and subcutaneous tissues with lidocaine with epinephrine so that sharp dissection can be performed through a relatively bloodless field. Hash marks may be used to facilitate realignment of the incision at the end of the case.

Step 3. The skin and platysma are divided sharply along the posterior border of the sternocleidomastoid. Subplatysmal flaps are raised sharply toward the midline. The external jugular vein and greater auricular nerve are left intact on the surface of the SCM.

Step 4. The transverse cutaneous nerve is identified, and the communicating branch is traced to the cervical branch of facial nerve. These branches of the transverse cutaneous nerve are sacrificed as they cross the anterior border of the SCM. Care is taken to identify and preserve the marginal mandibular branch of the facial nerve at the medial aspect of the subplatysmal flap. The tissue along the anterior border of the SCM is sharply opened along its entire length.

Step 5. The SCM is retracted laterally. **The superior boundary of the functional neck dissection is the fibrofatty node-bearing tissue just inferior and medial to the spinal accessory nerve (Level IIa).** Nodal tissue lateral and posterior to the spinal accessory nerve (Level IIb) is often spared in the functional neck dissection for thyroid cancer unless it is grossly involved. The spinal accessory nerve is identified as it crosses the jugular vein and passes into the posterior aspect of the SCM approximately 1 cm inferior to the posterior belly of the digastric muscle. Retract the node-bearing tissue caudad and dissect inferior to the spinal accessory nerve until the deep cervical fascia is encountered.

Step 6. The lateral boundary of the functional neck dissection is the posterior border of the SCM. Incise the tissue along the posterior border of the SCM from Level IIa down to the omohyoid muscle. Identify and spare the sensory branches of the cervical plexus if possible.

Step 7. The posterior boundary of the dissection is the deep cervical fascia overlying the scalene muscles. Stay on top of the cervical plexus and sharply dissect the fibrofatty bundle off the deep cervical fascia. The plane of dissection is anterior to the phrenic nerve. **The medial boundary of the dissection is the carotid sheath.** Continue the dissection medially to the carotid sheath and identify and preserve the vagus nerve. Roll the internal jugular vein medially to complete the dissection of the lymph nodes deep to the carotid sheath.

65

Step 8. Either divide the omohyoid or dissect it completely from the underlying cervical fascia. If the omohyoid is spared, then retract the muscle cephalad and pass the fibrofatty bundle underneath it. Identify and spare the transverse cervical artery and branches of the brachial plexus.

Step 9. The inferior boundary of the dissection is the clavicle. Dissection of the node-bearing tissue continues sharply along the deep cervical fascia overlying the scalene muscles until the clavicle is reached. On the left, at the junction of the internal jugular vein and subclavian vein, the thoracic duct will be encountered. The duct must not be inadvertently injured or the patient will develop a chyle leak. If the duct has been intentionally or inadvertently injured, it must be controlled with ties or clips.

Step 10. The node-bearing tissue is retracted medially and sharp dissection is carried over the carotid sheath, which should free the tissue from the dissection bed en bloc.

Step 11. The neck is then irrigated, and a closed suction drain is placed through a separate stab wound. The platysma and skin are then reapproximated.

Postop

A chest radiograph is performed in the recovery room to rule out pneumothorax or elevated hemidiaphragm. The closed suction drain should be left in place until the patient is eating a regular diet and output is less than 30 ml in 24 hours. Patients may be discharged with the drain in place.

Complications

The complications specific to functional neck dissection include possible temporary or permanent nerve injury, temporary or permanent hypoparathyroidism, chyle fistula, chylothorax, hematoma, and infection. The most important nerves at risk during the operation are the spinal accessory nerve, vagus nerve (and therefore recurrent laryngeal nerve), hypoglossal nerve, greater auricular nerve, marginal mandibular nerve, brachial plexus, and phrenic nerve. Patents should be told about the possibility of resecting the jugular vein. Injury to the transverse cervical artery could lead to hypoparathyroidism because it gives off the inferior thyroid artery. Pneumothorax is a rare complication.

Follow-Up

Physical therapy is given to patients who sustain injury to the spinal accessory nerve. Voice alterations are almost always temporary. Patients with significant voice alteration or any aspiration of thin liquids should undergo laryngoscopy and swallow evaluation. The surgical site should be evaluated within the first few weeks, and again between 3 to 6 months. Adjuvant treatment for thyroid cancer depends on tumor type, pathologic findings, and prior radiation exposure.

Acknowledgment

The editors and authors wish to acknowledge Peter Angelos and Jeffrey D. Wayne for contributing to the previous version of this chapter.

SECTION 3: SURGICAL ONCOLOGY
Section Editor: David M. Mahvi

Chapter 66

Transanal Excision of Rectal Tumor

Amy L. Halverson

Indications

Transanal excision is appropriate for benign lesions that are not amenable to endoscopic resection and for early-stage malignant tumors in select individuals. A tumor may be considered for transanal resection if it is less than 9 cm from the anal verge, less than 4 cm in length, mobile (there should be no suggestion of anal sphincter involvement), and involves less than one-third the circumference of the rectal wall. Malignant tumors should be well or moderately differentiated and have no lymph or vascular invasion. Extension beyond the rectal wall or lymph node involvement should be ruled out with preoperative endorectal ultrasound. Transanal excision may be used for palliation in patients with overt metastatic disease.

Preop

- Complete history, including family history
- Physical examination
- Endorectal ultrasound to assess depth of invasion and lymph node involvement
- Colonoscopy to evaluate the entire colon
- CT scan to rule out metastatic disease
- Complete bowel preparation
- The patient should be placed in the prone position if the lesion is anterior. The lithotomy position may be used for posterior lesions, although some surgeons prefer the prone position for posterior lesions as well.

Procedure

Step 1. Visualize the mass through an operating anoscope.

Step 2. Score the line of resection around the mass with electrocautery.

Step 3. Excise the lesion. Consider using a laparoscopic electrosugical instrument to facilitate reach and hemostasis. A clamp may be placed on the mass for retraction. For tumors that are malignant or suspicious for malignancy, full-thickness excision should be performed. The yellow perirectal fat will be visible with full-thickness resection. For benign lesions, the submucosa may be infiltrated with saline to aid in resection of the polyp leaving the muscular layer of the bowel wall intact.

Step 4. Wounds limited to within 3-4 cm of the dentate line may be closed or left open. Higher lesions should be closed in a transverse fashion with full-thickness absorbable suture.

Step 5. Orient the specimen for the pathologist. This may be done by securing the polyp to a piece of cardboard or using different sutures to mark proximal or distal and right or left.

Step 6. After completion, double check for hemostasis.

Northwestern Handbook of Surgical Procedures, 2nd Edition, edited by Nathaniel J. Soper and Dixon B. Kaufman. ©2011 Landes Bioscience.

Postop

Patients should take stool softeners for approximately one week, or more if they tend to have constipation. Bleeding, urinary retention, or recurrence occur in 20% of patients.

Follow-Up

For benign or malignant lesions, repeat proctoscopy in 6 weeks to detect residual disease or early recurrence. For malignant tumors, surveillance includes physical examination including proctoscopy every 3 months for 2 years.

66

Chapter 67

Abdominoperineal Resection

Steven J. Stryker

Indications

Indications for abdominoperineal resection (APR) of the rectum can be categorized into absolute and relative. Absolute indications include malignancy of the rectum with sphincter involvement, carcinoma of the anal canal in an individual with prior pelvic radiation for an unrelated malignancy, carcinoma of the anal canal that is persistent or has recurred after combined modality chemotherapy and radiation, and anorectal Crohn's disease with uncontrolled local septic complications. Relative indications for APR include malignancy of the rectum not involving the sphincter when continence is already impaired preoperatively, ulcerative colitis or Crohn's proctitis requiring surgical intervention in an individual not desiring a sphincter-preserving procedure, and radiation-induced proctitis not responding to nonoperative measures or fecal diversion alone.

Preop

Preoperative preparation consists of a thorough diagnostic workup to assess the extent and severity of the disease process. This may include, but is not limited to, colonoscopy, endorectal ultrasound, and computerized tomography. If studies demonstrate unilateral or bilateral hydronephrosis, cystoscopy with placement of ureteral catheters should be considered at the time of the APR. The patient should meet with an enterostomal therapist prior to surgery, both to mark the best-suited site for colostomy placement and to provide educational materials relative to stomal function and care. Perioperative intravenous antibiotics are administered to decrease the incidence of postoperative infectious complications. The advantages of oral antibiotics on the day prior to surgery, as well as a mechanical bowel lavage, are not as well-documented to decrease infection and are currently used much less frequently than in the past. All patients undergoing APR should have a type and crossmatch because of the risk, albeit low, of hemorrhage if the presacral venous plexus is disrupted intraoperatively. Pneumatic compression sleeves for the lower extremities are used to minimize the risk of deep venous thrombosis and, unless medically contraindicated, all patients receive postoperative subcutaneous heparin.

Procedure

Step 1. The patient is positioned in a dorsal lithotomy fashion allowing simultaneous access to the abdomen and perineum. It is preferable to position the patient awake to check for comfort in positioning with respect to the back, hips, and knees. Once the patient confirms that the positioning is comfortable, general anesthesia is induced. Care should be taken that there is no excessive pressure on the calves or the lateral aspect of the proximal leg after positioning to avoid compartment syndrome or peroneal nerve injury postoperatively.

Northwestern Handbook of Surgical Procedures, 2nd Edition, edited by Nathaniel J. Soper and Dixon B. Kaufman. ©2011 Landes Bioscience.

Step 2. The abdomen and perineum are widely prepped with a chlorhexidine prep. The preoperatively chosen stoma site should be scratched with an 18-gauge needle to facilitate intraoperative localization. The rectum is irrigated with a dilute povidone solution and then the anal verge is sewn shut with a heavy silk suture in a pursestring fashion.

Step 3. The abdomen and perineum are draped to provide wide access to these areas using the particular drape combinations available to the surgeon.

Step 4. A lower midline incision is made taking care to divide the midline fascia down to the pubic symphysis. A thorough abdominal exploration is undertaken to assess the extent of tumor involvement.

Step 5. The lateral and medial peritoneal reflections of the sigmoid colon are incised down to and across the rectovesical or rectovaginal reflection. The left ureter is identified and displaced laterally along with the gonadal vessels.

Step 6. The superior hemorrhoidal vessels and distal sigmoid vessels are ligated proximally, taking care to identify and avoid the left ureter throughout this maneuver.

Step 7. The rectum and mesorectum are sharply mobilized en bloc off the sacrum and lateral pelvic sidewalls, staying on the visceral aspect of the endopelvic fascia. The hypogastric nerves are visualized and preserved during the posterior mobilization. Waldeyer's fascia is divided posteriorly deep in the pelvis. In a male with malignant disease, Denonvillier's fascia is divided anteriorly as the rectum is separated off the seminal vesicles and prostate. In a female, the rectum is mobilized off the posterior vaginal wall. These planes of dissection are separated down to the levator musculature circumferentially. At this point, a cloth pack is placed deep in the pelvis between the rectum and coccyx.

Step 8. Following complete mobilization of the rectum, the proximal margin of transection is chosen, typically in the proximal one-third of the sigmoid. The colon is divided at this point with a linear stapling device.

Step 9. The operating surgeon relocates to the perineum at this point and makes a circumanal incision. For malignant disease, the incision is deepened, extending into the ischiorectal fossae bilaterally. (For benign disease, this incision can be continued proximally in the intersphincteric plane between the internal and external sphincter.) The anococcygeal raphe is divided in the posterior midline as the posterior three-quarters of the anus is mobilized in a cephalad direction using electrocautery. The inferior hemorrhoidal vessels are encountered in the anterolateral region of the ischiorectal fossae at the level of the upper anal canal. These vessels usually require separate ligation.

Step 10. When the levator ani muscles are reached from the perineal dissection, they are incised in the posterior midline, using the previously placed pack as a guide. After entering the pelvis from below, the levators are incised bilaterally from posterior to anterior until a large defect exists in the pelvic floor.

Step 11. The proximal end of the specimen is grasped from below and delivered through the perineal wound. The anterior portion of the perineal dissection is now completed by carefully incising the perineal body and continuing this dissection cephalad to the levators. In a male, this requires division of the rectourethralis ligament. In a female, the transverse perineal musculature is incised. When the pelvis is reached, the dissection is complete and the specimen is sent to pathology.

Step 12. The perineal wound is carefully inspected and hemostasis achieved. The wound is closed in layers, separately reapproximating the levators (if possible), the ischiorectal fat, and the perineal skin. In heavily irradiated tissues, a myocutaneous flap may be required for primary closure of the perineal wound.

Step 13. The pelvis is inspected once again from above and irrigated. A suction drain is placed in the presacral space and brought out through the anterior wall. No attempt is made to reapproximate the residual pelvic peritoneum.

Step 14. The previously marked colostomy site is excised and the proximal sigmoid end exteriorized through this transrectus opening. The colonic segment is sutured to the anterior abdominal wall from within the peritoneal cavity.

Step 15. The midline wound is closed and the colostomy is matured by excising the staple line and sewing the full thickness of the bowel wall to the dermis circumferentially.

Postop

Intravenous antibiotics are continued for 24 hours postoperatively. The patient ambulates on the evening of surgery. The pelvic drain is removed in 24-48 hours. A urinary catheter is left in for 2-3 days. Clear liquids are begun orally upon resumption of bowel activity.

Complications

The most common early postoperative complications encountered include atelectasis, urinary tract infection, abdominal or perineal wound infection, or prolonged ileus. Late complications include peristomal herniation and adhesive small bowel obstruction.

Follow-Up

Cancer patients are seen at 3-month intervals for the first 2 years, at 6-month intervals for the next 3 years, and annually thereafter.

Chapter 68

Right Hepatic Lobectomy

David M. Mahvi

Indications

Hepatic resection should be considered for either primary hepatic malignancies, or metastatic lesions confined to the right lobe of the liver. The right lobe includes Couinaud segments 5-8 which are anatomically to the right of Cantlie's line (an imaginary plane that connects the gallbladder and the right side of the vena cava). In addition to the general condition of the patient, assessment of the volume and condition of the residual liver is critical. The function of residual liver is affected by intrinsic conditions, such as cirrhosis and treatment prior to resection with cytotoxic chemotherapy. The treatment of patients with malignancy of the liver requires multidisciplinary input. In addition to a surgeon with extensive experience in hepatic resection these patients should be discussed in a conference attended by medical oncologists, interventional radiologists gastroenterologists, hepatologists and transplant surgeons.

Lobectomy is indicated in *noncirrhotic* patients in whom clearance of tumor can be obtained without compromising hepatic arterial and portal venous inflow or biliary drainage to the left lobe and in *cirrhotic* patients with adequate liver function (generally Child's class A).

Preop

- Preoperative imaging (either CT or MRI) to determine the location of the disease, native hepatic anatomy, and resectability.
- Preoperative transcutaneous biopsy of liver masses is not always advisable given (1) the clinical presumptive diagnosis, (2) availability of serum tumor markers, (3) accurate diagnostic imaging, and (4) risk of tumor dissemination.
- Correction of anemia and of coagulopathy and appropriate single dose of antibiotic prophylaxis (e.g., cephazolin). Patients with a history of cardiorespiratory disease and all patients over 65 years are submitted to full investigation.
- Type and crossmatch for two units of packed red blood cells or type and screen (depending on the ability of the institution to supply blood expeditiously should bleeding occur).
- Use appropriate deep venous thrombosis prophylaxis.
- Once general anesthesia is accomplished (supine position), place tube for gastric decompression during the procedure, Foley catheter, and invasive monitoring (arterial or central venous line). Prep skin from nipple level to pubis.

Northwestern Handbook of Surgical Procedures, 2nd Edition, edited by Nathaniel J. Soper and Dixon B. Kaufman. ©2011 Landes Bioscience.

Procedure

Step 1. Incision: right subcostal with a superior midline extension or bilateral subcostal extending across the left rectus muscle. Use a retractor system that provides elevation of the right costal margin to aid in right lobe mobilization and visualization of both the suprahepatic inferior vena cava (IVC) and retrohepatic IVC. The central venous pressure should be maintained as low as possible to reduce blood loss. This may require limiting fluid, epidural anaesthesia and vasodilators.

Step 2. Ligate and divide the ligamentum teres hepatis. Using diathermy, divide the falciform ligament up to the coronary ligament superiorly.

Step 3. Explore the abdomen thoroughly to exclude extrahepatic malignancy. This includes both manual exploration and intraoperative ultrasound.

Step 4. The line of resection of liver extends from the gallbladder fossa anteriorly to the vena cava posteriorly (main portal scisura or Cantlie's line). Ensure this line permits adequate resection margin (using bimanual palpation or intraoperative ultrasound).

Step 5. The gallbladder is removed after division of the cystic artery and duct(unless adjacent to tumor, where it should be rotated laterally and kept en bloc with the right lobe specimen).

Step 6. At the base of the gallbladder fossa, the right primary branches of the hepatic artery, portal vein, and bile ducts must be exposed and identified with great care to avoid injury to the structures serving the remnant liver (left lobe). Ligation of the hepatic artery is typically accomplished first followed by portal vein ligation and finally by division of the right hepatic duct. Because the hepatic duct bifurcates higher in the liver than the inflow vessels it is sometimes difficult to safely identify the right duct. In that case the duct can be divided during the parynchmal dissection. The right portal vein is best stapled (endo-GIA stapler with 2.8 mm vascular load) to prevent suture dislodgment or compromise of the portal vein bifurcation. A line of color demarcation will be seen on the liver corresponding to the anatomic division between the right and left lobes.

Step 7. Using diathermy, divide the remaining coronary ligament to expose the right hepatic vein as it enters the IVC at the base of segment 8 superiorly.

Step 8. Mobilize the right lobe by dividing the right triangular ligament.

Step 9. Continue by elevating the lobe medially, mobilizing the right lobe at the bare area (adherent to the diaphragm posteriorly). Use caution:

1. superiorly as the right hepatic vein is close to the capsule/diaphragmatic junction;
2. inferiorly as the adrenal gland will be elevated with the lobe until lowered; and
3. medially where the IVC meets the posterior aspect of the liver.

Step 10. Ligate and divide the short hepatic veins and caudate veins directly entering the retrohepatic IVC until the right lobe specimen is free medially. Divide the inferior vena caval ligament to expose the inferior margin of the right hepatic vein and aid in cava-hepatic separation. The right hepatic vein can then be divided with a vascular stapling device. This is possible in many but not all cases. The vein has a relatively long connection to the cava and cannot always be dissected free of the liver prior to parynchmal transection.

Step 11. Replace the right lobe and begin liver parenchymal transection in an anteroinferior to posterosuperior fashion. There are many different techniques to divide the parenchyma including:

1. sequential parenchymal fracture with hemostasis;
2. harmonic scalpel;
3. CUSA (Cavitron Ultrasonic Aspirator) in combination with Argon beam coagulation;

4. sequential parenchymal stapling;
5. sequential application of a coagulation device (radiofrequency and microwave devices are currently available) followed by transection; and
6. inflow occlusion (Pringle maneuver) in combination with above options.

There is no data suggesting that any technique is superior to another. Option 1 with inflow occlusion is the preference of the author. This allows individual ligation of hepatobiliary structures in a relatively bloodless field.

NOTE: As the transection advances deeper into the interlobar plane, be aware that the right anterior sector vein (draining segments 5 and 8 into the middle hepatic vein) crosses the transection plane and must be identified, dissected, and ligated definitively. This vein is large (5-10 mm) and is 2-3 cm deep, coursing obliquely into the inferior aspect of segment 4a.

Step 12. After completing transection, control and divide the right hepatic vein (if not done previously) either through clamp/suture or stapling, being careful to avoid injury to the IVC or middle hepatic vein.

Step 13. Where appropriate, send the right lobe specimen to surgical pathology to assess resection margin.

Step 14. Survey liver cut surface for both hemostasis and biliary leak. Control with sutures unless adjacent to middle hepatic vein and porta hepatis where caution is necessary.

Step 15. Irrigate cavity with saline and inspect cut surface once again. Biliary leaks (4% of liver resections of any magnitude) are subtle and difficult to identify.

Step 16. Closed suction drainage of the cavity is controversial. It is unnecessary in most cases (no reduction in complications) unless there is a leak that is difficult to control or in cases where a biliary reconstruction is necessary.

Step 17. Examine liver remnant for viability and risk of torsion into the right hepatic fossa. If torsion is possible, reconstruct falciform ligament to support left lobe using absorbable suture material.

Step 18. Close incision in layers with suture material of choice with particular attention to approximation of tissues at apex (midline, center).

Postop

Monitor for hemorrhage and hepatic decompensation in early postoperative period. Particular attention should be directed to phosphate repletion in the early postoperative period. A nasogastric tube is generally not necessary. Institute diet as tolerated. Continue deep venous thrombosis and pulmonary prophylaxis.

Complications

Fever and leukocytosis warrant imaging to exclude biloma which will typically require drainage. Drainage should be performed percutaneously. Continue drainage (90% will resolve) unless high output in which case ERCP-stent should be considered.

Follow-Up

The patient should be followed until fully recovered or, if applicable, bile leak subsides. All patients with liver malignancies should be evaluated by a medical oncologist.

Acknowledgment

The editors and author wish to acknowledge Alan J. Koffron for contributing to the previous version of this chapter.

Chapter 69

Axillary Lymphadenectomy

Kevin Bethke

Indications

Axillary lymphadenectomy is performed for the treatment of metastatic cancer (usually breast or melanoma) to the axilla as manifest by clinical adenopathy or a positive sentinel lymph node biopsy.

Preop

Preoperative lymphoscintigraphy may be necessary if dissection is performed for melanoma of the trunk and the nodal drainage pattern is ambiguous.

Procedure

Step 1. A soft roll is placed longitudinally beneath the shoulder and flank to displace the axilla anteriorly and allow easier access to the axilla. The arm is placed at slightly less than a 90° angle from the trunk. Care is taken to prevent hyperextension of the shoulder by elevating the arm on pillows and then securing it to an arm board in a stationary position.

Step 2. A 4-5 cm, gently curved incision is made at the inferior axillary hairline between the lateral border of the pectoralis major muscle and the anterior border of the latissimus dorsi muscle.

Step 3. The lateral border of the pectoralis major and anterior border of the latissimus dorsi muscles are exposed. Dissection proceeds along the latissimus dorsi muscle superiorly to the point where the axillary vein crosses the latissimus tendon. Care is taken to protect the intercostobrachial nerve which runs transversely through the mid-axilla.

Step 4. The axillary fascia is incised inferior to the axillary vein. Gentle blunt dissection using a Kittner dissector is utilized along with constant inferior retraction to dissect the fatty, node-bearing tissue off the axillary vein. The axillary vein should not be completely stripped of all tissue because this would disrupt lymphatic vessels which course along the vein and drain the arm, leading to an increased risk of lymphedema.

Step 5. The axillary tissue is bluntly dissected away from the serratus anterior muscle medially and the lattissimus dorsi muscle laterally. The long thoracic and thoracodorsal nerves are identified and protected. The medial pectoral nerve is identified along the superior lateral border of the pectoralis major muscle and protected.

Step 6. Once the long thoracic and thoracodorsal nerves have been identified, the Level II nodes posterior to the pectoralis minor muscle are retracted inferiorly and the axillary contents are swept inferiorly along the groove between the serratus anterior and lattissimus dorsi muscles. If the Level III nodes medial to the pectoralis minor muscle are grossly involved they are also dissected and swept inferiorly.

Northwestern Handbook of Surgical Procedures, 2nd Edition, edited by Nathaniel J. Soper and Dixon B. Kaufman. ©2011 Landes Bioscience.

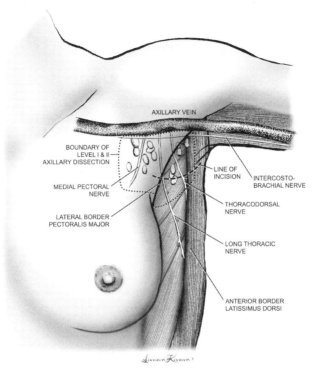

Figure 69.1. Axillary lymphadenectomy.

Step 7. The intercostobrachial sensory nerve, if not grossly involved with tumor, is dissected free of nodal tissue and retracted anteriorly. The axillary contents are then swept inferiorly and behind the nerve. Care is taken to avoid injury to the long thoracic nerve in the inferior axilla where the nerve enters the serratus anterior muscle. The larger veins draining the inferior axilla are ligated.

Step 8. A closed suction drain is placed in the axilla and brought out via a stab wound placed inferior to the incision and sutured in place.

Step 9. Bupivacaine is injected along the incision for prolonged postoperative pain relief and the wound is closed in two layers utilizing interrupted absorbable suture for the subcutaneous layers and a running 4-0 subcuticular stitch with steri-strips for the skin.

Postop

The patient is usually discharged 2-3 hours after the operation with a prescription for oral narcotic analgesia. Drain care instructions are given prior to discharge and include stripping the drain 2-3 times per day to help prevent clots from obstructing it. The patient will empty the suction bulb 2-3 times per day and record the output. When the output falls to 30-40 ml per 24 hours the drain is removed in the clinic. The patient may shower on postoperative day one and begin gentle arm and shoulder exercises.

Complications

Upper medial arm numbness can be avoided by careful dissection and preservation of the intercostobrachial nerve. Axillary seromas can be minimized by leaving the drain in until the output is down to 30-40 ml per 24 hours (usually 7-10 days). Injury to the long thoracic nerve causes dysfunction of the serratus anterior muscle and a "winged scapula." This is a very rare complication and can be minimized by adequate exposure and gentle blunt dissection of the axillary contents off the serratus muscle.

Clinical lymphedema occurs in 10-15% of patients. The more extensive the dissection performed, the greater the risk. Denuding the axillary vein during the superior axillary dissection will increase this risk, as will removal of the Level III nodes. Patients should be referred to an experienced physical therapy department at the earliest sign of lymphedema.

69

Follow-Up

The patient should be followed by the surgeon until the wound is healed and the drain is removed. Some form of adjuvant therapy (chemotherapy and/or radiation therapy) will likely be required after which the patient will need to be followed indefinitely for signs of recurrence.

Chapter 70

Inguinal Lymphadenectomy

Kevin Bethke

Indications

Inguinal lymphadenectomy is performed for the treatment of metastatic melanoma (and other cancers) to the groin as manifest by clinically involved lymph nodes without other evidence of distant metastasis or by a positive sentinel lymph node biopsy.

Preop

Preoperative lymphoscintigraphy is indicated if the primary cancer is located on the trunk and the nodal drainage pattern is ambiguous. All patients are at high risk for postoperative lower extremity lymphedema and should be measured preoperatively for thigh-high compression hose. Intravenous cefazolin is given 30 minutes prior to incision. Deep vein thrombosis prophylaxis with compression boots or heparin is provided according to the degree of patient's risk.

Procedure

Step 1. The patient is placed in a supine position and general endotracheal anesthesia is induced. A pillow is placed beneath the knee, and the leg is slightly externally rotated.

Step 2. After prepping and draping, a gently curved incision located parallel to and below the inguinal crease is made, extending from the anterior iliac spine to several centimeters below the pubic tubercle.

The extent of groin dissection required can be superficial or deep. Steps 3-6 describe a superficial groin disection; Step 7 describes a deep dissection.

Step 3. The superior skin flap is raised and the nodal tissue overlying the external oblique aponeurosis and inferior to the inguinal ligament is resected. Care must be taken to resect all nodal tissue, especially if the primary site is on the trunk, superior to the incision.

Step 4. With medial traction, dissection begins at the lateral border of the sartorius muscle and proceeds towards the femoral sheath. The femoral nerve and lateral femoral cutaneous nerve lie deep to the sartorius fascia and are protected.

Step 5. The dissection proceeds over the femoral vessels in a subadventitial plane. The saphenous vein is ligated at the saphenofemoral junction. With lateral traction, the adductor longus fascia is incised and dissection proceeds medially towards the femoral sheath.

Step 6. The saphenous vein is ligated inferiorly at the apex of the femoral triangle. The only remaining attachment is at the femoral ring where Cloquet's node resides, medial to the femoral vein. The fascia overlying the femoral ring is incised and Cloquet's node is dissected free from the properitoneal fat.

Northwestern Handbook of Surgical Procedures, 2nd Edition, edited by Nathaniel J. Soper and Dixon B. Kaufman. ©2011 Landes Bioscience.

70

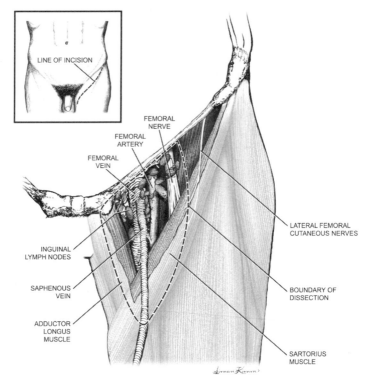

Figure 70.1. Inguinal lymphadenectomy.

Step 7. (Deep groin dissection). In some situations a more radical (ilioinguinal) groin dissection is indicated based on the status of Cloquet's node or significant inguinal adenopathy. Exposure for the deep dissection can be gained either by transecting the inguinal ligament as it crosses the femoral canal or by retracting the superior skin flap and making an incision through the abdominal wall superior and parallel to the inguinal ligament. The peritoneum is retracted superiorly and medially and the retroperitoneal node-bearing tissue is resected from the femoral ring (inferiorly), bladder (medially), genitofemoral nerve (laterally), and aortic bifurcation (superiorly). The inguinal ligament is then reapproximated (if transected) and/or the abdominal wall muscles are anatomically approximated. A drain is not necessary for the deep dissection.

Step 8. At the conclusion of dissection, the sartorius muscle is detached from the iliac spine and transposed over the femoral vessels. It is sutured to the inguinal ligament.

Step 9. A closed suction drain is placed in the groin wound and brought out from the abdominal wall superior to the incision in order to avoid compromised groin tissues.

Step 10. One centimeter of traumatized skin is excised from both wound edges and the wound is meticulously closed in two or three layers.

Postop

The leg is wrapped with an elastic bandage from the foot to the thigh. The leg is elevated on pillows during the hospital stay. Ambulation is encouraged on the first postoperative day. Several days after surgery the previously custom-fitted compression stocking is placed. The groin drain is left in place until the output is 30-40 ml over a 24-hour period (usually 2-3 weeks) and then removed in the clinic.

Complications

Wound problems are common, with skin ischemia, subsequent necrosis and wound infection a feared complication. The most severe complication is a femoral artery blowout in an infected groin, which fortunately is quite rare when the sartorius muscle is transposed over the vessels. Lymphedema is a common problem, especially with a combined superficial and deep dissection. Patients must be warned of this and encouraged to wear their support hose for at least 4-6 months after surgery.

Follow-Up

The patient should be followed by the surgeon until wounds are healed and any lymphedema problems are addressed. Either the surgeon or oncologist must follow the patient indefinitely for signs of recurrent tumor.

70

Chapter 71

Breast Biopsy after Needle Localization

Jacqueline Jeruss and Nora Hansen

Indications

Needle-localized breast biopsy is indicated for diagnosis of nonpalpable breast abnormalities considered suspicious for carcinoma (BIRADS 4 or 5) which are not suitable for image-guided breast biopsy. Needle biopsy may also be indicated for removal of lesions diagnosed as atypical hyperplasia or carcinoma by core biopsy.

Preop

On the day of surgery, a localization wire is placed into or within 1 cm of the suspicious lesion using mammographic or ultrasound guidance. The wire insertion is done in the radiology suite using local anesthesia. The biopsy is then performed in the operating room using local anesthesia with intravenous sedation.

Procedure

Step 1. Localization films (two views) are reviewed to determine the relationship of the wire to the lesion and the distance from skin entry to lesion. Proper incision placement is critical to success. The incision is placed in proximity to the known abnormality, *not* at the point the localization wire enters the skin.

Step 2. The patient is placed in the supine position. The skin incision is made in Langer's lines.

Step 3. Subcutaneous fat and superficial breast tissue are divided. The wire is identified within the breast parenchyma and pulled inward through the skin into the operative field.

Step 4. The breast tissue surrounding the wire is carefully divided with scissors or cautery until the area of the lesion is approached. Identification is facilitated by using a localization wire with a thickened distal segment, and that is placed through the lesion.

Step 5. At the level of the lesion, excision is carried out to remove the wire and the lesion (often with prior biopsy clip), surrounded by a margin of approximately 5 mm of normal breast tissue on all sides.

Step 6. The specimen is marked with a short suture in the superior margin and a long suture in the lateral margin and sent for specimen radiography to confirm removal of the target.

Step 7. After obtaining hemostasis, the cavity resulting from the excision is irrigated with saline. The four walls of the cavity are marked with clips.

Step 8. The deep dermis is closed with interrupted 3-0 Vicryl sutures, knots inverted. The skin is closed with a running subcuticular 4-0 Vicryl suture, steri-strips, and a light gauze dressing.

Northwestern Handbook of Surgical Procedures, 2nd Edition, edited by Nathaniel J. Soper and Dixon B. Kaufman. ©2011 Landes Bioscience.

Postop

This is an outpatient procedure. The dressing can be removed and the patient may shower in 24 hours.

Complications

Failure to remove target abnormality in 1-2%, hematoma, and infection.

Follow-Up

The patient should be examined in 1-2 weeks to ensure adequate healing. Other follow-up procedures are specific to pathology on biopsy.

Acknowledgment

The editors and authors wish to acknowledge Monica Morrow for contributing to the previous version of this chapter.

71

Chapter 72

Lymphatic Mapping and Sentinel Node Biopsy

Seema A. Khan

Indications

Lymphatic mapping and sentinel node biopsy are indicated for axillary nodal staging in patients with invasive breast carcinoma and a clinically negative axilla. In patients who have received neoadjuvant systemic therapy, sentinel node biopsy can be performed either prior to the initiation of systemic therapy or following its completion, at the time of definitive surgical treatment. If suspicious axillary lymphadenopathy is present at diagnosis however, the accuracy of sentinel node biopsy is suboptimal and Level I-II axillary dissection is advised. The same is true in patients with inflammatory carcinoma of the breast. Sentinel node staging may also be indicated in patients with ductal carcinoma in situ (DCIS) diagnosed by core needle biopsy when the likelihood of microscopic invasion in the excision specimen is high (e.g., palpable DCIS and large DCIS lesions).

Preop

Lymphatic mapping can be accomplished with either a radioactive tracer (technetium sulfur colloid) or a dye tracer (lymphazurin or methylene blue), or both. The radioactive tracer is usually injected in the nuclear medicine department one or more hours prior to surgery; the dye tracer is injected in the operating room by the surgeon. Radioactive tracer (RAT) injection can be peritumoral or intradermal, or a combination of these. Dye tracer is not injected intradermally to avoid tattooing of the skin. If using methylene blue, it is particularly important to avoid intradermal injection, since this may cause significant skin necrosis. If image-guided wire localization is needed for nonpalpable lesions, this is performed prior to the RAT injection. The sentinel node biopsy can be performed under sedation, with local anesthesia, but general anesthesia is often preferred.

Procedure

Step 1. The patient is placed in the supine position with the ipsilateral arm extended on an arm board at a nearly right angle to the operating table. Dye tracer can be injected prior to skin preparation, or following preparation and draping. Skin preparation includes the entire breast if wide excision of the breast carcinoma is being performed at the same time. The ipsilateral arm is prepared to the elbow or lower and draped into the field.

Step 2. Five ml of dye tracer is injected in a peritumoral or retroareolar fashion; (if methylene blue is being used, 2 ml should be diluted to a total volume of 5 ml). When using dual tracers, the dye tracer is often injected in the retroareolar region, and the radioactive tracer can be injected in a peritumoral location. (several studies have now shown that the entire breast shares the same sentinel node distribution and the location of the tracer injection is not crucial) The breast is then vigorously massaged for 5 minutes. If RAT mapping is also being used, the handheld gamma detector is draped and prepared for use.

Northwestern Handbook of Surgical Procedures, 2nd Edition, edited by Nathaniel J. Soper and Dixon B. Kaufman. ©2011 Landes Bioscience.

Step 3. For RAT mapping, the injection site is surveyed with the gamma detector, and the surgeon identifies the zone in the axillary tail where the radioactivity related to the injection site falls off to background levels. Then the axilla is surveyed, moving the gamma detector slowly. Once a hot spot is identified (counts greater than two-fold background), it must be confirmed that this is not "shine-through" from the injection site by angling the probe tip away from the injection site and pressing down with the probe tip. If a hot spot is present, counts will rise as the probe is pushed closer to it.

Step 4. For RAT lymphatic mapping, the proposed skin incision can be guided by the site of the hot spot. If dye tracer alone is being used, the proposed incision is marked at the inferior end of the hair-bearing area of the axilla, along a natural skin crease.

Step 5. The incision is made with a scalpel. It usually does not need to be more than 2 cm long. The dissection is deepened through the subcutaneous fat to the axillary fascia, which is opened. Blue-stained lymphatic channels encountered superficial to the fascia can be ignored. If RAT mapping is also being used, the probe is inserted into the axillary space and used to direct the dissection towards the hot spot. Again, care is taken not to point the probe towards the injection site.

Step 6. Once a blue-stained lymphatic channel is identified in the axillary fat, it is followed to a blue-stained lymph node. A node that is significantly replaced by tumor may not stain blue, even though a blue lymphatic channel leads directly to it. If RAT tracing is being used, the dissection is guided by the location of the hot spot. Digital palpation is also extremely helpful in final location of the sentinel node.

Step 7. Once the sentinel node is identified, it is excised by bovie, scissor or clamp dissection, applying clips or ligatures as needed for hemostasis. If RAT tracing was used, the excised node is scanned to determine that it is radioactive ex vivo.

Step 8. The axilla is surveyed with the gamma detector to confirm that no additional hot spots are present. If any are identified, they should be excised until the radioactivity of the axilla has been reduced to background levels. If only dye tracer was used, the axillary space is examined for additional blue-stained lymphatic channels or nodes. Any additional such sentinel nodes are excised. For radioactive mapping, nodes are removed only if the in situ count is 10% or more of the hottest node. There is marginal additional benefit from excising more than five sentinel nodes, and the procedure can usually be terminated if five nodes have been excised. Frozen section or touch prep examination of the sentinel nodes is optional, but is advisable if the sentinel node(s) are grossly suspicious or in the patient who is undergoing mastectomy with immediate reconstruction. If a microscopic diagnosis of sentinel node involvement with tumor is made intraoperatively and the matter has been discussed with the patient preoperatively, the procedure can be converted to a Level 1 and 2 axillary dissection.

Step 9. Hemostasis is obtained in the axillary space. The additional infiltration of 0.5% bupivicaine is optional at his point. The wound is closed in two layers, using absorbable interrupted sutures for the subcutaneous tissue and a subcuticular suture for the skin. Steri-strips and a dry dressing are applied.

Postop

Most patients are discharged on the day of surgery. Dressings can be removed and the patient allowed to shower within 1 or 2 days of surgery. Prescription analgesics are required.

Complications

Inability to identify a sentinel node requires conversion to standard axillary dissection. Although rare (failure to map occurs in less than 2% of cases in most recent series), the risk for this is higher in elderly and obese women. Nerve injury is possible during sentinel node biopsy, particularly to the intercostobrachial nerve. Hematoma and infection are extremely rare.

Follow-Up

Follow-up care may include adjuvant radiation and systemic therapy, as indicated by final pathologic stage of the tumor.

Chapter 73

Partial Mastectomy and Axillary Dissection

Jacqueline Jeruss and Nora Hansen

Indications

The majority of patients with intraductal carcinoma or Stage I and II invasive cancer are candidates for a partial mastectomy. Contraindications to the procedure are multicentric carcinoma, prior irradiation to the breast region, first or second trimester pregnancy, diffuse indeterminate microcalcifications on a mammogram, and the inability to achieve negative margins after an adequate number of surgical attempts. Relative contraindications are a large tumor-to-breast ratio and scleroderma or systemic lupus. Patients with invasive carcinoma who are clinically node negative should have their axillary nodal status determined by sentinel node biopsy and axillary dissection if the sentinel node is positive. Those who present with node positive disease require axillary dissection.

73

Preop

General anesthesia is often the method of choice, though some institutions are now also using regional blocks. Long-acting muscle relaxants should be avoided to allow intraoperative nerve identification. Prophylactic antibiotics are used. Deep vein thrombosis prophylaxis with sequential compression devices or subcutaneous heparin is used according to patient risk factors.

Procedure

Step 1. The patient is positioned supine with the ipsilateral arm at nearly a right angle on an armboard. The arm should be circumferentially prepped and draped in stockinette to allow for intraoperative movement if needed.

Step 2. If the lesion is palpable, the incision is placed over the tumor in the breast. In the superior half of the breast, incisions are in skin creases. In the inferior breast, radial incisions are used. Circumareolar incisions should be reserved for tumors in proximity to the areola since they should not encompass more than half the areolar circumference.

Step 3. The subcutaneous fat overlying the tumor is divided and preserved. This will help maintain breast contour. Subcutaneous fat needs to be removed only for tumors approaching the dermis.

Step 4. The depth of the tumor is determined by palpation, and the breast tissue overlying the tumor is divided to a depth of 1.0-1.5 cm anterior to the tumor.

Step 5. Double-pronged skin hooks or small Richardson retractors are placed, and using the knife or electrocautery, flaps are raised to allow the tumor to be excised surrounded by approximately 1.0-1.5 cm of normal breast tissue. Raising the flaps at the depth of the tumor helps to maintain the breast contour.

Northwestern Handbook of Surgical Procedures, 2nd Edition, edited by Nathaniel J. Soper and Dixon B. Kaufman. ©2011 Landes Bioscience.

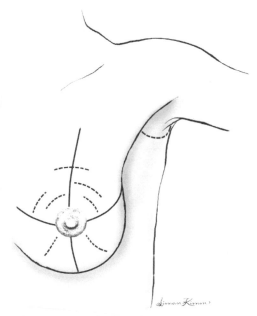

Figure 73.1. Partial mastectomy and axillary dissection.

Step 6. The tumor is controlled and retracted with the nondominant hand while it is sharply excised. This reduces the risk of accidentally cutting into the tumor. Grasping the tumor and surrounding breast tissue with clamps should be avoided as it is difficult to precisely determine the tumor location within the mass of tissue.

Step 7. The specimen is marked with orienting sutures (short suture superior, long suture lateral) and examined to ensure that the tumor is covered on all surfaces by a margin of normal breast tissue. If a margin appears inadequate, an additional specimen is excised with a knife or electrocautery, marked with an orienting suture or surgical clip to indicate the new margin surface, and sent as a separate specimen. Routine frozen sections of margins are not employed.

Step 8. Hemostasis is obtained with cautery, and the walls of the cavity are marked with hemoclips. The wound is packed with a sponge while axillary dissection is carried out if intraoperative evaluation reveals the sentinel node to be positive.

Step 9. The axillary incision is made at the edge of the hairline in a skin crease and extends from the pectoralis major anteriorly to the latissimus dorsi posteriorly. In patients with a very narrow axillary space, the ends of the incision should be curved superiorly in a U-shaped configuration to provide exposure.

Step 10. Using cautery, the subcutaneous fat is divided. Double-pronged skin hooks are placed and flaps are raised superiorly to the level of the axillary vein, medially to expose the free edge of the pectoralis major, inferiorly to the junction of the breast tissue and the axilla, and laterally to expose the latissimus dorsi.

Step 11. The axillary investing (clavipectoral) fascia is opened medially along the edge of the pectoral muscles, with care being taken not to injure the medial pectoral neurovascular bundle.

Step 12. The latissimus dorsi is cleared of fat along the anterior surface until it turns tendinous. The axillary vein will be identified at this point. Approximately two-thirds of the way to the vein a large branch of the intercostobrachial nerve is encountered. This should be preserved.

Step 13. The axillary vein is cleared of overlying fat from lateral to medial on its anterior surface with care being taken not to strip the vein.

Step 14. The first assistant retracts the axillary contents inferiorly. Dissection 0.5-1.0 cm inferior to the vein is carried out from lateral to medial and venous branches are ligated with 3-0 ties and surgical clips.

Step 15. The thoracodorsal bundle is identified and preserved.

Step 16. The axillary fat medial to the thoracodorsal bundle is retracted laterally. Two fingers are inserted into the fat adjacent to the chest wall immediately below the axillary vein and spread in a cranial and caudal direction to expose the long thoracic nerve. The nerve is sharply dissected free from the axillary contents, taking care to keep it out of harm's way. The medial end of the intercostobrachial nerve is usually identified at this time.

Step 17. The intercostobrachial nerve is dissected free from the specimen, retracted superiorly and spared if possible.

Step 18. The fat between the long thoracic and the thoracodorsal nerves is encircled with a right-angle clamp, divided, and tied proximally. The distal fat is bluntly swept down with a sponge.

Step 19. Small branches of the thoracodorsal vessels entering the specimen are clipped, and the specimen is dissected free of its inferior attachments to the chest wall and sent to pathology.

Step 20. Hemostasis is obtained with clips, ties, and cautery as required. The lumpectomy site is reinspected for hemostasis and both wounds irrigated with saline. A #19 flat closed suction drain is placed in the axilla.

Step 21. Both wounds are closed with an interrupted layer of 3-0 absorbable sutures, knots inverted, in the deep dermis, and 4-0 subcuticular skin closure. No attempt is made to approximate the cavity in the breast. A light gauze dressing is used.

Postop

Patients are discharged within 24 hours of surgery after instruction in drain care. The drain is removed when the output is less than 30 ml per 24 hours for 2 days. Stretching exercises to maintain shoulder mobility are begun on postoperative day 2.

Complications

Complications include infection, hematoma, seroma, anesthesia in distribution of the intercostobrachial nerve, and damage to thoracodorsal and long thoracic nerves.

Follow-Up

An examination for arm mobility, seroma formation, and wound healing should be done 1-2 weeks postoperatively. Adjuvant radiation therapy and possibly systemic therapy will be required. The patient will require lifelong surveillance for ipsilateral and/or contralateral tumor recurrence.

Acknowledgment

The editors and authors wish to acknowledge Monica Morrow for contributing to the previous version of this chapter.

73

Chapter 74

Modified Radical Mastectomy

Jacqueline Jeruss and Nora Hansen

Indications

Modified radical mastectomy is indicated for the treatment of invasive breast cancer when breast conserving therapy (BCT) is contraindicated or when the patient chooses mastectomy over BCT.

Preop

General anesthesia or regional block is indicated. The patient is positioned in the supine position with the arms extended at nearly 90° on armboards. The ipsilateral operative arm is prepped into the field. Standard perioperative antibiotic is administered prior to incision. Long-acting muscle relaxants should be avoided. Deep venous thrombosis prophylaxis with sequential compression devices should be used.

Procedure

Step 1. The subcutaneous tissue of the breast and into the axilla is infiltrated with 500-1500 ml tumescent solution (1 L lactated Ringers' solution with 30 ml of 1% lidocaine solution with epinephrine at 1:1000). Alternatively, if electrocautery is to be employed for the dissection, tumescent solution should not be used.

Step 2. An incision is chosen which includes excision of the nipple-areolar complex and incorporates any previous biopsy incision. Remaining skin may be spared if immediate reconstruction is planned, otherwise an elliptical incision is chosen.

Step 3. Skin flaps are raised sharply with a scalpel or electrocautery, extending superiorly to the clavicle, medially to the lateral border of the sternum, inferiorly to the superior aspect of the rectus sheath, and laterally to the latissimus dorsi muscle.

Step 4. The pectoralis major fascia is incised, controlling internal mammary perforators (medially) with ties.

Step 5. The breast and pectoralis fascia are excised with knife or cautery. The breast is left attached inferolaterally to provide traction.

Step 6. The latissimus dorsi muscle edge is followed superiorly along its anterior surface using Richter scissors or cautery. Care is taken to preserve the intercostobrachial nerves as encountered. As the muscle becomes tendinous, the axillary vein will be encountered crossing superior to it.

Step 7. The axillary vein is cleared on its anterior surface in a layer-by-layer, lateral-to-medial fashion from the latissimus muscle to the chest wall, taking care not to strip the vein.

Step 8. Dissection is then continued along the axillary vein about 5 mm inferior to the vein, again in a layer-by-layer, lateral-to-medial fashion from the latissimus muscle to the chest wall. There is generally an anterior thoracic branch of the vein that should be controlled, ligated, and divided.

Northwestern Handbook of Surgical Procedures, 2nd Edition, edited by Nathaniel J. Soper and Dixon B. Kaufman. ©2011 Landes Bioscience.

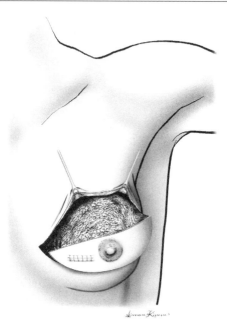

Figure 74.1. Modified radical mastectomy. Incision.

Step 9. The thoracodorsal bundle will come into view during the dissection along the inferior edge of the axillary vein. The nerve is usually slightly posteromedial to the vein. The nerve can be tested to confirm contraction of the latissimus dorsi muscle and preserved.

Step 10. Tissues along the chest wall are spread with blunt finger dissection until the long thoracic nerve comes into view. The nerve can be tested to confirm contraction of the serratus anterior muscle and preserved.

Step 11. The tissue lying between the nerves is surrounded with a right-angle clamp, taking care not to catch the nerves in the clamp. The tissue is divided and tied at its superior aspect to control lymphatics and small vessels.

Step 12. The tissue remaining between the nerves is swept inferiorly with an open gauze sponge over the first two fingers.

Step 13. The intercostobrachial nerves are separated from the specimen and retracted out of the way when possible.

Step 14. The thoracodorsal bundle is followed inferiorly along its anterior surface, clipping and dividing small vein branches as they arise.

Step 15. When the thoracodorsal bundle turns and dives into the latissimus dorsi muscle and the long thoracic nerve turns and enters the chest wall, the remainder of the specimen is removed by electrocautery where it remains attached inferolaterally, in such a way that the breast and axillary portions of the specimen remain intact.

Step 16. A #19 round Jackson-Pratt drain is placed beneath the inferior skin flap and brought out through a separate stab incision. A second drain is placed into the axilla, taking care that it does not contact the axillary vein. The drains are secured to the skin with nylon drain stitches.

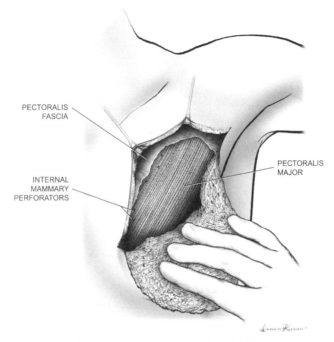

PECTORALIS
FASCIA

PECTORALIS
MAJOR

INTERNAL
MAMMARY
PERFORATORS

74

Figure 74.2. Modified radical mastectomy. Extent of depth of dissection.

Step 17. The deep dermal layer is closed with interrupted 3-0 absorbable sutures and the skin with running subcuticular 4-0 absorbable suture.

Postop
The Jackson-Pratt drains should be stripped periodically and output recorded. The drains can be removed when output is less than 30 ml/day for 2 days. Range-of-motion exercises should be initiated about 2 days after surgery.

Complications
The most common complications associated with this procedure are hematoma beneath flaps, skin necrosis, lymphedema, and/or arm numbness.

Follow-Up
The patient is referred for systemic therapy, and/or radiation therapy as needed according to the stage and biologic markers of tumor. The patient should be monitored for local recurrence and have bi-annual examinations and annual mammography to look for primary disease in the contralateral breast.

Acknowledgment
The editors and authors wish to acknowledge Valerie L. Staradub for contributing to the previous version of this chapter.

Chapter 75

Simple Mastectomy

Jacqueline Jeruss and Nora Hansen

Indications

Simple mastectomy is indicated for extensive ductal carcinoma in situ (DCIS) of the breast, DCIS in a patient who is not a candidate for radiation therapy, or DCIS in a patient who elects mastectomy over breast conservation therapy. Simple mastectomy may be combined with sentinel node biopsy in selected cases of invasive carcinoma.

Preop

General orotracheal anesthesia is preferred. Intravenous access should be on the nonsurgical side. Prophylactic antibiotic is given immediately prior to incision. Deep venous thrombosis prophylaxis with sequential compression devices and/or subcutaneous heparin is used depending on patient risk factors.

Procedure

Step 1. The patient is placed in supine position with the arms extended nearly 90° on armboards. The subcutaneous tissue of the breast is infiltrated with 500-1500 ml of tumescent solution (1 L lactated Ringers' solution with 30 ml of 1% lidocaine solution with epinephrine at 1:1000). Alternatively, if electrocautery is to be employed for the dissection, tumescent solution should not be used.

Step 2. An elliptical incision is made to include excision of the nipple-areolar complex and any recent biopsy incision. Other skin may be preserved if immediate reconstruction is planned, otherwise excess breast skin is excised.

Step 3. Skin flaps are raised sharply with a knife superiorly to the clavicle, medially to the lateral border of the sternum, inferiorly to the superior aspect of the rectus sheath, and laterally to the latissimus dorsi muscle.

Step 4. The pectoralis major fascia is incised, controlling internal mammary perforators (medially), with ties.

Step 5. The breast and pectoralis fascia are removed from superior to inferior with knife or cautery.

Step 6. Laterally, the latissimus dorsi muscle is followed superiorly until the axillary investing fascia is entered, to ensure removal of the axillary tail of breast tissue. The axillary tail is marked with a stitch for orientation of the specimen.

Step 7. A #19 round closed Jackson-Pratt suction drain is placed beneath the inferior skin flap, brought out through a separate stab incision, and secured with a nylon drain stitch.

Step 8. The deep dermal layer is closed with interrupted 3-0 absorbable sutures. The skin is closed with a running subcuticular 4-0 suture.

Northwestern Handbook of Surgical Procedures, 2nd Edition, edited by Nathaniel J. Soper and Dixon B. Kaufman. ©2011 Landes Bioscience.

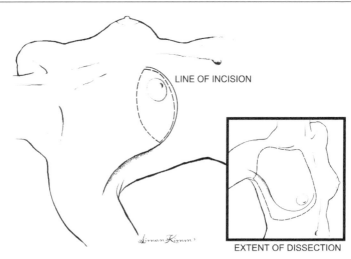

LINE OF INCISION

EXTENT OF DISSECTION

75

Figure 75.1. Simple mastectomy.

Postop
The Jackson-Pratt drain should be stripped periodically and output recorded. The drain can be removed when output is less than 30 ml/day for 2 days.

Complications
Most common complications include hematoma beneath the flap and skin necrosis.

Follow-Up
The patient should be monitored indefinitely for local recurrence or contralateral disease. Antiestrogen therapy for contralateral risk reduction should be considered.

Acknowledgment
The editors and authors wish to acknowledge Valerie L. Staradub for contributing to the previous version of this chapter.

Chapter 76

Major Excision and Repair/Graft for Skin Neoplasms

Jeffrey D. Wayne

Indications

Until recently, all but the thinnest of invasive melanomas of the trunk and proximal extremities were treated with radical resections, with 3-5 cm margins. While rates of local recurrence were low, most of these excisions required a skin graft for closure. Four randomized, prospective surgical trials including the WHO and Intergroup trials challenged such wide margins of excision and established the treatment standards employed in the management of malignant melanoma today. Namely, in situ lesions are routinely resected with a margin of 0.5-1.0 cm. T1 melanomas, with a Breslow's depth of less than or equal to 1.0 mm, are excised with 1 cm margins. T2 lesions, 1.01-2 mm in depth can be safely excised in all locations with a 1-2 cm margins. T3 lesions (2.01-4 mm in depth) should be excised with 2 cm margins. Finally, while they may have a higher propensity to local recurrence, thick (>4.0 mm) lesions are also excised with a 2.0 cm margins.

76

Preop

Most often, the primary lesion has been resected either by a punch or excisional biopsy. For lesions on the trunk, or proximal extremity, primary closure is the norm. Most patients can undergo elliptical excision of the scar and tumor bed under local anesthetic, with or without intravenous sedation. In general, a proper excision involves resection of the scar and tumor bed, with an appropriate margin of normal tissue, and an en bloc resection of the subcutaneous tissues down to the underlying muscular fascia. There is no evidence that removal of the underlying fascia improves local control or survival, and thus it is usually spared.

This procedure is usually performed on an outpatient basis. The patient is admitted to the operating room on the morning of surgery. If intravenous sedation or a general anesthetic is to be used, the patient is instructed to take nothing by mouth after midnight. If a previous biopsy has been performed, a single dose of a first-generation cephalosporin is given in the holding area, at least 45-60 minutes prior to the planned incision. If the patient has an allergy to penicillin, clindamycin may be used as an alternative.

Northwestern Handbook of Surgical Procedures, 2nd Edition, edited by Nathaniel J. Soper and Dixon B. Kaufman. ©2011 Landes Bioscience.

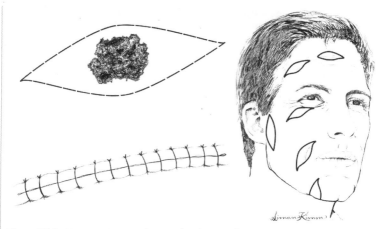

Figure 76.1. Major excision and repair for skin neoplasms.

Procedure

76

Step 1. To start the procedure, the patient is prepped and draped in the usual sterile fashion. A time out is then performed, and the side and site of surgery are verified. A sterile ruler is then used to measure the desired (0.5-2.0 cm) margin from all edges of the primary lesion or biopsy scar. This usually results in a circular or oval area to be excised. The incision is extended in either direction to form an ellipse. In general the ellipse should be oriented along Langer's lines, or the long axis of the extremity, to allow for maximal utilization of available local skin. To achieve a closure with minimal tension, the length of the incision must be at least three times the measured width of the planned excision. Local anesthetic (usually 1% lidocaine with or without epinephrine, mixed in a 1:1 concentration with 0.5% bupivicaine) is then instilled into the area to be resected. A number 15 blade is used to create the incision. The incision is carried down to the level of the underlying muscular fascia, in a perpendicular fashion, using electrocautery. Once the boundaries of the excision have been defined, the specimen is dissected off the underlying fascia using electrocautery.

Step 2. After the specimen has been excised, it is oriented for the pathologist with a marking stitch of 2-0 or 3-0 silk and sent for permanent sections. The wound is then irrigated with sterile saline, and a meticulous check for hemostasis is undertaken. Final hemostasis is usually achieved with electrocautery. Single skin hooks are then used to apply slight traction to ends of the incision and a marking pen is used to place lines across the incision at 8-10 mm intervals to help with suture alignment. Alternatively, a series of towel clips can be used to align the wound for closure. Depending on the size of the resultant soft-tissue defect, a closed-suction drain may be placed in the wound bed prior to closure. The deep dermis is then re-approximated with an interrupted layer of 3-0 undyed, absorbable sutures. The skin may be closed with either interrupted 3-0 nylon sutures or a running 4-0 monofilament subcuticular stitch, depending on the tension on, anatomic location of, and anticipated mobility of the individual incision.

Step 3. If the incision will not close in a primary fashion, local advancement flaps may often be used. Alternatively, a split-thickness or full-thickness skin graft may be employed. For lesions on the extremities, the donor site is usually chosen on the opposite extremity. A Zimmer dermatome is used to harvest a graft of 0.16 mm in thickness. A mesher or scalpel is then used to create holes in the graft and allow for the seepage of serum from the wound. Should a seroma or hematoma form under the graft, the take will be poor.

The wound bed is prepared by imbricating the edges of the wound with absorbable sutures. This minimizes the area to be covered and allows for a smooth transition from the wound to the surrounding skin. The graft is secured in place with 4-0 absorbable sutures and then a nonadherent (petroleum gauze) dressing is placed over the graft. A bolster of cotton balls soaked in glycerin is then placed over the petroleum gauze dressing and held in place using 3-0 silk sutures placed at the periphery of the wound. The donor site may be dressed with petroleum gauze or simply a Tegaderm™ or other occlusive dressing. The area is immobilized for approximately 7 days, at which time the dressings are taken down and take of the graft is assessed.

Postop

As above, most patients are discharged to home on the day of surgery. If a closed suction drain has been placed, the patient is instructed on care of the drain, asked to empty the drain twice daily, and to record the output. The patient is brought back to the clinic for drain removal when the total output for 24 hours is less than 30 ml for 2 days in a row. This is usually the case by postoperative day 7. Sutures are removed 10-14 days after the operation and steri-strips are placed.

Complications

Cellulitis or minor wound infections are the most common complications. The first line of treatment is often oral or intravenous antibiotics, with elevation of the extremity in question. Only rarely will sutures have to be removed and the wound allowed to close by secondary intention. Wound-edge separation with the formation of a wide scar is another complication. This often results from the use of a subcuticular closure in a high-tension wound or from early suture removal.

Follow-Up

Current guidelines call for all patients to be examined, with a full head-to-toe skin survey and palpation of all lymph node basins, at least twice a year. Many physicians who care for patients with melanoma will also follow chest X-rays and LDH levels; however, this is considered optional. Any new symptoms should prompt radiologic imaging as appropriate.

76

Chapter 77

Sentinel Lymph Node Biopsy for Melanoma

Jeffrey D. Wayne

Indications

Although therapeutic lymph node dissection remains the procedure of choice for grossly involved regional lymph nodes in melanoma patients, elective lymph node dissection (ELND) has been abandoned as treatment for the clinically negative lymph node basin. Sentinel lymph node biopsy (SLNB) is a more accurate and less morbid way to assess the regional lymphatics in patients with negative physical findings. The operative concept is that individual areas of skin have specific patterns of drainage, not only to a regional basin, but to a specific node or nodes within that basin. This sentinel lymph node may be identified at the time of operation using a vital blue dye, a radio-labeled colloid, or both.

Thin melanomas (Breslow's depth <1.0 mm) have a low propensity to metastasize to regional lymph nodes (<5%). Thus, patients with thin lesions are usually not offered a SLNB unless they have a lesion with one or more adverse prognostic indicators. Such factors are: Clark's level >III, ulceration, evidence of regression, or mitoses>0. All patients with intermediate thickness and thick lesions, with clinically negative nodes (clinical Stage IB and II), are appropriate candidates for SLNB.

Preop

Successful SLNB is predicated upon identifying the lymph node basin(s) at risk. While this is a relatively straightforward task in extremity lesions, there is often ambiguous drainage from lesions located on the trunk, head, and neck. Thus, preoperative lymphoscintigraphy is crucial to operative planning. Furthermore, preoperative lymphoscintography helps identify not only the number and relative location of the sentinel lymph nodes within a particular basin, but it also allows for the identification of "in transit" sentinel lymph nodes. These nodes, often found in the epitrochlear or popliteal spaces, may be sources of regional or distant failure if not recognized and properly excised.

The highest reported sentinel node identification rates (in excess of 97%) are found when both 1% isosulfan blue dye and ^{99m}technetium sulfur colloid are used for localization. Sentinel lymph node biopsy is ideally performed at the same operative setting as excision of the primary lesion. It is an outpatient procedure, and patients can often be done under local anesthesia with intravenous sedation.

The patients are seen in the nuclear medicine department upon arrival to the hospital, and the preoperative lymphoscintigraphy is performed. A four-point intradermal injection technique is used to administer 0.5-1 mCi of ^{99m}technetium sulfur colloid. Early and frequent images using a gamma camera are obtained in anterior and lateral views until all sentinel lymph nodes are identified.

The patient is then transported to the preoperative holding area, where a handheld gamma probe is used to verify the location of the nodes transcutaneously.

Northwestern Handbook of Surgical Procedures, 2nd Edition, edited by Nathaniel J. Soper and Dixon B. Kaufman. ©2011 Landes Bioscience.

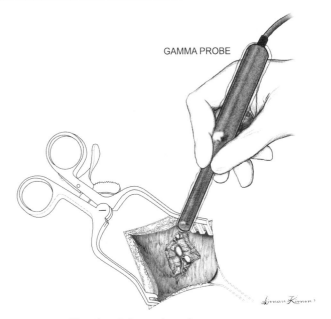

GAMMA PROBE

Figure 77.1. Sentinel lymph node biopsy for melanoma.

Procedure

Step 1. The patient is positioned so as to allow access to the primary site and any lymph node basins harboring a sentinel node. This sometimes requires the use of a beanbag and the lateral decubitus position. Prior to prepping, 1-3 ml of 1% isosulfan blue dye is injected sterilely in the area around the primary lesion or biopsy scar. The patient is then prepped and draped in the usual sterile fashion.

Step 2. A time out is performed and the side and sites of surgery are verified. Attention is first turned to the sentinel lymph node basin(s). A handheld gamma probe is brought on the field in a sterile sheath and used to localize the node in question. A #15 blade is then used to make a focused 2-3 cm incision directly over the highest point of radioactivity. Care must be taken to align the incision such that it can be easily excised should a completion lymph node dissection be necessary.

Step 3. A self-retaining retractor may then be placed to facilitate dissection. Careful blunt and sharp dissection with meticulous hemostasis is then used until afferent and efferent blue lymphatic channels are identified. These channels are traced to the first blue node which is isolated and excised. Afferent and efferent channels as well as small vessels may be controlled using electrocautery and 3-0 or 4-0 absorbable ties, as necessary. The use of hemoclips is discouraged, as they may interfere with a completion dissection. The sentinel lymph node is then scanned ex vivo, and the maximal count is recorded. The basin is then rescanned, and all blue nodes or any node with an in vivo count within 10% of the hottest node should be removed. Nodes are individually identified and sent to pathology for permanent histology and immunohistochemistry. Markers frequently employed to confirm the diagnosis of metastatic melanoma are HMB-45, MELAN-A, and S-100.

Step 4. Once the background count in the basin is less than 10% of the hottest node, sterile moist gauze is placed in the wound, and attention is turned to the primary lesion. Using a second set of instruments, a wide local excision and primary closure is performed, as described previously. Gloves and instruments are again changed, and the lymph node basin and all intervening tissues are rescanned to rule out any additional in transit nodes. The sentinel node wound is then closed in two layers using an absorbable suture to close the deep dermis and a monofilament, absorbable suture to close the skin in a subcuticular fashion. Steri-strips or Dermabond® are then placed.

Postop

Patients may be discharged on the day of surgery with oral analgesics.

Complications

Unlike with ELND or therapeutic lymph node dissection, complication rates with SLNB are low. The most common complications are cellulitis, minor wound infections, and seroma formation. Large seromas may be drained in the office using a sterile technique and 18-gauge needle. Lymphedema is rarely observed.

Follow-Up

Should a positive sentinel lymph node be found, completion lymph node dissection is usually scheduled within 2-3 weeks. Alternatively, patients may consider entry onto the MSLT-II trial, which randomizes patients to either completion nodal dissection or observation with ultrasound at 4-month intervals. The patient is also referred to a medical oncologist for consideration of treatment with interferon α-2β, observation or a clinical trial.

77

Chapter 78

Radical Excision of Soft Tissue Tumor (Sarcoma)

Jeffrey D. Wayne

Indications

Soft tissue sarcomas (STS) are a rare and diverse group of tumors. It is estimated that there were be 10,520 new cases diagnosed, and 3,920 deaths from STS in the United States in 2010. Most of these tumors are derived from the embryonic meso-derm or ectoderm, and individual sarcomas take their name from the parent tissue of origin. The mainstay of treatment for most sarcomas is surgical. However, in an effort to facilitate limb salvage, radiation therapy is frequently employed in either the neoadjuvant or adjuvant setting. This is especially true if the individual lesion is high grade, large (T2 or ≥5 cm in greatest dimension), or deep to muscular fascia. A typical course consists of 50 Gy given over 25 fractions. Sarcomas of the extremity typically present as a painless mass. An MRI or CT scan of the affected limb usually suggests the diagnosis. However, a confirmatory biopsy is often required, especially if a neoadjuvant treatment strategy is to be employed. In this day and age, the diagnosis is typically secured via core-needle biopsy. This is an office-based procedure which may be accomplished under a local anesthetic. Deeper lesions may undergo core bi-opsy in interventional radiology under ultrasound or CT guidance. Rarely, an open (incisional) biopsy is required. If this technique is employed, care must be taken to orient the incision along the long axis of the limb so as to facilitate future attempts at limb salvage. To illustrate some of the concepts of radical excision for STS, we will outline the en bloc excision of a sarcoma of the anterior compartment of the thigh.

78

Preop

Testing is dictated by the location, size, and grade of the sarcoma. For a small, superficial (T1, ≤5 cm) sarcoma, a chest X-ray and CBC may be all that is required. For larger (T2, >5 cm) or high-grade lesions CT scan of the chest, abdomen and pelvis (if a lower extremity sarcoma), are routinely obtained to rule out disseminated disease. If a previous biopsy has been performed, A single dose of a first-generation cephalosporin is administerd prior to making an incision. If the patient is penicillin allergic, clindamycin may be substituted.

Procedure

Step 1. The patient is taken to the operating suite and placed in the supine position with the affected limb slightly abducted and externally rotated. A sterile bump of folded towels provides adequate support beneath the knee. Care must also be taken to provide adequate padding under the lateral malleolus. A general anesthetic is usually employed, as an epidural catheter may interfere with neurologic monitoring of the extremity. Neuromuscular blocking agents may be given for induction, but are not typically re-dosed. The limb is prepped from the knee to a level above the anterior superior iliac spine. A circumferential prep is helpful to allow for repositioning of the leg as necessary.

Northwestern Handbook of Surgical Procedures, 2nd Edition, edited by Nathaniel J. Soper and Dixon B. Kaufman. ©2011 Landes Bioscience.

Step 2. The operation begins by making an elliptical incision around any previous biopsy incision or scar, with 1-2 cm margins. This incision is extended proximally and distally over the course of the longitudinal extent of the lesion. Medial and lateral flaps are then raised with the assistance of skin rakes.

Step 3. In the case of large tumors, it is often helpful to expose major anatomic structures, such as vessels and nerves proximal and distal to the lesion. This will facilitate their salvage if oncologically feasible. Thus, in the case of a deep thigh sarcoma, the common femoral artery, vein, and nerve are identified cephalad to the tumor, and encircled with vessel loops. Similarly, the superficial femoral vessels are located at the medial aspect of the distal thigh and likewise encircled with vessel loops.

Step 4. Deep dissection is then begun at whichever aspect of the mass that allows for maximal dissection with the least amount of resistance. Care should be taken to stay outside the tumor "capsule" which is a misnomer in most cases of sarcoma. This is actually a pseudocapsule composed of compressed tumor cells at the periphery of the mass. A bovie electrocautery, 10 blade, or Metzenbaum scissors are used as appropriate for the thickness and location of the tissue. A harmonic scalpel or advanced bipolar sealing device (e.g., EnSeal®) may also be employed. Any muscle fibers, for example from the rectus femoris, adductor magnus, or vastus medialis, are taken en bloc with the specimen. Only rarely must an entire muscle belly be sacrificed.

Step 5. The deep aspect of the tumor is addressed last. If the mass can be removed in a margin-negative fashion without sacrifice of the vessels, then this is optimal. However, vascular reconstruction with reversed saphenous vein or polytetrafluoroethylene (PTFE) is an option, if necessary. If a positive margin is unavoidable in the vicinity of bone, or a major motor nerve, thought should be given to the placement of brachytherapy (after loading) catheters in the wound bed. In any instance, it is always advisable to place large, titanium hemoclips in the resection bed to facilitate adjuvant external-beam radiation.

Step 6. Finally, the wound is irrigated with copious amounts of warm saline solution and a meticulous check of hemostasis is made. A flat, closed-suction drain is placed in the resection bed and brought out through an inferiorly placed stab incision, in line with the primary incision. The deep dermis is then reapproximated with a layer of 3-0 absorbable sutures in an airtight fashion and then the skin is closed with interrupted nylon sutures in an interrupted vertical mattress fashion.

Postop

In the absence of vascular or plastic surgery reconstruction, the patient is typically kept in the hospital overnight. The leg is elevated, and the patient is instructed in drain management. Patients are asked to return to the clinic when the drain output is less than 30 ml for two consecutive 24-hour periods. The patient is seen by the physical therapist on postoperative day one for instruction in weight bearing as tolerated. The patient is given a prescription for an oral narcotic to take home (e.g., Norco®). Sutures are removed at 2 weeks.

Complications

Inadvertent nerve or vascular injury may be minimized by utilizing the techniques outlined above. Specifically, dissecting out major anatomic structures prior to any attempt at removal of the main tumor mass will allow for these structures to be followed along their course and preserved, as they are dissected sharply away from the mass. Wound infection rates in the range of 25-30% are reported when external beam radiation is employed in a neoadjuvant fashion. Liberal use of autologous tissue flaps are advised in this instance.

Follow-Up

Patients are seen at intervals of 3-6 months for the first 2-3 years, depending on the size and grade of the lesion. At each visit a physical exam, chest imaging, and site-specific imaging (CT, MRI, or ultrasound) is performed. Follow-up is continued for 10 years due to the small, but definable instance of late recurrence.

78

SECTION 4: PLASTIC SURGERY

Section Editor: Thomas Mustoe

Chapter 79

Burn Debridement and/or Grafting

John Kim

Indications

First degree—Superficial partial-thickness burns. Erythematous. Superficial epidermal burns, such as from sunburn, heal with topical therapy alone.

Second degree—Deep partial-thickness burns. Bullae often present. Less pain, deep dermis exposure with or without hair follicle exposure.

Third degree—Full-thickness burns. Relatively painless, dry, leather-like appearance.

Fourth degree—Muscle or bone exposure.

Preop

The surgeon must assess and manage (A)irway, (B)reathing, and (C)irculation, as with any form of trauma. Estimate the total body surface area (TBSA) burn using the rule of nines. Intravenous fluid replacement is done using the Parkland formula, 4 ml lactated Ringers/kg body wt/%TBSA burn, given over the first 24 hours: one-half in the first 8 hours. Continue IV fluids to maintain a urine output of 0.5 ml/kg wt/hours. Avoid systemic antibiotics. Topical antimicrobial agents—silvadene, sulfamyalon, or silver nitrate—are used liberally. Silvadene is the agent of first choice; silver nitrate is used in patients with sulfa allergy; and sulfamyalon is used for patients who require eschar penetration. Treat associated trauma.

Procedure

Step 1. Bring patient to room with temperature greater than 75° F.

Step 2. Anesthesia with appropriate monitoring.

Step 3. Prep and drape all involved areas.

Step 4. Debride all nonviable tissue. Tangential excision is preferred using a burn knife or dermatome.

Step 5. Avoid skin grafting directly onto adipose tissue.

Step 6. Using a dermatome (Zimmer), obtain a skin graft from an uninjured area. Set to thickness of 12-15/1000 in.

Step 7. Consider meshing skin to increase surface area coverage while minimizing the donor area. Avoid meshed grafts for the hand.

Step 8. Hemostasis using topical thrombin and cautery.

Step 9. Apply meshed skin graft with staples or sutures.

Step 10. Cover skin graft with Adaptic™ or Owens gauze to avoid skin graft disruption from the removal of the dressings. May use VAC sponge as an alternative.

Step 11. Apply mineral oil-soaked cotton or gauze to the Adaptic™ dressing.

Step 12. Cover donor site Xeroform™, Biobrane™ or Opsite™ dressing.

Step 13. Apply extremity splints as necessary to immobilize grafted areas and prevent shear.

Step 14. Do not graft more than 25% TBSA in one sitting.

Northwestern Handbook of Surgical Procedures, 2nd Edition, edited by Nathaniel J. Soper and Dixon B. Kaufman. ©2011 Landes Bioscience.

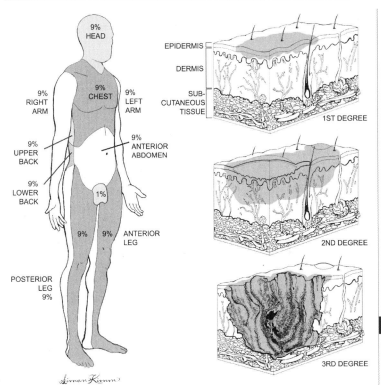

Figure 79.1. Burn debridement and/or grafting.

Postop
Control temperature. The dressings are kept in place for 3-5 days. Fluid and nutritional support. Begin physical therapy at 7-10 days.

Complications
Hematoma/seroma, infection, graft failure, scarring/deformity, and need for additional surgery.

Follow-Up
Aggressive physical therapy. Compressive garments. Secondary reconstruction. Scar/deformity management.

79

Chapter 80

Split-Thickness Skin Grafts

Robert D. Galiano

Indications

Split-thickness skin graft closure is useful for wounds with a stable granulating or vascular base adequate for plasma diffusion/neovascular growth. Split-thickness skin grafts (STSG) are very reliable but require a donor area that will heal by secondary means. STSGs tend to contract over time and should not be used in wounds which will not tolerate contracture (i.e., hand).

Preop

The desired skin graft site should have a stable, clean base of granulation tissue or muscle/fascia to allow for successful plasma diffusion/neovascularization. Wound biopsy should be done if the wound is suspected to have greater than 500,000 microorganisms. Repeated wound debridement may be necessary to achieve a clean base for grafting.

Procedure

Step 1. The patient is positioned to allow access to the wound (recipient site) and the donor site.

Step 2. The donor area (typically the lateral thigh) is prepared by removing betadine and applying mineral oil to reduce dermatome friction.

Step 3. An air hydraulic dermatome or knife is set 12-16/1000th inch thickness, depending on wound location and/or type.

Step 4. Using four-point traction on the donor area, the dermatome is placed at a 45° angle and advanced with uniform pressure to harvest desired skin.

Step 5. The skin graft can be meshed to increase surface area and/or to permit fluid egress, which may hinder graft take.

Step 6. The wound (recipient site) is cleaned/debrided to healthy, bleeding tissue.

Step 7. The skin graft is applied to the wound, dermis side down, and secured with sutures or staples. ALWAYS CHECK DERMIS SIDE DOWN.

Step 8. Adaptic™, Owens, or Xeroform™ gauze is applied to the secure skin graft to prevent skin adhesion to the dressing.

Step 9. Mineral oil-soaked cotton balls or Reston foam is applied to the adaptic.

Step 10. A second Reston foam pad is sewn or stapled to keep shear movement to a minimum, with moderate compression. Ace wrap can be used on the extremities.

Step 11. The donor site is checked for hemostasis, then a Tegaderm™ dressing is applied.

Northwestern Handbook of Surgical Procedures, 2nd Edition, edited by Nathaniel J. Soper and Dixon B. Kaufman. ©2011 Landes Bioscience.

Postop

Prophylactic antibiotics may be used. The area should be elevated and kept dry for 5-7 days. The recipient dressing is removed at 5-7 days. The donor dressing is left on until it falls off or shows signs of delayed healing. The donor area should re-epithelialize by 7-14 days. Activity should be limited while the skin graft dressing is on.

Complications

Bleeding, infection, neurovascular injury, delayed wound healing, skin graft loss, scarring, and need for additional surgery.

Follow-Up

Once the grafted area is exposed, it should be kept clean and lubricated. After washing, aloe or vitamin A & D ointment can be used daily. Split-thickness skin grafts lack dermal support structures and are at risk for dessication.

80

Chapter 81

Debride/Suture Major Peripheral Wounds

Gregory Dumanian

Indications

Major acute wounds of the head and neck, chest, abdomen, back, and extremities should be closed promptly to restore patient homeostasis. Chronic wounds, in contradistinction, should be assessed for etiology, reasons for failure to close, and appropriateness for closure.

Preop

As all major acute wounds should be closed, all that is left to decide is the method of wound closure and its timing. The method of wound closure depends on the importance of the soft tissues exposed. Exposed fractures and orthopedic hardware, tendons, open joints, recent incision lines in blood vessels or bowel, and prosthetic material all require soft tissue flaps to aid in wound healing. Surface wounds without critical underlying exposed tissue often heal with split-thickness skin grafts. Timing of major reconstructive surgery is an interplay between overall patient status, the ability of the underlying tissue to remain "open," and the complexity of the reconstructive procedure. For extremity wounds, assessment of distal limb nerve, artery, and muscle function is critical.

Procedure

Step 1. Patient positioning should be appropriate for exposing the major wound and any other incision for flap and skin graft harvest. The greater the patient exposure, the better.

Step 2. "Create the defect" by first sharply debriding devitalized tissue. Aim to cut back to normal-appearing tissue in order to reduce bacterial loads and to allow for primary wound healing.

Step 3. Pulse-lavage tissues which are critical, contaminated, and cannot be removed during the sharp debridement process.

Step 4. Assess the wound and the ability of tissues to either be closed primarily, to receive a skin graft, or those wounds which will need flap coverage.

Step 5. Wounds which can be closed primarily should not have undue tension, should not require more than a 2-0 nylon to appose the tissue, and should not put undue pressure on underlying tissue. Drain the underlying soft tissues liberally. Keep deep foreign material in the wound to a minimum to help avoid postoperative infections.

Step 6. Split-thickness skin grafts are best harvested from the posterolateral thigh with a mechanical dermatome at 12-14/1000th inch. Donor sites are dressed with a semipermeable membrane dressing. The grafts are usually meshed and held in place on the wound with stent dressings.

Northwestern Handbook of Surgical Procedures, 2nd Edition, edited by Nathaniel J. Soper and Dixon B. Kaufman. ©2011 Landes Bioscience.

Step 7. Pedicled flaps from adjacent tissue require a knowledge of flap and blood vessel anatomy. Ensure that the flap blood vessels have not been damaged by the agent which caused the wound. "Creating the defect" first before flap elevation ensures proper flap dimensions for wound coverage. Elevate the flap, close the donor site, and then insert the flap on the wound. Liberally drain the donor site and the flap recipient site.

Step 8. For major wounds requiring free flap transfer, the flap donor site will be removed spatially from the wound. Blood vessels near the wound will need to be dissected out for eventual anastomosis to the flap blood vessels. Donor site closure and flap inset will be the same as for pedicled flaps.

Step 9. Immobilization of major wounds with postoperative dressings is a critical feature of these procedures. Plaster splints made in the operating room are often used to immobilize soft tissues and help to prevent unwanted movement and potential late joint contractures.

Postop

Elevation, immobilization, and pressure relief of recently treated wounds will decrease edema and help wound healing. Skin graft dressings can be removed 3-5 days after placement. Flaps need to be checked for vascularity.

Complications

Wound dehiscence, wound infection, hematoma, partial or total skin graft loss, partial or total pedicled flap loss, free flap arterial or venous thrombosis.

Follow-Up

Initial acute wound management may only be the first phase in eventual reconstruction. Extremity injuries frequently require physical or occupational therapy to restore function.

81

Chapter 82

Repairing Minor Wounds

Thomas Mustoe

Indications

Indications include lacerations penetrating the superficial dermis with some gaping of the wound; blunt trauma injuries causing splitting of the skin with some gaping; and combination laceration/abrasion injuries where skin can be reapproximated even in the setting of surrounding abrasions.

Preop

Anesthetize the area if sensitive to allow proper cleaning. Utilize xylocaine with epinephrine in the head and neck to allow sufficient anesthesia time to complete the repair and minimize bleeding. Avoid epinephrine in the hands or feet, particularly in the digits. Arterial spasm with eventual digit necrosis is a well-known complication of utilizing epinephrine in those regions. For young children consider an oral or intramuscular sedative, although it may not be necessary. Sedatives can make the experience less traumatic and allow a more meticulous repair. For infants, immobilize the baby in a papoose or blanket. Prep the area with betadine and utilize a simple drape. Two towels work fine or optimally a single sheet with a cut-out circle.

Procedure

Step 1. Irrigate the wound thoroughly. Irrigation under adequate pressure is much more effective than soaking or low-pressure irrigation. Utilizing a 10 ml syringe with a 20-gauge needle at close range will deliver adequate force. For a small wound 30-40 ml of saline should suffice unless the wound is very dirty and there is ground-in foreign material. With abrasions, it is critical to remove all foreign material to prevent tattooing. That may necessitate extensive irrigation.

Step 2. Excise ragged or extensively traumatized wound edges, as long as the wound can be brought together with minimal tension. The eventual scar will be superior. An easy method is to score the 1-2 mm excision at the edge and then complete the excision of the traumatized edge with scissors.

Step 3. If the wound is gaping substantially and penetrates the entire dermis, utilize buried intradermal sutures with the knot on the deep surface of the tied ligature. Absorbable sutures such as polygalactic acid should be utilized.

Step 4. For optimal epidermal approximation, cutaneous sutures with either a permanent suture or an absorbable suture can be utilized. In the head and neck a permanent suture such as 6-0 or 5-0 nylon works well. Alternatively, for children or when follow-up is uncertain, a 5-0 fast-break plain gut or a 6-0 mild chromic suture can be utilized. The disadvantage of absorbable sutures is an increased risk of inflammation with potential deleterious effects on eventual scar outcome. In general interrupted

Northwestern Handbook of Surgical Procedures, 2nd Edition, edited by Nathaniel J. Soper and Dixon B. Kaufman. ©2011 Landes Bioscience.

sutures are safer in the case of infection in a dirty wound, but in a clean wound with good dermal approximation, a running suture is very acceptable and saves time. In the hands and feet, a one-layer closure with 5-0 nylon is preferred.

Step 5. Occlusive dressings minimize inflammation and optimize the eventual scar outcome. In routine cases this is best accomplished by steri-strips and an adhesive such as benzoin or gum mastic (Mastisol®). If significant exudate is expected, or there are surrounding abrasions, then an occlusive ointment applied repeatedly is a good alternative. Antibiotic ointments, particularly those that contain neomycin, run the risk of allergy. If the wounds are kept clean of exudate in the postoperative period, then a nonantibiotic ointment such as A & D ointment will minimize the risk of allergic dermatitis.

Postop

The major principle to recognize is that bacterial growth in the eschar of and serum on the top of an incision will lead to inflammation, delayed epithelization, and a suboptimal scar. Therefore if the wound is clean and postoperative exudate is presumed to be minimal, then steri-strips will protect the wound and prevent bacterial overgrowth. There is no need to keep the area dry in the shower. If the wound is dressed without an occlusive dressing, ointment should be applied and the area should be washed free of dried blood and protein exudate at least daily to prevent bacterial overgrowth in the microenvironment of the wound, with application of an appropriate ointment as described above.

Complications

If the wound has been properly irrigated, with removal of foreign body and devitalized tissue, wound infection should be very unusual (1-3%). If interrupted sutures are used, then a suture can be removed to allow drainage of a superficial abscess in the postoperative period.

82

Follow-Up

In cosmetically sensitive areas such as the face, neck, or upper chest, periodic follow-up is necessary to observe for the possibility of hypertrophic scar or keloid. If poor scarring develops, intralesional steroids or topical silicone gel sheeting should be instituted. If these steps are not successful or if the physician is inexperienced with these methods, appropriate referral to a plastic surgeon should be considered.

Chapter 83

Removal of Moles and Small Skin Tumors

Thomas Mustoe

Indications

Growing lesions, changing lesions, painful, itchy or bleeding lesions, darkening of pigmentation, elevation of lesion, or patient preference, with the knowledge that there will be a scar.

Preop

This is typically an office-based procedure but may be done under local or in an ambulatory operating room setting, or as part of a more significant operation. It can be prudent to stop aspirin and anticoagulants, but often these procedures are so minor that even this is not necessary.

Procedure

Step 1. Prep the skin with alcohol.

Step 2. Mark the lesion for excision along the skin tension lines. Lenticular excisions are usually preferred. When in doubt, excise the lesion as a circle and determine later which orientation to close.

Step 3. Inject with lidocaine and epinephrine.

Step 4. Allow 5-10 minutes to pass between injection and incision.

Step 5. Prep with betadine and sterile towels.

Step 6. Excise the lesion with a 15 blade. Extend into the subcutaneous tissue. On lesions for which you have not previously decided upon orientation, a circular excision will have been done and at this stage pull gently with tooth forceps to see which direction the skin closes most easily. Once this is determined, place a deep buried suture in the middle, and then remove the resultant dog ears as necessary. In general closing an angle of more than 30° will cause a dog ear (elevation of end of incision), which will help determine length of excision. It is necessary to take a margin on benign lesions, and in general if there is a suspicion of malignancy an excisional biopsy can preparation for a more definitive procedure.

Step 7. Finish closing with further buried sutures and if necessary external sutures; then cover with skin tape. Internal buried sutures will take tension off the external closure and may lead to a finer scar. This must be balanced against the potential for a buried suture to extrude and cause inflammation.

Step 8. Send all suspected skin cancers and all pigmented lesions for pathology.

Northwestern Handbook of Surgical Procedures, 2nd Edition, edited by Nathaniel J. Soper and Dixon B. Kaufman. ©2011 Landes Bioscience.

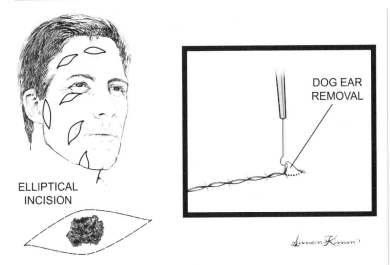

Figure 83.1. Removal of skin moles, small tumors, etc.

Additional Comments

Many variations of suture techniques exist. My preference is to use internal clear nylon sutures to provide long-lasting support to the scar. External clear nylon sutures are then used as necessary to get adequate skin coaptation. Using a single suture for both the internal and external suture cuts down on the number of sutures that are necessary for these minor procedures. I typically will use a 5-0 suture on the face and 4-0 sutures elsewhere on the body. I have found that a single layer of 4-0 polypropylene typically works well on the scalp, and the blue suture is visible, and makes suture removal easier. A small disposable stapler also works well in the scalp. For postoperative care, pressure is held on the excision for 15-20 minutes to reduce bruising. Patients may shower the next day.

Complications

Bleeding and infection are always possible with any excision, but they are quite rare with simple skin excisions. They are especially rare in the face. Antibiotics are not routinely used.

Follow-Up

Pathology results must be transmitted to the patient. In tumors, it is important to maintain adequate follow-up for both recurrence and development of new tumors. This is often done by a dermatologist or a primary care physician.

Acknowledgment

The editors and author wish to acknowledge Neil A. Fine for contributing to the previous version of this chapter.

83

Chapter 84

Removal of Subcutaneous Small Tumors, Cysts and Foreign Bodies

Gregory Dumanian

Indications

Indications include lesions that are growing in size or painful, or are suspicious for malignancy due to their hard size, irregularity, or fixation to the overlying skin. Patients may desire removal of lesions for cosmetic reasons but should understand that a scar is inevitable.

Preop

Discuss with the patient the nature of the scar and the unpredictability of scars. Patients who take aspirin or anticoagulants are advised to stop them. Often however, if the lesions are small this is not necessary. This procedure would commonly be performed in an office or in an ambulatory care setting using local anesthesia or as a conjunctive procedure as part of a more involved procedure requiring some other form of anesthesia.

Procedure

Step 1. Position the patient so the cyst or lesion is visible.

Step 2. Cleanse the skin with alcohol.

Step 3. Mark the overlying skin in the form of an ellipse if the lesion is protuberant. This will allow for excision of excess skin due to the stretching of the protuberant cystic lesion.

Step 4. Inject with lidocaine with epinephrine. Allow at least 10 minutes to pass prior to incision of the skin. Prep the skin with chlorhexidine or an iodine compound, and place sterile towels. Incise the skin with a 15 blade and switch to scissors or continue with the 15 blade down to the surface of the cyst or foreign body. Once at the level of the cyst or tumor, it is most advantageous to spread and dissect with scissors. Next, when the cyst or tumor is freed circumferentially, try to remove in one piece. If it is not possible to remove in one piece, special care should be taken in the case of a cyst to ensure that all fragments of the true cyst itself are removed to decrease recurrence. Small bleeding points can be tied or sutured with small absorbable sutures. Alternatively, handheld cautery units may be utilized. Often due to the vascular constrictive effects of epinephrine, neither of these is necessary. The skin is closed in layers with deep buried suture followed by an external skin suture if necessary and then skin tapes.

84

Northwestern Handbook of Surgical Procedures, 2nd Edition, edited by Nathaniel J. Soper and Dixon B. Kaufman. ©2011 Landes Bioscience.

Comments

Before the start of any skin lesion excision, the anatomic region of the area should be assessed for the presence of nerves that can be injured during the procedure. Some key nerves to remember are the spinal accessory nerve in the posterior triangle, and the superifical radial nerve at the wrist. Injury to these nerves or other superficial sensory nerves can cause disability that is profound.

The patient should be warned that scars on the upper back, chest, and shoulders are particularly prone to forming hypertrophic scars or keloids. Any suspicious or unusual lesions need to be sent to pathology.

The more mobile the mass under the skin preoperatively, the easier will be the removal. Cysts that have previously ruptured will have uncertain soft tissue planes, and some subcutaneous fat will need to be removed with the cyst in these instances. Lipomas that are mobile can often be "squeezed" out of an incision half the size of the lesion. This technique is useful in its protection of surrounding subcutaneous nerves.

A key to removal of sebaceous cysts is the removal of the punctum or the connection of the skin to the epidermis. Excision of an ellipse of skin over the cyst helps to ensure the removal of the punctum and in addition decreases the chance of rupturing the cyst.

While absolutely sterile technique is the goal, small breaks in sterile technique do not seem to correlate with postprocedure infections or complications.

Postop

Pressure is held overlying the area for 15-20 minutes to decrease bruising. Patients are allowed to shower the next day.

Complications

Bleeding and infection are always possible with any excision, but they are quite rare with simple skin excisions. They are especially rare in the face. Antibiotics are not routinely used.

Follow-Up

84

Pathology results must be transmitted to the patient. In tumors, it is important to maintain adequate follow-up for both recurrence and development of new tumors. This is often done by a dermatologist or a primary care physician.

Acknowledgment

The editors and author wish to acknowledge Neil A. Fine for contributing to the previous version of this chapter.

SECTION 5: CARDIOTHORACIC SURGERY

Section Editor: Malcolm DeCamp

Chapter 85

Esophagectomy: Ivor-Lewis

Alberto de Hoyos and Malcolm DeCamp

Indications

1. Esophageal malignancies involving the esophagogastric junction or mid-lower esophagus (at or below the azygos vein level).
2. Barrett's esophagus with high-grade dysplasia.
3. End-stage benign esophageal diseases (typically after multiple failed operations for gastroesophageal reflux or motor disorders).

Preop

1. Accurate diagnosis of the esophageal disease is accomplished by a detailed history, barium esophagogram, pH and manometric evaluation and endoscopy with biopsy.
2. In resections for esophageal cancer, rigorous tumor staging is necessary to determine that the lesion is localized and therefore amenable to a curative resection. This is accomplished by clinical evaluation (history and physical examination) and by computed tomographic (CT) scan of the chest and abdomen. Endoscopic ultrasound (EUS) is routinely used to assess the depth of penetration of the esophageal wall by the tumor (T stage). Additionally, EUS is used to assess nodal stage by using echo characteristics and directed fine needle aspiration (N stage). PET scanning is also commonly used to identify distant metastases (M stage). Preoperative surgical staging using thoracoscopy and laparoscopy has also been described but has not been universally accepted.
3. Fiberoptic bronchoscopy is necessary to rule out the possibility of airway involvement by tumors of the mid and upper esophagus.
4. Standard preoperative hematological and biochemical blood work includes electrolytes, liver functions, complete blood count with platelets, coagulation profile; urinalysis; electrocardiogram (EKG); type and screen. If the patient has documented cardiac disease or an abnormal EKG, dyspnea or chest pain with exertion, or is unable to exercise, a formal cardiology assessment should be conducted. If the patient has documented pulmonary disease, a significant smoking history, or if there is any significant breathing impairment, formal pulmonary function tests (PFT) and arterial blood gas (ABG) should be obtained.
5. Many centers advocate the use of preoperative (neoadjuvant) chemo-radiotherapy in esophageal carcinoma although there are insufficient data to support this approach as the standard of care in this disease. Patients with T_3 or N_1 disease should be considered for induction therapy protocols.
6. Optimal pain control is provided by intravenous patient-controlled analgesics (PCA) or an epidural catheter.

Northwestern Handbook of Surgical Procedures, 2nd Edition, edited by Nathaniel J. Soper and Dixon B. Kaufman. ©2011 Landes Bioscience.

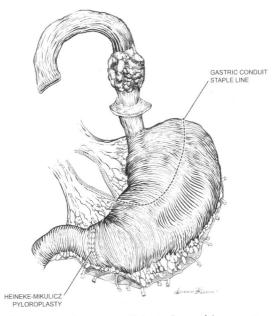

GASTRIC CONDUIT
STAPLE LINE

HEINEKE-MIKULICZ
PYLOROPLASTY

Figure 85.1. Ivor-Lewis esophagectomy. Abdominal part of the operation.

Procedure

Step 1. The first phase of the operation is performed with the patient positioned supine with arms tucked at the sides. Monitoring includes intravenous access, radial arterial line, Foley catheter, and central venous line inserted in the right side of the neck.

Step 2. The abdomen is prepared and draped as a large sterile field.

Step 3. An upper midline incision from the xiphoid to umbilicus is made. The xiphisternum may be excised to improve exposure. The falciform ligament is divided, and a table-mounted retractor is inserted to exert strong upward and lateral pull on the costal arches bilaterally. A Balfour retractor is also inserted to improve the abdominal exposure.

Step 4. In resections for carcinoma, a full inspection of the abdomen and pelvis is performed to rule out the possibility of metastatic disease. Careful attention is paid to the liver, the primary tumor, and the upper abdominal lymph nodes. Once this has proven to be satisfactory, the resection and gastric pull-up are performed.

Step 5. The proximal stomach and distal esophagus are mobilized in anticipation of stomach being used as the replacement conduit for swallowing. The stomach will ultimately be perfused by the right gastroepiploic and right gastric arterial systems so extreme care must be taken throughout the dissection not to damage these vessels. Gastric mobilization starts with the upper stomach and gastroesophageal junction. The left lateral segment of the liver is retracted to provide exposure of the hiatus. The lesser sac is entered through the lesser omentum. The gastrohepatic ligament is palpated to exclude the presence of a replaced left hepatic artery. The lesser omentum is then divided to the right crus. The distal esophagus is dissected circumferentially and encircled with a Penrose drain. The left phrenic vein is divided and the diaphragm opened anteriorly to provide a four finger hiatus.

Step 6. The lesser and greater curvatures are fully mobilized. The gastrocolic ligament is divided with cautery to enter the lesser sac, staying a generous distance away from the gastroepiploic arcade. The posterior stomach wall and the transverse mesocolon must be accurately visualized. The greater curve can be mobilized with cautery, with the harmonic scalpel being reserved for larger vessels (e.g., short gastric vessels). The mobilization is taken to the right just beyond the pylorus. The duodenum is fully Kocherized using cautery, giving enough mobility to allow the pylorus to lie near the esophageal hiatus after the stomach is transposed to the chest. All of the short gastric vessels are divided, to arrive at the left crus of the diaphragm. The large posterior vascular pedicle containing the left gastric vessels and lymph nodes is isolated at the upper border of the pancreas and divided using sequential applications of the linear vascular stapler. Alternatively, the vascular pedicle may be dissected sweeping nodal tissue toward the lesser curvature. The skeletonized left gastric vessels are then ligated and divided. Dissection may be done by elevating the stomach and isolating the pedicle posteriorly or through the lesser omentum superiorly and anteriorly.

Step 7. Pyloroplasty is performed using the Heinecke-Mikulicz technique. Stay sutures of 3-0 silk are placed at the superior and inferior margins of the pylorus. The pylorus is then opened longitudinally using cautery, slightly onto the duodenal and gastric sides, and ensuring that all of the pyloric muscle has been transected. The opening is then closed transversely, using a series of full-thickness sutures of 3-0 silk.

Step 8. The esophagus is dissected into the mediastinum. A long sweetheart retractor and sponge sticks are used to retract tissues around the esophagus. The esophagus is freed up well above the esophageal hiatus with cautery or the harmonic scalpel. The operating table is tipped into Trendelenburg position to improve the exposure up into the mediastinum. Care is taken to provide adequate hemostasis.

Step 9. The gastric tube is created. The lesser curvature is resected along with the esophagus, using sequential firings of the linear stapler to leave a gastric tube consisting of the greater curvature. The conduit width should be 8 cm. The lesser curve staple line is inverted using a continuous Lembert suture of 4-0 polypropylene. The tubularized stomach is tacked to the distal esophagus using two 3-0 prolene sutures for retrieval in the chest. Care is taken to maintain conduit orientation.

Step 10. A feeding jejunostomy catheter is placed, using a modified (short) Witzel technique and a 12 F red rubber catheter. The ligament of Treitz is identified on the inferior surface of the transverse mesocolon, and the catheter is typically inserted about 20-40 cm distal to this point. Two pursestring sutures of 4-0 silk are used in the antimesenteric border of the bowel to secure the catheter at the entry site. A series of interrupted 4-0 silk Lembert sutures are used to bury the catheter in the bowel wall over a short distance and to secure the bowel to the parietal peritoneum. A tacking stitch is placed proximal to the site where the j-tube penetrates the bowel to prevent torsion or angulation and obstruction of the jejunum at the site of the j-tube. The catheter is brought out through a small stab wound in the left midabdomen and secured at the skin with a 3-0 nylon suture.

Step 11. The laparotomy incision is closed with continuous #2 Maxon™ suture in the fascia and staples in the skin. Sterile dressings are applied, and the drapes are taken down.

Step 12. The right thoracotomy is performed. A left-sided double-lumen tube, or alternately a right-sided bronchial blocker, must be placed by the anesthesiologist between the abdominal and thoracic phases of the operation. After this is complete, the patient is repositioned in the left lateral decubitus position. The right chest is prepared and draped as a sterile field. A right posterolateral thoracotomy incision is made, entering the chest in the 5th (or preferably 4th) intercostal space. The serratus anterior muscle

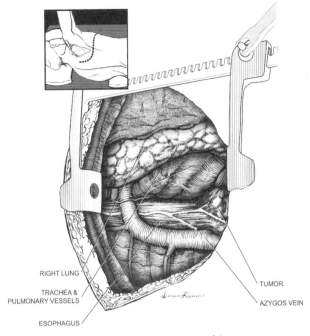

RIGHT LUNG

TRACHEA &
PULMONARY VESSELS

ESOPHAGUS

TUMOR

AZYGOS VEIN

Figure 85.2. Ivor-Lewis esophagectomy. Thoracic part of the operation.

can be spared during creation of the thoracotomy. Subperiosteal removal of a 2 cm segment of the posterior portion of the inferior rib (shingling) will substantially improve exposure. Retractors are inserted at right angles to one another to provide the exposure. The right lung is deflated and packed anteriorly with moist laparotomy sponges.

Step 13. The thoracic esophagus is mobilized, the esophagectomy is completed, and the gastric tube is prepared. The azygos vein arch is dissected out, and the vein is divided between heavy silk ligatures or with a vascular stapler. The esophagus is dissected out and encircled with a large Penrose drain. The esophagus is then fully mobilized in the chest to the desired level of transection [usually near the thoracic inlet]. Below the carina, the segmental blood supply to the esophagus is isolated as a series of small stalks and divided distal to clips or silk ligatures. This method is more secure than dividing the stalks with cautery alone, as an unrecognized lymphatic injury can result in postoperative chylothorax. When mobilizing the esophagus above the carina, it is vital to keep the dissection immediately on the adventitia of the esophagus to minimize the chance of operative injury to the right recurrent laryngeal nerve and membranous airway. Traction is applied to the distal esophagus to deliver the stomach into the chest, taking care to maintain the proper orientation (lesser curvature facing towards the patient's right side). The esophagus is transected proximally with a linear stapler usually several cm above the divided azygos vein. The esophagogastrectomy specimen is passed off the field. If resection is for tumor or Barrett's dysplasia, a frozen section is obtained to verify that the proximal margin is clear of tumor or residual Barrett's mucosa which would mandate ongoing surveillance.

Step 14. The end-to-side esophagogastrostomy is created, usually midway between the carina and the thoracic inlet. It is desirable to create the anastomosis as high as reasonably possible since this will decrease the chance of troublesome postoperative bile reflux and associated complications (e.g., esophagitis or stricture). The anastomosis begins with excision of the staple line from the esophagus and creation of a small gastrotomy on the upper greater curve (staying away from the lesser curve suture line). A common method of hand-sewn anastomosis involves two layers of simple interrupted 4-0 silk sutures. Sutures of the inner layer incorporate the esophageal mucosa to the full thickness of the stomach and are placed so that knots are tied on the inside to maximize mucosal inversion. The outer row of sutures approximates the esophageal muscle to the seromuscular layer of the stomach and buries the inner suture line. An alternate technique involves the use of the linear stapler to create a functional end-to-end anastomosis. The endo-GIA stapler, with a 45 mm long blue cartridge (3.5 mm staples) is inserted into both the esophagus and stomach, which have been laid in apposition. The stapler is fired, creating most of the anastomosis and leaving a common opening into the esophagus and stomach. This common opening is closed using a continuous 4-0 monofilament absorbable suture, with a second layer of 4-0 silk sutures (approximating the esophageal muscle to the seromuscular layer of the stomach) used to bury this suture line. In either technique, a nasogastric (NG) tube must be passed by the anesthesiologist, guided across the anastomosis, and then advanced into the lower portion of the gastric tube before completion of the anastomosis.

Step 15. Two 28 F chest tubes are placed, brought out through separate stab wounds inferior to the main incision, secured to the skin with heavy polypropylene suture, and connected to a drainage apparatus. One tube lies anterior to the lung. The other is positioned posteriorly and lies adjacent to the anastomosis. It is secured here with a 3-0 chromic catgut suture to the pleura to prevent it from moving. This tube is intended to provide external drainage in case an anastomotic leak develops later.

Step 16. The right lung is manually reinflated and the thoracotomy is closed, using #2 Vicryl sutures as pericostal sutures, #1 Vicryl in the extracostal fascial layers, 2-0 Vicryl in subcutaneous tissues, and staples in the skin. Absorbent gauze dressings are applied and the drapes are taken down.

Postop

1. *Intravenous fluids.* Generous volume resuscitation is required in the first 24 hours postoperatively (e.g., normal saline at 150-175 ml/h) since the anticipated 3rd space volume loss will be substantial. Urine output is carefully monitored, with 0.5-1.0 ml/kg/hour considered adequate. ABGs should be obtained periodically to monitor PaO_2 and $PaCO_2$ (during ventilator weaning; see below), and to monitor the base deficit, another useful parameter to assess the adequacy of the circulation. Maintenance of satisfactory circulation is vital, since the upper portion of the gastric conduit is perfused mostly by intramural collateral blood supply. This is best achieved by generous use of crystalloids (as described above), and transfusion of packed red cells (as dictated by low hemoglobin values). About 24-36 hours after surgery, the 3rd space volume loss subsides, and the extravascular volume is returned to the intravascular space. This is heralded by diuresis and signals the appropriate timing to reduce volume administration to maintenance fluids and judicious use of diuretics (e.g., furosemide 20 mg IV Q 8-12 h). Diuretics are continued until the patient has reached their preoperative weight.

85

2. *Respiratory status.* The patient is extubated in the operating room or maintained on a ventilator postoperatively and weaned and extubated the following morning. Supplemental oxygen is then given by face mask and ultimately by nasal prongs as indicated by peripheral pulse oximetry. ABGs are checked during the phase of ventilator weaning and immediately after extubation. If the patient is stable thereafter, pulse oximetry is sufficient to guide the administration of oxygen. During the 48 hours following extubation, nebulized bronchodilators and chest physical therapy are used preemptively to minimize the development of sputum retention and pneumonia. Ambulation is begun on the morning of postoperative day two. Deep vein thrombosis (DVT) prophylaxis (heparin 5000 U sc TID and sequential compression devices [SCDs]) is started prior to induction of anesthesia. Heparin is started immediately postoperatively. Once ambulation is well-established, the SCDs can be removed and subcutaneous heparin is continued until discharge from the hospital. Daily chest X-rays are obtained for several days to monitor for development of significant atelectasis, consolidation, pleural effusion, and position of the gastric conduit.

3. *Pain management.* A thoracic epidural catheter is inserted preoperatively and maintained for 3 days for optimal pain relief. Subsequently, a patient-controlled analgesia (PCA) pump is utilized. After oral intake is resumed, elixir can be taken by mouth or given by j-tube and is prescribed for home use after hospital discharge.

4. *GI tract management.* The patient is kept NPO with a sumping nasogastric (NG) tube on continuous suction until at least postoperative day 3 or until outputs decrease to less than 300 ml/24 hours. If the clinical progress is satisfactory, then a contrast swallow radiological evaluation is performed on postoperative day 5-7 to assess anastomotic integrity. Water-soluble contrast material is swallowed first, and if X-rays do not demonstrate a leak, then they are repeated as thin barium is swallowed. If this sequence of X-rays is negative for anastomotic leak, oral intake of water is started on postoperative day 5. The subsequent advancement of oral intake is dependent on the patient's clinical progress, but the general plan is as follows: postoperative day 6, clear liquids; postoperative day 7, full liquids; postoperative day 8, mechanical soft diet. Five percent dextrose and water (D5W) is started through the jejunostomy catheter on postoperative day 2, and tube feeds are started on postoperative day 3 and gradually advanced to the target rate over the ensuing 36 hours. If the patient's oral intake of soft food is sufficient by postoperative day 9, they may be discharged from the hospital without home jejunostomy feeds. Patients are instructed to simply flush the catheter with tap water BID until their return visit to the outpatient clinic at which time the catheter is removed. If the patient's oral intake of protein and calories is insufficient, the jejunostomy catheter may be used for home tube feeds for a few days/weeks, until oral intake has improved.

5. *Antibiotics.* Cefuroxime 1.5 g is given 30 minutes prior to the incision and repeated Q 8 h for two doses.

6. *Wound care.* Skin staples are removed at the first postoperative visit. Chest tubes are initially kept on 20 cm water suction and switched to water seal when the patient begins to ambulate. The anterior chest tube may be removed by postoperative day 2. The posterior chest tube draining the region of the anastomosis is removed after the patient's oral intake has progressed to soft solids without evidence of a leak.

85

Complications

1. Medical complications following major surgery include: atelectasis, pneumonia, DVT and pulmonary embolism (PE), atrial fibrillation, myocardial infarction, and stroke.
2. General surgical complications of such major surgery include bleeding (requiring transfusion) and surgical site infection(s).
3. Complications pertinent to Ivor-Lewis esophagectomy include ischemic necrosis of the gastric conduit, with or without mediastinitis or empyema; anastomotic leak, with or without mediastinitis or empyema; right recurrent laryngeal nerve injury; injury to thoracic duct with ensuing chylothorax; splenic injury necessitating splenectomy; and pleural injury causing left pneumothorax and/or pleural effusion.
4. Since the thoracic esophagus is dissected free under direct vision using a right thoracotomy, the risk of intraoperative injury to the membranous airway or major vessels (e.g., aorta) should be very low.

Follow-Up

1. Postoperative follow-up at 3 days to 2 weeks after discharge entails clinical examination and chest X-ray. Removal of jejunostomy catheter is recommended 4-6 weeks later.
2. If postoperative adjuvant therapy (chemotherapy, radiation therapy) is recommended, it can be started between 4 and 8 weeks postoperatively.
3. Subsequent follow-up is at 3-6 monthly intervals that includes clinical examination and chest X-ray. If esophagectomy was performed for carcinoma, follow-up must be conducted for a total of 5 years after conclusion of treatment. CT scans may be obtained to monitor for recurrence at 6-12 month intervals, assuming that clinical progress is otherwise favorable.

Acknowledgment

The editors and authors wish to acknowledge Sudhir Sundaresan for contributing to the previous version of this chapter.

85

Chapter 86

Esophagectomy: Left Transthoracic

Alberto de Hoyos and Malcolm DeCamp

Indications

Left transthoracic esophagectomy is indicated for resection of esophageal malignancies involving the distal esophagus or esophagogastric junction, particularly when the tumor involves a substantial part of the proximal stomach, thus limiting the amount of stomach available to pull up. It is also the best approach for the rare esophageal tumor resection that would include a portion of diaphragm.

Preop

1. Accurate diagnosis of the esophageal disease is accomplished by a detailed history and endoscopy with biopsy. Barium esophagram may be helpful in defining anatomy but does not routinely need to be obtained.
2. In resection for esophageal cancer, rigorous tumor staging is necessary to determine that the lesion is localized and therefore amenable to a curative resection. This is accomplished by clinical evaluation (history and physical examination) and by computed tomographic (CT) scan of the chest and abdomen, endoscopic ultrasound (to assess the degree of penetration of the esophageal wall by the tumor and the status of the locoregional lymph nodes), and positron emission tomography (PET scanning) to identify distant bloodborne metastases. Preoperative surgical staging using thoracoscopy and laparoscopy has also been described but has not been universally accepted.
3. Standard preoperative hematological and biochemical blood work includes electrolytes, liver functions, complete blood count with platelets, coagulation profile, urinalysis, electrocardiogram (EKG), and type and screen. If the patient has documented cardiac disease or if the EKG is abnormal, a formal cardiology assessment should be conducted. If the patient has documented pulmonary disease, a significant smoking history, or if there is any significant breathing impairment, formal pulmonary function tests (PFT) and arterial blood gas (ABG) should be obtained.
4. Many centers advocate the use of preoperative (neoadjuvant) chemo-radiotherapy in esophageal carcinoma, although there are insufficient data to support this approach as the standard of care in this disease. Patients with T_3 or N_1 disease should be considered for induction therapy protocols.
5. Optimal pain control is provided by intravenous patient-controlled analgesics (PCA) or an epidural catheter.

Procedure

The operation can be performed in one of three ways.

Option 1

The first option entails separate midline laparotomy and left thoracotomy approaches. This obviously requires repositioning of the patient and creation of a separate sterile field

Northwestern Handbook of Surgical Procedures, 2nd Edition, edited by Nathaniel J. Soper and Dixon B. Kaufman. ©2011 Landes Bioscience.

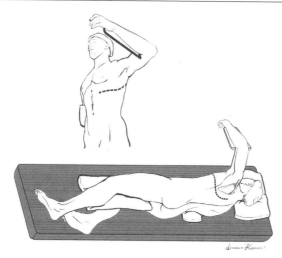

Figure 86.1. Left transthoracic esophagectomy through a left posterolateral thoracotomy.

for the left thoracic portion. In this scenario, the operation is virtually identical to the Ivor-Lewis resection and begins with Steps 1-12 already described for that procedure, except that a left thoracotomy is utilized, entering the chest typically in the 6th intercostal space. The subsequent conduct of the thoracic portion is also similar to Steps 13 and 14 for the Ivor-Lewis resection, except that the esophagus is mobilized through a left thoracotomy to just below the arch of the aorta. The esophagus must be carefully freed from the pericardium, inferior pulmonary vein, and left main bronchus, and attention is paid to carefully securing the segmental blood supply as already described. The left vagus nerve must be separated from the esophagus below the origin of the recurrent branch in order to preserve vocal cord function. The anastomotic technique, chest tube use, and thoracotomy closure are identical to that described for the Ivor-Lewis resection (Steps 14-16).

Option 2

The second option entails a left posterolateral thoracotomy alone, entering the chest in the 6th intercostal space. The periphery of the left diaphragm is then opened as follows: Paired stay sutures of 3-0 polypropylene are placed on either side of the proposed phrenotomy. Cautery is used to make a full-thickness opening, conforming to the curvature of the periphery of the diaphragm and leaving a small rim attached to the chest wall to facilitate its subsequent reconstitution. The stay sutures are useful to retract the diaphragm and to guide reapproximation during diaphragm closure. The exposure of the upper abdomen through this technique is superb and will facilitate mobilization of the entire stomach (Ivor-Lewis resection, Steps 5-10) except for the Kocher maneuver and pyloroplasty. Since the gastric tube is required to reach only as high as the aortic arch, the elimination of the Kocher maneuver here is acceptable. The delivery of the stomach via the hiatus into the chest, completion of the resection, preparation of the gastric tube, creation of the anastomosis, chest tube placement, and closure are as described previously. The only additional step is the closure of the phrenotomy. This is done using a series of horizontal mattress sutures of #1 polypropylene, followed by a continuous #1 polypropylene suture line ("vest-over-pants" closure).

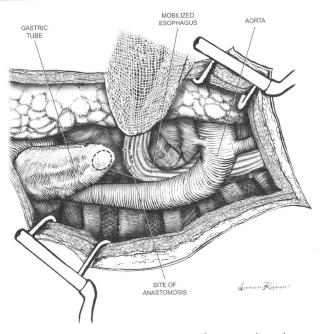

GASTRIC TUBE

MOBILIZED ESOPHAGUS

AORTA

SITE OF ANASTOMOSIS

Figure 86.2. Left transthoracic esophagectomy. Site of gastroesophageal anastomosis.

Option 3

The third option entails a left thoracoabdominal incision. This approach is less favored since the incision involves division of the costal arch, which is associated with considerable late morbidity (pain). The main advantage of this approach (and the reason that many surgeons use this approach preferentially for total esophagectomy with neck anastomosis) is that it improves the abdominal exposure, particularly to the right of the midline, and will therefore facilitate a complete Kocher maneuver and pyloroplasty. The remaining steps described below will detail this approach.

Step 1. The patient is positioned in full right lateral decubitus position as for left thoracotomy but the hips are allowed to rotate slightly posteriorly to provide improved abdominal midline exposure. A sterile field is created that includes the chest and abdomen slightly to the right of the midline and below the umbilicus.

Step 2. With the left lung deflated, a 6th or 7th intercostal space thoracotomy is created and taken obliquely across the costal arch, ending near the midline of the epigastrium, midway between the xiphisternum and the umbilicus. The costal arch is divided sharply with a scalpel. The abdominal obliques and the anterior and posterior layers of the left rectus sheath are also divided, although the rectus abdominis muscle is simply retracted to the right. A small circumferential opening in the periphery of the left diaphragm is also necessary to provide exposure (see Option 2). A chest retractor (for the thoracotomy) and a Balfour retractor (for the abdomen) will provide superb exposure. The process of gastric mobilization, pyloroplasty, Kocher maneuver, jejunostomy, esophagogastrectomy, preparation and transposition of the gastric tube, and esophagogastric anastomosis have already been described.

86

Step 3. Once the operation is complete, the chest tube insertion and wound closure is performed as already described. The abdomen is closed with #1 polypropylene in the anterior and posterior layers of the rectus sheath, in the obliques, and in the diaphragm (diaphragm closure described above). A short plug of cartilage is excised at the costal arch to eliminate any overriding of the divided ends, and a figure-of-eight suture of heavy polypropylene is used to secure the arch at this point, in addition to the remaining pericostal sutures. The technique of chest closure has already been described. The left lung is manually reinflated prior to completion of closure. Absorbent gauze dressings are applied to the incision and the drapes are taken down.

Postop

1. *Intravenous fluids.* Generous volume resuscitation is required in the first 24 hours postoperatively (e.g., normal saline at 150-175 ml/h) since the anticipated 3rd space volume loss will be substantial. Urine output is carefully monitored, with 0.5-1.0 ml/kg/h considered adequate. ABGs should be obtained periodically to monitor PaO_2 and $PaCO_2$ (during ventilator weaning; see below) and to monitor the base deficit, another useful parameter to assess the adequacy of the circulation. Maintenance of satisfactory circulation is vital, since the upper portion of the gastric conduit is perfused mostly by intramural blood supply. This is best achieved by generous use of crystalloids (as described above), and transfusion of packed red cells (as dictated by low hemoglobin values). About 24-36 hours after surgery, the 3rd space volume loss subsides, and the extravascular volume is returned to the intravascular space. This is heralded by diuresis, and signals the appropriate time to reduce volume administration to maintenance fluids and judicious use of diuretics (e.g., furosemide 20 mg IV Q 8-12 h). Diuretics are continued until the patient has reached their preoperative weight.

2. *Respiratory status.* The patient is maintained on a ventilator postoperatively and weaned and extubated the following morning. Supplemental oxygen is then given by face mask and ultimately by nasal prongs until it is stopped. ABGs are checked during the phase of ventilator weaning and immediately after extubation. If the patient is stable thereafter, pulse oximetry is sufficient to guide the administration of oxygen. During the 48 hours following extubation, nebulized bronchodilators and chest physical therapy are used preemptively to minimize the development of serious sputum retention and pneumonia. Ambulation is begun on the morning of the 2nd postoperative day. Deep vein thrombosis (DVT) prophylaxis (heparin 5000 U sc TID, and sequential compression devices [SCDs]) are started prior to induction of anesthesia. Once ambulation is well-established, the SCDs can be removed and subcutaneous heparin is continued until discharge from the hospital. Daily chest X-rays are obtained for several days to monitor for development of significant atelectasis, consolidation, pleural effusion, and position of the gastric conduit.

3. *Pain management.* A thoracic epidural catheter is inserted preoperatively and maintained for 3 days for optimal pain relief. Subsequently, a patient-controlled analgesia (PCA) pump is utilized. After oral intake is resumed, elixir can be taken by mouth or given by j-tube and is prescribed for home use after hospital discharge.

4. *GI tract management.* The patient is kept NPO with a sumping nasogastric (NG) tube on continuous suction until at least postoperative day 3 or until outputs decrease to less than 300 ml/24 hours. If the clinical progress is satisfactory, a contrast swallow radiological evaluation is performed on postoperative day 5-7 to assess anastomotic integrity. Water-soluble contrast material is swallowed

first, and if X-rays are negative for leak, they are repeated as thin barium is swallowed. If this sequence of X-rays is negative for anastomotic leak, the NG tube is removed and oral intake of water is started on postoperative day 5. The subsequent advancement of oral intake is dependent on the patient's clinical progress, but the general plan is as follows: postoperative day 6, clear liquids; postoperative day 7, full liquids; postoperative day 8, mechanical soft diet. Five percent dextrose and water (D5W) is started through the jejunostomy catheter on postoperative day 2, and tube feeds are started on postoperative day 3 and gradually advanced to the target rate over the ensuing 36 hours. If the patient's oral intake of soft food is sufficient by postoperative day 9, they may be discharged from the hospital without home jejunostomy feeds. Patients are instructed to simply flush the catheter with tap water BID until their return visit to the outpatient clinic at which time the catheter is removed. If the patient's oral intake of protein and calories is insufficient, then the jejunostomy catheter may be used for home tube feeds for a few days/weeks until oral intake has improved.

5. *Antibiotics.* Cefuroxime 1.5 g is given 30 minutes prior to the incision and repeated Q 8 h for two doses.

6. *Wound care.* Skin staples are removed on the day of discharge. Chest tubes are initially kept on 20 cm water suction. The anterior chest tube may be removed by postoperative day 2. The posterior chest tube (draining the region of the anastomosis) is removed after the patient's oral intake has progressed to soft solids.

Complications

1. Medical complications following major surgery include: atelectasis, pneumonia, DVT and pulmonary embolism (PE), atrial fibrillation, myocardial infarction, and stroke.

2. General surgical complications of such major surgery include bleeding (requiring transfusion) and wound infection(s).

3. Complications pertinent to left transthoracic esophagectomy include ischemic necrosis of the gastric conduit, with or without mediastinitis or empyema; anastomotic leak, with or without mediastinitis or empyema; left recurrent laryngeal nerve injury; injury to thoracic duct with ensuing chylothorax; splenic injury necessitating splenectomy; and pleural injury causing right pneumothorax and/or pleural effusion.

4. Since the thoracic esophagus is dissected free under direct vision using a left thoracotomy, the risk of intraoperative injury to the membranous airway or major vessels (e.g., aorta) should be very low.

86

Follow-Up

1. Postoperative follow-up at 2-3 weeks entails clinical examination and chest X-ray. Removal of jejunostomy catheter is recommended 4-6 weeks later.

2. If postoperative adjuvant therapy (chemotherapy, radiation therapy) is recommended, it can be started between 4 and 8 weeks postoperatively.

3. Subsequent follow-up is at 3-6 month intervals and includes clinical examination and chest X-ray. If esophagectomy was performed for carcinoma, follow-up must be conducted for a total of 5 years after conclusion of treatment. CT scans may be obtained to monitor for recurrence at 6-12 month intervals, assuming that clinical progress is otherwise favorable.

Acknowledgment

The editors and authors wish to acknowledge Sudhir Sundaresan for contributing to the previous version of this chapter.

Chapter 87

Esophagectomy: Transhiatal

Alberto de Hoyos and Malcolm DeCamp

Indications

1. Esophageal malignancies involving the esophagogastric junction, lower third of the esophagus at or below the inferior pulmonary vein level, or upper esophagus at or above the thoracic inlet. In the latter situation, the tumor resection usually requires pharyngolaryngoesophagectomy with a permanent tracheostomy in order to achieve a satisfactory proximal margin and incorporates a transhiatal esophagectomy with gastric pull-up to the hypopharynx.
2. Barrett's esophagus with high-grade dysplasia.
3. End-stage benign esophageal diseases (typically after multiple failed operations for gastroesophageal reflux or motor disorders). Transhiatal esophagectomy may be contraindicated in some cases for the resection of midesophageal tumors lying adjacent to the membranous trachea or left main bronchus.

Preop

1. Accurate diagnosis of the esophageal disease is accomplished by a detailed history, barium esophagogram, and endoscopy with biopsy paying particular attention to the proximal and distal extent of the tumor.
2. In resections for esophageal cancer, rigorous tumor staging is necessary to determine that the lesion is localized and therefore amenable to a curative resection. This is accomplished by clinical evaluation (history and physical examination), computed tomographic (CT) scan of the chest and abdomen. Esophageal ultrasound (EUS) is routinely used to assess the degree of penetration of the esophageal wall by the tumor (T stage). Additionally, EUS is used to assess nodal stage by using echo characteristics and directed fine needle aspiration (N stage). Positron emission tomography (PET) scanning is also commonly used to confirm or refute N stage and to identify distant bloodborne metastases (M stage). Preoperative surgical staging using thoracoscopy and laparoscopy has also been described but has not been universally accepted.
3. For midesophageal tumors fiberoptic and rigid bronchoscopy are necessary to rule out the possibility of membranous airway involvement by the tumor.
4. Standard preoperative hematological and biochemical blood work includes electrolytes, liver functions, complete blood count with platelets, and coagulation profile; urinalysis; electrocardiogram (EKG); type and cross. If the patient has documented cardiac disease or if the EKG is abnormal, a formal cardiology assessment should be conducted. If the patient has documented pulmonary disease, a significant smoking history, or if there is any significant breathing impairment, formal pulmonary function tests (PFT) and arterial blood gas (ABG) should be obtained.

Northwestern Handbook of Surgical Procedures, 2nd Edition, edited by Nathaniel J. Soper and Dixon B. Kaufman. ©2011 Landes Bioscience.

5. Many centers advocate the use of preoperative neoadjuvant chemo-radiotherapy in esophageal carcinoma. There are insufficient data to support this approach as the standard of care in this disease. Patients with T3 or N1 disease should be considered for induction therapy protocols.

6. Optimal pain control is provided by intravenous patient-controlled analgesics (PCA) or an epidural catheter.

Procedure

Step 1. The patient is positioned supine with arms tucked at the sides. An inflatable mattress is placed beneath the shoulders and inflated to achieve neck extension, with the head turned slightly to the right. Monitoring includes intravenous access, radial arterial line, Foley catheter, and optionally a central venous line inserted in the right side of the neck.

Step 2. The neck, chest, and abdomen are prepared and draped as one large sterile field.

Step 3. An upper midline laparotomy is made from the xiphoid to the umbilicus. The xiphisternum is excised. The falciform ligament is divided, and a table-mounted retractor is positioned to exert strong upward and lateral pull on the costal arches bilaterally. A Balfour retractor is also inserted to improve the abdominal exposure.

Step 4. In resections for carcinoma, full inspection of the abdomen and pelvis is performed to rule out the possibility of metastatic disease. Careful attention is paid to the liver, the primary tumor, and the upper abdominal lymph nodes. Suspicious areas are biopsied and sent for frozen section. Once this has proven to be satisfactory, the resection and gastric pull-up are performed.

Step 5. The stomach is fully mobilized in anticipation of being used as the replacement conduit for swallowing. The stomach will ultimately be perfused by the right gastroepiploic and right gastric arterial systems, so extreme care must be taken throughout the dissection not to damage these vessels and corresponding veins. Gastric mobilization begins with division of the lesser omentum (with cautery) starting at the mid-lesser curvature of the stomach and ending at the right crus of the diaphragm. The distal esophagus is dissected out circumferentially and encircled with a Penrose drain to provide traction.

Step 6. The greater curvature is fully mobilized. The gastrocolic ligament is divided with cautery to enter the lesser sac, staying a generous distance away from the gastroepiploic arcade. The posterior stomach wall and the transverse mesocolon must be accurately visualized. The greater curve can be mobilized with cautery, with the harmonic scalpel being reserved for larger vessels (e.g., short gastric vessels). Mobilization is taken to the right just beyond the pylorus. The duodenum is fully kocherized (the inferior vena cava can be exposed) using cautery and blunt dissection, giving enough mobility to allow the pylorus to lie near the esophageal hiatus after the stomach is transposed to the neck. All of the short gastric vessels are divided to arrive at the left crus of the diaphragm. The large posterior vascular pedicle containing the left gastric vessels and lymph nodes is isolated at the upper border of the pancreas, thinned into two or three smaller stalks, and divided between right angle clamps or using sequential applications of the linear vascular stapler. The stapler can be applied either from the lesser curvature or from behind the stomach. The stomach is now fully mobile, and it must be verified that it will reach the neck without limitation. Throughout the dissection and mobilization, extreme care is exercised to protect the gastroepiploic arcade.

87

Step 7. Pyloroplasty is performed using the Heinecke-Mikulicz technique. Stay sutures of 3-0 silk are placed at the superior and inferior margins of the pylorus. The pylorus is then opened longitudinally along the axis of the stomach and duodenum using low cautery, slightly onto the duodenal and gastric sides, and ensuring that all of the pyloric muscle has been divided. The opening is then closed transversely, using a series of full-thickness sutures of 2-0 silk, taking care to include bites of the mucosa.

Step 8. The left neck is opened to permit mobilization of the cervical esophagus. An oblique incision is made along the anterior border of the left sternocleidomastoid (SCM) muscle and deepened with cautery through the platysma. The incision starts at the level of the brow of the thyroid cartilage and includes a short hockey-stick extension into the suprasternal notch. The SCM is mobilized laterally to expose the omohyoid muscle which is divided. The fascia along the lateral border of the strap muscles is divided, and the strap muscles are mobilized and divided with cautery near their sternal attachments. Gentle digital pressure is applied to retract the trachea to the right. The middle thyroid vein and inferior thyroid artery are identified and divided between 3-0 silk ligatures or small clips. Dissection is then carried out on the adventitia of the esophagus to isolate it and encircle it with a Penrose drain. Care is taken to avoid the left recurrent laryngeal nerve in the tracheoesophageal groove and preserve it; there must be no cautery or pressure from metal instruments applied to the nerve.

Step 9. The mediastinal portion of the esophagus is fully mobilized. Working from the neck, blunt dissection (using the finger tip or a sponge forcep) is used to free the esophagus to the level of the carina. The success of this mobilization depends on accurately isolating the esophagus in the correct plane (Step 8). This maneuver is safe since there is no segmental esophageal blood supply at this level. Keeping the blunt dissection immediately on the longitudinal muscle layer of the esophagus minimizes the chance of injuries to the recurrent nerve, membranous trachea, or adjacent blood vessels. From the abdominal aspect, the esophageal hiatus must be enlarged. This is done using cautery after dividing the phrenic vein which is routinely identified crossing anterior to the hiatus. The operating table is tipped into Trendelenburg position to improve the exposure up into the mediastinum. The esophagus is then freed up using cautery or the harmonic scalpel. The last portion of mobilization at the carinal level requires blunt digital dissection, taking care to gently disrupt the esophageal attachments close to the esophageal wall. Temporary hypotension may ensue from manual compression of the heart during this maneuver. Communication with the anesthesiologist is essential to avoid prolonged periods of hypotension. Entry into the pleural space is avoided but if entry occurs, a chest tube is inserted in the corresponding side.

Step 10. The esophagectomy is completed and the gastric tube is prepared. Strong traction is applied to the cervical esophagus to deliver the esophagus maximally out of the neck incision. The esophagus is transected as distal as possible with a linear stapler, with a long length of umbilical tape fastened to the distal stump. The thoracic esophagus is then delivered into the abdomen, leaving the umbilical tape traversing the mediastinum along the future path of transposition of the gastric tube. The lesser curvature is then resected along with the esophagus, using sequential firings of the linear stapler constructing a gastric tube 8 cm wide. The esophagogastrectomy specimen is passed off the field. If resection is for tumor, a frozen section is obtained to verify that the proximal mucosal margin is clear of tumor and Barrett's esophagus. The staple line along the lesser curvature is then inverted using a continuous Lembert suture of 4-0 polypropylene if desired.

Step 11. The gastric tube is transposed to the neck. After assuring hemostasis of the mediastinal dissection plane, a laparoscopy camera bag is snugged around the stomach (like an accordion) after first drying the serosa of the stomach with a lap sponge. Drying the serosa is important since the plastic camera bag will drag the gastric tube to the neck by virtue of the friction it exerts on the stomach. The umbilical tape is tied to the upper end of the plastic bag. Constant firm tension is applied to the tape from the neck. The tip of the bag (containing the leading edge of the stomach tube) is gently guided into the esophageal hiatus, and the outside of the bag is lubricated by irrigating it with saline. As the bag emerges from the neck incision, the tip of the gastric fundus will be seen. At this point the bag can be cut away, pulled through, and discarded. The gastric tip is securely grasped with a Babcock clamp in preparation for the anastomosis. Care is taken at this juncture to verify that there has been no rotation of the stomach during the transposition maneuver; the greater curvature should be oriented to the patient's left, and the lesser curvature suture line should be facing the right, both in the neck and in the abdomen. The pyloroplasty should be just below the esophageal hiatus.

Step 12. The cervical esophagogastrostomy is created. It begins with excision of the staple line from the esophagus and creation of a small gastrotomy on the upper greater curve (staying away from the lesser curve suture line). A circumferential myotomy is created in the esophagus 2 cm proximal to the staple line taking care to preserve the mucosa intact. The muscularis is swept distally toward the staple line. The mucosa is then transected close to the staple line creating a mucosal tube. A common method of hand-sewn anastomosis involves two layers of simple interrupted 4-0 silk sutures. Sutures of the inner layer incorporate the tube of esophageal mucosa to the full thickness of the stomach and are placed so that knots are tied on the inside, to maximize mucosal inversion. The outer row of sutures approximates the esophageal muscle to the seromuscular layer of the stomach and buries the inner suture line.

An alternate technique involves the use of the linear stapler to create a functional end-to-end anastomosis. The endo-GIA stapler, with a 45 mm long blue cartridge (3.5 mm staples) is inserted into both the esophagus and stomach, which have been laid in apposition. The stapler is fired, creating most of the anastomosis and leaving a common opening into the esophagus and stomach. This common opening is closed using a continuous 4-0 monofilament absorbable suture, with a second layer of 4-0 silk sutures (approximating the esophageal muscle to the seromuscular layer of the stomach) used to bury this suture line. Alternatively, a total mechanical anastomosis can be constructed by closing the defect with a TA-stapler. In either technique, a nasogastric (NG) tube must be passed by the anesthesiologist, guided across the anastomosis, and then advanced into the lower portion of the gastric tube, before completion of the anastomosis. The anastomosis, along with the redundant esophagus and excess stomach, are all returned to the mediastinum below the thoracic inlet.

Step 13. A feeding jejunostomy catheter is placed, using a standard Witzel technique and a 12 F red rubber catheter. The ligament of Treitz is identified on the inferior surface of the transverse mesocolon, and the catheter is typically inserted about 30 to 40 cm distal to this point. A pursestring suture of 3-0 silk is used in the antimesenteric border of the bowel to secure the catheter at the entry site. A series of interrupted 3-0 silk Lembert sutures is used to bury the catheter in the bowel wall over a short distance and to secure the bowel to the parietal peritoneum. The catheter is brought out through a small stab wound in the left midabdomen and secured at the skin with a 3-0 nylon suture.

Step 14. The incisions are closed. A Penrose or closed suction drain is guided down into the mediastinum next to the anastomosis from the lower portion of the neck incision. The platysma layer is closed with continuous 2-0 polyglactan suture, deliberately leaving a generous defect at the Penrose site. This is done to facilitate establishment of a cutaneous fistula in the event of an anastomotic leak. Skin is closed with staples. The laparotomy incision is closed with continuous #2 polypropylene suture in the fascia and staples in the skin. Absorbent gauze dressings are placed to both incisions.

Postop

1. *Intravenous fluids.* Generous volume resuscitation is required in the first 24 hours postoperatively (e.g., normal saline at 150-175 ml/h) since the anticipated 3rd space volume loss will be substantial. Urine output is carefully monitored, with 0.5-1.0 ml/kg/h considered adequate. ABGs should be obtained periodically to monitor PaO_2 and $PaCO_2$ (during ventilator weaning; see below) and to monitor the base deficit as an indicator of the adequacy of the circulation. Maintenance of satisfactory circulation is vital since the upper portion of the gastric conduit is perfused mostly by the intramural network of collateral flow. This is best achieved by generous use of crystalloids (as described above) and transfusion of packed red cells (as dictated by low hemoglobin values). Twenty-four to 36 hours after surgery, the 3rd space volume loss subsides, and the extravascular volume is returned to the intravascular space. This is heralded by diuresis and signals the appropriate timing to reduce volume administration to maintenance fluids and the judicious use of diuretics (e.g., furosemide 20 mg IV Q 8-12 h). Diuretics are continued until the patient has reached her preoperative weight.

2. *Respiratory status.* The patient is maintained on a ventilator postoperatively and extubated when they are normothermic and hemodynamically stable. However, most patients can be extubated in the operating room or shortly after arriving in the ICU. For those undergoing long procedures or who are requiring resuscitation it is prudent to maintain endotracheal intubation overnight. Once extubated, supplemental oxygen is given by face mask and ultimately by nasal prongs as indicated by peripheral pulse oximetry. ABGs are checked during ventilator weaning and immediately after extubation. If the patient is stable thereafter, pulse oximetry is sufficient to guide the administration of oxygen. For 48 hours following extubation, nebulized bronchodilators and chest physical therapy are used to minimize the development of sputum retention, atelectasis and pneumonia. Ideally, patients are out of bed on the first postoperative day and ambulation is begun on the morning of the second day. Deep vein thrombosis (DVT) prophylaxis consists of heparin 5000 U TID sc and sequential compression devices [SCDs). Once ambulation is well-established, the SCDs can be removed and sc heparin is continued until discharge from the hospital. Chest X-rays are obtained as needed to monitor for development of significant atelectasis, consolidation, pleural effusion, and position of the gastric conduit.

3. *Pain management.* A thoracic epidural catheter is inserted preoperatively and maintained for 3 days for optimal pain relief. Subsequently, a patient-controlled analgesia (PCA) pump is utilized. After oral intake is resumed, elixir narcotic analgesics can be taken by mouth and is prescribed for home use after hospital discharge.

4. *GI tract management.* The patient is kept NPO with a sumping nasogastric (NG) tube on continuous or low intermittent suction until postoperative day 3 to 5. If the clinical progress is satisfactory, a contrast swallow radiological evaluation is performed to assess anastomotic integrity between postoperative day 5 and 7. Water-soluble contrast material is swallowed first, and if X-rays are negative for leak, then they are repeated with a thin barium swallow. If this sequence of X-rays is negative for anastomotic leak, oral intake of water is started. The subsequent advancement of oral intake is dependent on the patient's clinical progress, but the general plan is as follows: postoperative day 6 and 7 full liquids advancing to mechanical soft diet by day 8. Five percent dextrose and water (D5W) is started through the jejunostomy catheter on postoperative day 2 at 20 ml/hour, and tube feeds are started on postoperative day 3 and gradually advanced to the target rate over the ensuing 36 hours. Cycling tube feeds (6 PM to 9 AM) at a rate that will meet protein and caloric requirements, allow patients to be disconnected from the pump for most of the day. If the patient's oral intake of soft food is sufficient by postoperative day 9, they may be discharged from the hospital without home jejunostomy feeds. Patients are instructed to simply flush the catheter with tap water twice a day until their return visit to the outpatient clinic, at which time the catheter is removed. If the patient's oral intake of protein and calories is insufficient, the jejunostomy catheter may be used for home tube feeds for a few days/weeks until oral intake has improved.
5. *Antibiotics.* Cefuroxime 1.5 g is given 30 minutes prior to the incision and repeated Q 8 h for two doses.
6. *Wound care.* The neck drain can be gradually shortened starting on postoperative day 6. The drain and skin staples are removed on the day of discharge.

Complications

1. Medical complications following major surgery include: atelectasis, pneumonia, DVT and pulmonary embolism (PE), atrial fibrillation, myocardial infarction, and stroke.
2. General surgical complications of such major surgery include bleeding (requiring transfusion) and surgical site infection(s).
3. Complications pertinent to transhiatal esophagectomy include ischemic necrosis of the gastric conduit, with or without mediastinitis or empyema; anastomotic leak, with or without mediastinitis or empyema; anastomotic stricture; left recurrent laryngeal nerve injury; injury to the thoracic duct with ensuing chylothorax; splenic injury necessitating splenectomy; mediastinal bleeding and pleural injury causing pneumothorax and/or pleural effusion.
4. Potentially catastrophic intraoperative injuries include those to the membranous airway or major vessels (e.g., aorta, azygos vein).

Follow-Up

1. Postoperative follow-up at 2-3 weeks entails clinical examination and chest X-ray. Removal of the jejunostomy catheter is recommended 4-6 weeks later.
2. If postoperative adjuvant therapy (chemotherapy, radiation therapy) is recommended, it can be started at about 4 weeks postoperatively.

87

3. Subsequent follow-up is at 3-month intervals and includes clinical examination and chest X-ray. If esophagectomy was performed for carcinoma, follow-up must be conducted for a total of 5 years after conclusion of treatment. CT scans may be obtained to monitor for recurrence at 6-12 month intervals, assuming that clinical progress is otherwise favorable.

Acknowledgment

The editors and authors wish to acknowledge Sudhir Sundaresan for contributing to the previous version of this chapter.

87

Chapter 88

Mediastinoscopy: Cervical

Alberto de Hoyos and Malcolm DeCamp

Indications

The objectives of cervical mediastinoscopy are to establish a histologic diagnosis of Level 2, 4, and 7 lymph nodes and to assess the operability of central lung cancers. The indication for the operation is middle mediastinal adenopathy suspected from clinical history, chest X-ray, or CT scan and to rule out unsuspected lymph node metastases from lung cancer. The more common etiologies of middle mediastinal adenopathy include primary malignancies, e.g., primary bronchogenic carcinoma and lymphoma; granuloma-producing diseases of infectious etiology, such as *Mycobacterium tuberculosis* and *Histoplasma capsulatum*; and granuloma-producing diseases of uncertain etiology, such as sarcoidosis.

Preop

A patient undergoing cervical mediastinoscopy requires no specific preoperative preparation. The operation is usually performed with a concomitant bronchoscopic examination, which is performed immediately prior to the cervical mediastinoscopy.

Contraindications to cervical mediastinoscopy include proven or suspected ascending or arch aortic aneurysm or dissection, and severe cervical spine disease that limits adequate neck extension. Because dissection is along the avascular pretracheal plane, previous sternotomy and superior vena cava syndrome generally do not preclude safe mediastinoscopy.

Procedure

Step 1. The patient is placed in the supine position on the operating table. The neck is extended. An inflatable support or towel roll placed beneath the shoulders facilitates neck extension. A pulse oximeter or arterial line is placed on the right upper extremity.

Step 2. Sterile preparation extends from chin to umbilicus in the unlikely event sternotomy would be required. Sterile drapes are applied to expose the anterior aspect of the neck and upper chest.

Step 3. A 1.5-2.5 cm transverse curvilinear skin incision is made 1 cm above the suprasternal notch. Dissection is carried transversely through the platysma.

Step 4. The raphe of the strap muscles is vertically incised and the muscles separated with blunt dissection. Intermittent palpation of the trachea helps to guide the surgeon directly toward that structure. Occasionally division of a few millimeters of the inferior portion of the thyroid gland is required as the surgeon approaches the trachea. Anterior jugular veins should be ligated if necessary.

Step 5. The pretracheal fascia is transversely divided.

Northwestern Handbook of Surgical Procedures, 2nd Edition, edited by Nathaniel J. Soper and Dixon B. Kaufman. ©2011 Landes Bioscience.

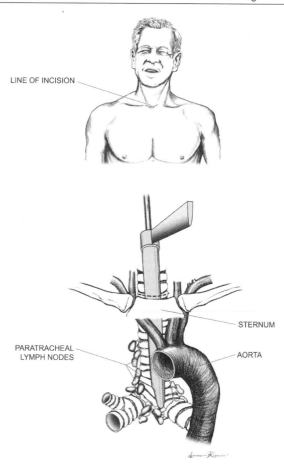

Figure 88.1. Cervical mediastinoscopy.

Step 6. The tip of the index finger is placed on the exposed tracheal rings and is advanced toward the carina in direct contact with the distal tracheal rings. When fully inserted, the finger will occupy a position between the trachea and the ascending aorta-innominate artery junction. The arterial pulsation should be readily felt when the finger is in this position. In addition, pathologic Level 2-R, 2-L and Level 4-R lymph nodes are occasionally palpable with the tip of the finger.

Step 7. After the finger is removed, the beveled end of the mediastinoscope is placed over the exposed upper tracheal rings, which are visualized through the scope.

Step 8. The mediastinoscope is advanced under direct vision along the anterior tracheal wall into the same space created by the previously inserted finger. Dampening of the pulse oximeter or arterial waveform during the procedure indicates compression of the innominate artery and should prompt repositioning of the mediastinoscope to avoid right carotid hypoperfusion.

Step 9. Through the mediastinoscope, right and left Level 2, 4 and Level 7 (sub-carinal) lymph nodes are identified by blunt dissection with a small-caliber rigid suction dissector whose rim is equipped with electrocautery. Electrocautery is not used along the left paratracheal space as the recurrent laryngeal nerve (which runs very close to the trachea) may be injured.

Step 10. Cup-tipped or alligator-tipped biopsy instruments are passed down the lumen of the mediastinoscope and are used to biopsy dissected lymph nodes.

Occasionally the appearance of the azygous vein or the superior vena cava will mimic the appearance of an anthracotic lymph node. Biopsy of either venous structure can produce massive bleeding. This complication usually can be prevented by inserting a 22-gauge spinal needle into the putative lymph node and aspirating. Biopsy is not performed if blood is aspirated. If the wall of a large vein is biopsied, the bleeding is usually controlled by firmly packing the pretracheal space with Surgicel® and a gauze sponge for 10-30 minutes. The same is true for injuries to the pulmonary artery since this is also a low-pressure system. If no bleeding occurs following removal of the sponge, the mediastinoscopy incision is closed. If bleeding persists, the mediastinum is repacked and preparations for an open procedure are made with anesthesia (better IV access, lung isolation), the blood bank and the OR nursing staff. The bleeding is controlled via either a median sternotomy or a right thoracotomy determined by the patient's pathological anatomy, the presumed location of the vascular injury and clinical status.

Step 11. The strap muscles are reapproximated with simple 3-0 polyglactan sutures, the platysma and the subcuticular layer approximated with running absorbable suture.

Postop

The patient usually can be extubated in the operating room. A chest X-ray is performed in the recovery room to assess the mediastinum and pleural spaces for pneumothorax or hemithorax. The patient can resume preoperative diet within an hour or two and usually can be discharged several hours later.

Complications

- Bleeding requiring sternotomy or thoracotomy
- Recurrent nerve injury (hoarseness, swallowing difficulties)

Follow-Up

An outpatient office visit that includes a PA and lateral chest X-ray is scheduled 1-2 weeks following the operation.

Acknowledgment

The editors and authors wish to acknowledge James W. Frederiksen for contributing to the previous version of this chapter.

88

Chapter 89

Lung Biopsy: Thoracoscopic

Alberto de Hoyos and Malcolm DeCamp

Indications
The objectives of the operation are:
1. to establish a histologic diagnosis in a patient with clinical and/or radiographic evidence of interstitial lung disease or diffuse air space disease;
2. to establish a histologic diagnosis of a peripheral lung nodule; and/or
3. to obtain lung tissue for culture of aerobic and anaerobic organisms, acid-fast organisms, fungi, viruses, and parasites in a patient with a suspected pulmonary infection.

The indications for the operation include suspected interstitial lung disease, peripheral lung nodule(s) identified on chest-X-ray or computed tomography, and suspected pulmonary infection in a patient in whom culture or other assessment of sputum, tracheal aspirate, bronchoalveolar lavage fluid, or transbronchial biopsy is unlikely or has failed to yield a diagnosis.

Preop
The surgeon should carefully assess the ventilatory status and the coagulation profile of each patient in whom the operation is contemplated. Patients likely to become hypoxic during single-lung ventilation should be prepared for open lung biopsy.

Significant prolongation of the PT (INR > 1.5), any prolongation of the PTT, and significant thrombocytopenia (platelet count <50,000) increase the risk of serious intra- and postoperative bleeding. Coagulopathic patients who require thoracoscopic lung biopsy may need pre- and intraoperative transfusions of fresh frozen plasma and/or platelets.

The possibility of co-existent pulmonary hypertension should be assessed and included in the risk-benefit analysis for surgical lung biopsy as its presence also increases the bleeding risk.

Procedure
Step 1. The patient is anesthetized while lying supine. Single-lung ventilation is achieved either through use of a double-lumen endotracheal tube or insertion of a bronchial blocker. Subsequently the patient is turned to the lateral decubitus position. Flexing the table slightly downward prevents the ipsilateral hip from limiting full excursion of the video camera.

Step 2. The hemithorax is draped in a manner suitable for a full thoracotomy. Rarely during a thoracoscopic biopsy operation, hemorrhage that is not controllable with thoracoscopic instruments occurs. Hence, the operative field should be prepared and draped widely enough for a thoracotomy, and instruments required to perform a thoracotomy should be readily available during this and all other thoracoscopic operations.

Northwestern Handbook of Surgical Procedures, 2nd Edition, edited by Nathaniel J. Soper and Dixon B. Kaufman. ©2011 Landes Bioscience.

89

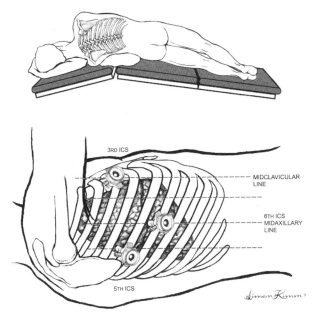

Figure 89.1. Thoracoscopic lung biopsy. Patient position and port placement.

Step 3. The anesthesiologist allows the lung to deflate as the surgeon creates two to three 2 cm thoracoscopic port incisions: one in the 6th, 7th, or 8th intercostal space in the midaxillary line, one in the 3rd or 4th intercostal space between the midclavicular line and the anterior axillary line, and one in approximately the 5th intercostal space posterior to the lateral border of the scapula. The skin and intercostal spaces are anesthetized with 0.25% bupivicaine prior to incision.

Step 4. The surgeon inserts the video camera—thoracoscope—through the midaxillary line port.

Step 5. Both thoracoscopic and standard instruments can be used. The surgeon thoroughly examines the lung, the pleura, and the mediastinum with the thoracoscope. To facilitate the examination, it is often necessary to retract the lung with endoscopic clamps, to divide adhesive bands with endoscopic scissors or electrocautery, and to insert the thoracoscope through one or both of the more superior port incisions. The surgeon can feel a small portion of the lung surface by inserting a finger through a port incision. Nodules should be palpated to assess depth and size prior to biopsy so that adequate margins can be obtained.

Step 6. Guided by preoperative imaging studies and by visual inspection, the surgeon identifies areas of abnormal and normal appearing lung.

Step 7. To establish a histologic diagnosis of a diffuse disease process, or to obtain tissue for culture, the surgeon usually excises a small "wedge" of abnormal appearing lung from each lobe with an endoscopic stapling instrument. The surgeon holds steady the lung tissue intended for removal by applying an endothoracic clamp or a ring forceps to that tissue as the stapler is being positioned and fired. If normal-appearing lung tissue is present, the surgeon should include some of this tissue in at least one of the specimens.

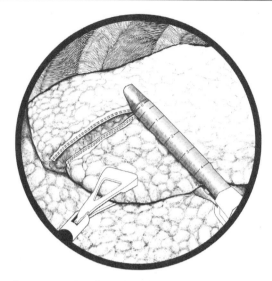

Figure 89.2. Thoracoscopic lung biopsy. Wedge resection.

Step 8. The surgeon sends at least one specimen containing abnormal lung tissue to the pathologist for frozen section examination. A brief summary of the patient's history and the suspected diagnosis should accompany the specimen so that the pathologist can prepare the tissue for special stains or for electron microscopic examination if appropriate. The pathologist is asked to examine a frozen section of a biopsy specimen, not necessarily to make a definitive diagnosis but to be certain that the tissue is lesional insuring a diagnosis can be made from permanent sections. If the patient is immunocompromised, or if the referring pulmonologist or infectious disease specialist suspects an infectious process, the surgeon sends tissue samples for appropriate cultures and microbiologic stains.

Step 9. If one or more pulmonary nodules are to be biopsied, the surgeon confirms the presence of the nodule by visual inspection.

Step 10. With a stapling device the surgeon excises the nodule, along with a surrounding 1- to 2-cm margin of normal lung. A resected specimen that contains a suspected malignant nodule is removed in a closed plastic bag via a port incision to help prevent malignant cells from being deposited along the incision during removal.

Step 11. Following removal of the biopsy specimen(s), the surgeon makes certain that each staple line has no significant air leak and is hemostatic. Intercostal nerve block is performed by injecting 3 ml of bupivicaine 0.25% into each intercostals space from T4 to T9. The surgeon inserts a 20-28 F chest tube via the midaxillary line port incision and advances the tip to the apex of the hemithorax. Visualizing the chest tube's insertion and verifying its apical position can be facilitated by inserting the thoracoscope via one of the more superior port incisions as the chest tube is inserted.

Step 12. The anesthesiologist inflates the lung as the surgeon secures the chest tube with a heavy suture and closes the other port incisions with absorbable suture. The chest tube is connected to a closed drainage device with an underwater seal set at -20 cm water suction.

89

Postop

The patient usually can be extubated in the operating room or shortly following transfer to the recovery room. A chest X-ray is obtained in the recovery room to assess the expansion of the lungs and the position of the chest tube. The drainage device is maintained at -20 cm water suction for 24 hours. Thereafter suction is discontinued, as long as the patient's lung has remained fully expanded. The chest tube is removed when significant drainage has ceased and the lung has remained satisfactorily expanded. On occasion, in patients that are stable with minimal evidence of abnormal pulmonary function, the chest tube can be removed several hours after the procedure as long as there is no air leak and the lung is fully expanded on the postoperative chest X-ray.

Complications

Bleeding, prolonged air leak and respiratory failure are the most common early postoperative complications.

Follow-Up

An outpatient visit that includes a chest X-ray is scheduled for approximately 2 weeks following operation.

Acknowledgment

The editors and authors wish to acknowledge James W. Frederiksen for contributing to the previous version of this chapter.

89

Chapter 90

Pulmonary Lobectomy: Open

Alberto de Hoyos and Malcolm DeCamp

Indications

The indications for pulmonary lobectomy include:

1. removal of lung cancer;
2. removal of deep seated isolated metastasis;
3. removal of lung destroyed by infection, especially tuberculosis; and
4. removal of lung abscess refractory to medical therapy.

Preop

1. Preoperative assessment is focused on whether or not the patient has the pulmonary reserve to withstand removal of the lobe. Two tests are used for this purpose: pulmonary function tests (PFTs) and the arterial blood gas (ABG). Based upon preoperative PFTs, the patient must generally have a postoperative predicted forced expiratory volume in 1 second (FEV1.0) or diffusing capacity (DLCO) of at least 40% of the corresponding value according to age, height and gender.

   Predicted postoperative function = preoperative function x [1- (Functional segments to be resected/Total number of functional segments)]

   A quick rule of thumb assigns a loss of function of 5.26% per functional segment removed. Patients with predicted postoperative DLCO or FEV1 <40% are at higher risk for pulmonary failure and death.

2. A preoperative ABG demonstrating an arterial pCO_2 greater than 45 to 50 mm Hg indicates high risk but does not necessarily prohibit resection. In selected patients, maximal oxygen uptake (VO_2 max) or quantitative ventilation perfusion scanning is indicated in order to avoid denying an operation to patients who do poorly on the basic testing.

3. Clinical staging dictates the best therapeutic approach. Staging tools include computed tomography (CT) scan of the chest and upper abdomen, bronchoscopy, positron emission tomography and mediastinoscopy or endobronchial ultrasound for biopsy of mediastinal lymph nodes.

4. Standard preoperative hematological and biochemical blood work includes electrolytes, complete blood count with platelets, coagulation profile; electrocardiogram (EKG); type and screen. If the patient has documented cardiac disease or an abnormal EKG, dyspnea or chest pain with exertion, or is unable to exercise, a formal cardiology assessment should be conducted.

Northwestern Handbook of Surgical Procedures, 2nd Edition, edited by Nathaniel J. Soper and Dixon B. Kaufman. ©2011 Landes Bioscience.

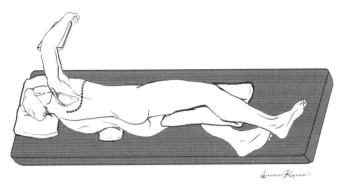

Figure 90.1. Open lobectomy. Patient position.

Procedure

Step 1. The procedure is performed under general anesthesia. The anesthesiologist intubates the patient with a double-lumen endotracheal tube or uses a bronchial blocker to permit one-lung ventilation.

Step 2. The patient is placed in the lateral decubitus position. The lower leg is partially flexed and the upper leg is kept on extension. The knees are appropriately padded to prevent injury to the peroneal nerve. A rolled towel is placed under the axilla to prevent pressure injury to the axillary structures. The ipsilateral arm is supported with an airplane splint to permit exposure to the posterolateral chest wall.

Step 3. The incision is initiated posteriorly, midway between the medial scapular border and the spinous processes. It extends anteriorly, in a curvilinear fashion, one or two fingerbreadths below the scapular tip. The incision extends to approximately the mid to anterior axillary line, one or two fingerbreadths below the level of the nipple. Some surgeons use a vertical axillary or lateral axillary muscle-sparing incision for routine lobectomy.

Step 4. The lattisimus dorsi muscle is divided with electrocautery, taking care to preserve the serratus muscle intact, which is retracted anteriorly. A hand is gently placed under the scapula and the ribs are counted. The second rib is clearly identified by the insertion of the scalene muscle. The 5th intercostal space is entered above the sixth rib. The intercostal muscles are divided with cautery and the ribs spread gently with a mechanical retractor.

Step 5. The lung is deflated as the anesthesiologist employs one-lung ventilation. The lung is reflected posteriorly, anteriorly and superiorly to perform a complete examination of the lung and pleural surfaces to identify any unexpected pathologic process. The inferior pulmonary ligament is exposed and divided with cautery to the level of the inferior pulmonary vein, taking care to biopsy lymph nodes frequently encountered in this location. Gentle palpation of the lung is performed to localize the tumor.

Step 6. The corresponding pulmonary venous drainage is identified: upper lobes and right middle lobe or lingula drain into the superior veins and lower lobes drain into the inferior veins. The appropriate vein is sharply dissected out, double ligated with 2-0 silk and divided. Alternatively, a vascular stapler is utilized (either linear or TA). When the vein is divided intrapericardially, the vein stump is oversewn with 4-0 polypropylene suture.

90

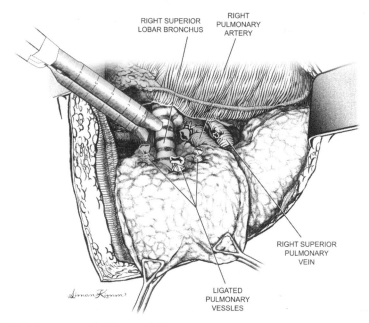

Fig 90.2. Resection of right upper lobe.

Step 7. The pulmonary artery is identified in the pulmonary hilum behind the divided vein. The pulmonary arterial blood supply to the appropriate lobe is sharply dissected with scissors taking care to open the plane of Leriche. Gentle handling is paramount to avoid traction injury with subsequent subadventitial hematoma and bleeding. A Kittner dissector is then utilized to complete full mobilization and the branches double ligated and divided. Alternatively, a vascular stapler is utilized as with the vein. If the arterial dissection proves difficult to complete, the corresponding main pulmonary artery should be dissected proximally, either outside or inside the pericardium and a tourniquet placed. The tourniquet can be snugged in the event of hemorrhage.

Step 8. The lobar bronchus is identified and sharply dissected out and stapled with an automatic stapling device (TA green load, 4.5 mm staple height). The bronchus is divided distal to the staple line and the lobe removed. Care is taken to sweep the lymph nodes upward toward the specimen before the application of the stapler.

Step 9. The anesthesiologist inflates the remaining lung and holds inflation at 20 cm H_2O pressure. The bronchial staple line is submerged in saline, and an air leak ruled out. If an air leak is identified, repair sutures of 4-0 polyglactan are utilized.

Step 10. Lymph node dissection is performed, taking care to sample at least 3 to 5 stations and harvest a minimum of 10 lymph nodes if possible. Particular attention is given to subcarinal lymph nodes and lower paratracheal lymph nodes. The overlying pleura is opened and the lymph nodes dissected and exposed. Caution is taken to achieve complete hemostasis and to avoid the recurrent laryngeal nerve on the left at the aorto-pulmonary window.

90

Step 11. Anterior and posterior chest tubes are placed through separate incisions.

Step 12. The intercostal space is closed with paracostal sutures. The musculature of the chest wall, subcutaneous tissue, and skin are then closed in appropriate layers.

Postop

1. *Chest tubes.* The chest tubes should be kept on suction for the first 24-36 h and removed when the patient has no air leak and the tubes are draining less than 300 ml per 24 hours. Chest tubes are placed on water seal on postoperative day 2.

2. *Respiratory status.* Supplemental oxygen is given by face mask and ultimately by nasal prongs as indicated by peripheral pulse oximetry. Nebulized bronchodilators and chest physical therapy are used preemptively to minimize the development of sputum retention and pneumonia. Ambulation is begun on the morning of postoperative day 1. Deep vein thrombosis (DVT) prophylaxis consists of heparin 5000 U sc TID and sequential compression devices (SCDs). SCDs are placed prior to induction of anesthesia. Heparin is started immediately postoperatively. Once ambulation is well-established, the SCDs can be removed and subcutaneous heparin is continued until discharge from the hospital. Daily chest X-rays are obtained for several days to monitor for development of significant atelectasis, consolidation, pneumothorax and pleural effusion.

3. *Pain management.* A thoracic epidural catheter is inserted preoperatively and maintained for 3 days for optimal pain relief. Subsequently, a patient-controlled analgesia (PCA) pump is utilized. Alternatively, ketorolac and oral narcotics are administered.

Complications

1. Prolonged air leak
2. Atelectasis
3. Empyema
4. Surgical site infection
5. Pneumonia
6. Bronchial stump dehiscence
7. Hemorrhage

Follow-Up

Pathologic staging determines if adjuvant therapy is required. In the absence of nodal disease or higher tumor stage (IIA-III), adjuvant therapy is not recommended. Patients undergoing resection for lung cancer are followed every 3-4 months for the first 2 years and yearly thereafter. CT scans of the chest and upper abdomen are obtained every 6 months.

90

Acknowledgment

The editors and authors wish to acknowledge David Fullerton for contributing to the previous version of this chapter.

Chapter 91

Pneumonectomy

Malcolm DeCamp

Indications

Carcinoma of the lung involving a main bronchus, the main pulmonary artery, or multiple lobes of the lung comprise the common indications for pneumonectomy. Occasionally pneumonectomy is indicated for a chronic longstanding infectious process that essentially destroyed the functional capacity of one lung.

Preop

General endotracheal anesthesia with double lumen tube or bronchial blocker, epidural catheter, arterial line, Foley catheter.

Procedure

Step 1. Appropriate lateral decubitus position.

Step 2. Posterior lateral thoracotomy entering the chest through the 4th or 5th intercostal space.

Step 3. Free any adhesions between the lung and chest wall and assess resectability.

Step 4. Deflate lung and retract it posteriorly and inferiorly to expose the hilum.

Step 5. Incise the mediastinal pleura anteriorly around the hilum.

Step 6. Sample Levels 5, 6, 7, 8 and 9 mediastinal lymph nodes in cancer cases.

Step 7. Identify the pulmonary artery and superior pulmonary vein.

Step 8. Dissect down to the adventitia of the pulmonary artery and mobilize and encircle it. Test clamp the pulmonary artery to confirm cardiovascular stability.

Step 9. Ligate the main trunk of the pulmonary artery and its first branches, then transect the artery. Alternatively the artery may be divided between rows of vascular staples.

Step 10. Identify the superior pulmonary vein, mobilize and either ligate or staple it and transect the vein.

Step 11. Displace the lung anteriorly and superiorly to expose the pulmonary ligament.

Step 12. Incise the pulmonary ligament from the diaphragm towards the hilum, exposing the inferior pulmonary vein.

Step 13. The inferior pulmonary vein is moblized, ligated or stapled and divided.

Step 14. The main bronchus is identified and lymph nodes are mobilized towards the lung.

Step 15. The bronchus is stapled and transected.

Step 16. Irrigate the pleural cavity with warm saline. Confirm airtight closure of bronchial stump by pressurizing the airway to 30 cm H_2O pressure.

Northwestern Handbook of Surgical Procedures, 2nd Edition, edited by Nathaniel J. Soper and Dixon B. Kaufman. ©2011 Landes Bioscience.

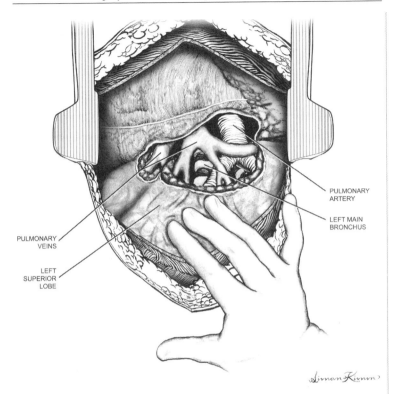

PULMONARY
ARTERY

LEFT MAIN
BRONCHUS

PULMONARY
VEINS

LEFT
SUPERIOR
LOBE

Figure 91.1. Left pneumonectomy. Anatomy.

Addsteps. Reinforce bronchial closure with vascularized tissue such as pleural flap, intercostal muscle flap or pericardial fat pad. Close the chest in layers with absorbable sutures. A chest tube/drain may or may not be used. If pleural drainage is performed clamp chest tube or place to water seal. Note applying suction to the drainage tube after a pneumonectomy will cause significant ipsilateral mediastinal shift and cardiovascular collapse secondary to impaired venous blood return.

Postop
Monitor oxygen saturation, position of the mediastinum, and level of accumulation of fluid within the pleural space.

Complications
Complications include respiratory failure, disruption of bronchial closure, infection within pleural space, mediastinal shift with cardiovascular instability, arrhythmias, and postpneumonectomy pulmonary edema.

91

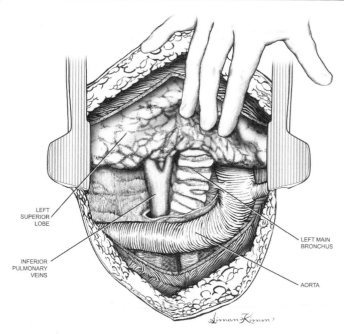

LEFT
SUPERIOR
LOBE

INFERIOR
PULMONARY
VEINS

LEFT MAIN
BRONCHUS

AORTA

Figure 91.2. Left pneumonectomy. Anatomy.

Follow-Up

Serial chest X-rays until pleural space is stable.

Acknowledgment

The editors and author wish to acknowledge Robert Vanecko for contributing to the previous version of this chapter.

Chapter 92

Pleurodesis: Thoracoscopic

Malcolm DeCamp

Indications

The purpose of the operation is to obliterate the pleural space to prevent reaccumulation of fluid or air. The indications are for treatment of recurrent spontaneous pneumothorax, recurrent benign pleural effusions and malignant pleural effusion.

Preop

The procedure is performed under general anesthesia. Single-lung ventilation is accomplished with either a double-lumen endotracheal tube or a bronchial blocker. The patient is placed in the lateral decubitus position. The knees are padded to prevent peroneal nerve injury. A rolled towel or gel pad is placed under the axilla to prevent pressure injury to the brachial plexus. The ipsilateral arm is suspended on an airplane splint or placed in the "praying position with appropriate padding.

Procedure

Step 1. Two 2 cm thoracoscopic port incisions are made, one in the 5th intercostal space near the lateral border of the scapula and one in the midaxillary line in the 5th intercostal space.

Step 2. The thoracoscope is inserted into the midaxillary port. A long clamp holding a folded sponge or bovie scratch pad is inserted through the other port.

Step 3. The parietal pleural surface along the chest wall is mechanically abraded with the folded sponge/pad. Care is taken to avoid injury to the subclavian vessels and brachial plexus. The diaphragm and pericardial surfaces are not abraded.

Step 4. After the mechanical pleurodesis has been completed, a solution of 100 ml of doxycycline (1,000 mg) or 2 grams or aerosolized sterile talc powder is placed into the pleural space to add a chemical pleurodesis.

Step 5. A chest tube is placed through the midaxillary port incision.

Step 6. The posterior port is closed in with a 2-0 polyglactan in the muscular layer and a 4-0 polyglactan subcuticular suture.

Postop

The patient should be extubated in the operating room. The chest tube remains in place until the patient has no air leak and the chest tube drainage has decreased to less than 100 ml per shift.

Complications

1. Pneumonia
2. Wound infection
3. Empyema
4. Pneumonitis

Northwestern Handbook of Surgical Procedures, 2nd Edition, edited by Nathaniel J. Soper and Dixon B. Kaufman. ©2011 Landes Bioscience.

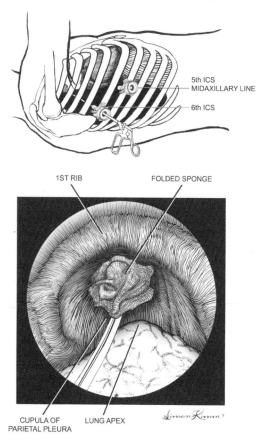

5th ICS
MIDAXILLARY LINE

6th ICS

1ST RIB   FOLDED SPONGE

CUPULA OF   LUNG APEX
PARIETAL PLEURA

Figure 92.1. Thoracoscopic pleurodesis

Follow-Up

Chest X-rays should be followed for approximately the first month after the procedure. Thereafter, chest radiography is indicated for recurrent symptoms.

Acknowledgment

The editors and author wish to acknowledge David Fullerton for contributing to the previous version of this chapter.

92

Chapter 93

Tracheostomy

Alberto de Hoyos

Indications

Tracheostomy has been performed since antiquity for the emergency management of upper airway obstruction. Presently the principal indications include:

1. relief of upper airway obstruction;
2. long-term ventilatory support for patients with respiratory failure; and
3. control of secretions in patients prone to aspirate.

Tracheostomies most commonly are performed for respiratory insufficiency requiring prolonged mechanical ventilation. Typically, these patients have had an orotracheal or nasotracheal tube prior to their tracheostomy. While there is no firm rule regarding the length of time a patient may be managed with an endotracheal tube, tracheostomies are usually performed after 7 to 14 days of intubation. It may be performed earlier in patients who are unlikely to be extubated during the ensuing 1 to 2 weeks. Originally, this time frame was shorter due to the potential for tracheal stenosis from the circumferential cuff on an endotracheal tube. While this complication has decreased significantly with the use of high-volume, low-pressure cuffs, there is evidence that long-term endotracheal intubation can lead to laryngotracheal stenosis. This can be avoided by timely performance of a tracheostomy.

Preop

While tracheostomy may be done with local anesthesia, with the patient in a supine position and the neck hyperextended, due to the fact that an endotracheal tube is in place, it is typically done with a general inhalational or intravenous anesthesic. While the procedure may be done at the bedside in the intensive care unit with proper sterile technique and adequate lighting, we prefer the operating room to allow optimal operating conditions. Trachesotomy can be performed as a traditional open procedure or as a percutaneous bronchoscopic guided technique, utilizing a commercially available kit.

Procedure

Step 1. With the patient supine and the neck hyperextended using a shoulder roll, the patient is sterilely prepped and draped. Care is taken to support the head appropriately.

Step 2. The following landmarks are palpated: thyroid cartilage, cricothyroid membrane, cricoid cartilage and sternal notch. It is important to remember that the tracheotomy is performed in relation to the cricoid cartilage, whereas the skin incision to gain access to the trachea is relative to the sternal notch.

Northwestern Handbook of Surgical Procedures, 2nd Edition, edited by Nathaniel J. Soper and Dixon B. Kaufman. ©2011 Landes Bioscience.

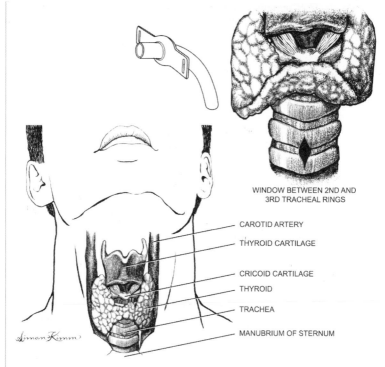

WINDOW BETWEEN 2ND AND
3RD TRACHEAL RINGS

CAROTID ARTERY

THYROID CARTILAGE

CRICOID CARTILAGE

THYROID

TRACHEA

MANUBRIUM OF STERNUM

Figure 93.1. Tracheostomy.

Step 3. A horizontal incision, approximately 3-5 cm in length, is made approximately 1 fingerbreadth above the sternal notch and is carried through the platysma muscle with electrocautery.

Step 4. The strap muscles are separated vertically in the midline to avoid bleeding.

Step 5. The thyroid isthmus is identified and dissection is carried out below it and below the lower border of the cricoid cartilage.

Step 6. The thyroid isthmus may need to be divided and if so, hemostasis is critical. Once divided, the edges are oversewn with a continuous 2-0 polyglactan suture, starting on one end, running it to the other end and back to the starting point. The needle is removed and the suture is used for retraction and exposure.

Step 7. More commonly, the thyroid gland can be dissected free and retracted cranially.

Step 8. The cartilaginous rings must be identified and the exact number of the ring must be double-checked.

Step 9. It is imperative that the first cartilage be left intact so that pressure by the tracheostomy tube will not erode the first ring or the cricoid cartilage. Injury to the cricoid cartilage will result in collapse of the cartilaginous support once the trachesotomy tube is removed.

93

Step 10. Traction sutures should then be placed around the 3rd cartilage laterally. Typically, this is a 2-0 or 3-0 monofilament suture. Preoxygenation with 100% oxygen will ensure optimal saturation and prevent desaturation during the remainder of the procedure.

Step 11. The 2nd and 3rd cartilages and potentially part of the 4th cartilage, if necessary, are incised at the midline. Alternatively, a transverse incision is made between the second and third cartilages and the incision extended down at the sides to create a "trap door".

Step 12. Small flaps may be created to enlarge the opening, but care should be taken to avoid making too large an opening, even with a flap that is left as a "trap door." Any opening will heal by scarring, and the larger the opening the more likely it is for the ensuing scar to narrow the trachea.

Step 13. Care should be taken to avoid damage to the cuff of the endotracheal tube when incising the trachea. The endotracheal tube should be visible through the opening and the cuff should clearly be below the stoma site.

Step 14. Once the tracheostomy opening has been created, the cuff of the endotracheal tube is deflated and the tube is pulled back to the subglottic space just above the stoma site. The stoma is occluded with a finger and ventilation resumed to ensure proper oxygenation prior to attempting introduction of the tracheostomy tube.

Step 15. Under direct vision, using the traction sutures, the stoma is gently dilated if necessary with a tracheal dilator and the tracheostomy tube is then slid into position. The obturator is removed and the inner cannula inserted.

Step 16. The cuff of the tracheostomy tube should then be inflated and checked to make sure that it is intact.

Step 17. The tracheostomy is connected to the anesthesia/ventilator circuit.

Step 18. Exhaled carbon dioxide should be noted once connected to insure correct placement.

Step 19. The platysma and skin can then be closed with widely spaced simple sutures which allow drainage from the wound. .

Step 20. A gauze dressing is placed at the level of the skin to protect the skin from the external flange and collar that helps hold the tracheostomy tube in place.

Step 21. This flange should initially be sutured to the skin, but once a tract is formed, the sutures may be removed on postoperative day 5 to 7. Care is taken to avoid too much traction on the tube by the ventilator circuit.

Postop

An arterial blood gas and a chest X-ray should be performed to insure the proper functioning and position of the tracheostomy tube. Additional care is similar to an endotracheal tube with regard to suctioning to control secretions and maintain airway patency.

Complications

There are principally four complications of tracheostomy: infection; hemorrhage; airway obstruction, and dislodgment. Additional complications include tracheoesophageal fistula or persistence of the stoma. As would be expected, the longer a tracheostomy is in place, the more likely it is that complications will occur.

93

1. *Infection.* All tracheostomies are contaminated and frequently will grow numerous bacteria. Therefore, sterile care and cleansing of the stoma site and appropriate maintenance of the respiratory equipment is necessary to minimize the possibility of a wound or lower airway infection.

2. *Hemorrhage*. Occasionally, the tracheostomy tube may erode into the innominate artery, and massive hemorrhage can occur. There may be a brief, sentinel bleed that signals the fistulization. In such a case, immediate tamponade of the arterial leak by finger pressure and prompt surgical treatment are required. The tracheostomy may need to be removed and an endotracheal tube placed. However, as this is done in an emergency setting, care must be taken to avoid loss of control of the airway. The injured artery should be resected and both ends sutured closed. Prosthetic graft material should not be used in this contaminated field. The tracheal innominate fistula is fortunately a rare, but frequently a lethal complication. In the few patients who have been successfully treated, neurologic sequelae have not been noted as a result of resecting and suturing the innominate artery. Bleeding from granulation tissue or the skin is far more common but is usually less massive and can be treated with local measures.

3. *Airway obstruction*. Despite having a tracheostomy tube in place, airway obstruction can occur for a variety of reasons. The cuff may prolapse over the end of the tracheostomy due to overdistention, but this should be avoidable. Crusting of secretions and obstruction of the tube can be addressed by the use of an inner cannula, humidification, and suctioning. Occasionally, granulations may form at the end of the tracheostomy, which may need to be removed.

The most troubling type of obstruction is due to tracheal stenosis, either at the stoma site or at the cuff site. This can be prevented by having lightweight connectors that allow for pivoting around the connection site, as opposed to torquing on the tracheostomy. Meticulous surgical technique and creation of an appropriate-sized stoma will prevent stenosis later as the stoma heals. Large-volume, low-pressure cuffs that are typically employed today, conformed to the shape of the trachea rather than distending it, are key to preventing pressure necrosis and ensuing tracheal stenosis.

4. *Dislodgment*. If the tracheostomy tube becomes dislodged during the first 1-3 days following its insertion, no attempt should be made to reintroduce the tube through the stoma. The patient should be intubated orotracheally and the tracheostomy reinserted electively under more favorable conditions in the operating room. Attempts at reinsertion through the stoma will likely result in a false passage and loss of airway control.

Follow-Up

Long-term follow-up is primarily removal of the tracheostomy once the patient's respiratory failure has abated. Prior to removing the tracheostomy tube, and once off the ventilator, a smaller uncuffed tracheostomy tube can be inserted to allow air to pass around the tube while plugged to permit the patient to talk. Decannulation is performed by sequential downsizing of the tubes. The stoma site will heal usually within a few weeks. If shortness of breath or noisy breathing develop, the patient should be investigated for possible laryngeal or tracheal stenosis.

93

Acknowledgment

The editors and author wish to acknowledge Keith A. Horvath for contributing to the previous version of this chapter.

Chapter 94

Cricothyrotomy

Alberto de Hoyos and Malcolm DeCamp

Indications

Cricothyrotomy (cricothyroidotomy) is an emergency procedure recommended for urgent, short-term airway control. It is the procedure of choice in patients who have a strong indication for intubation and in whom the trachea cannot be intubated for any reason. Common indications for an emergent surgical airway include:
1. failure to accomplish orotracheal intubation;
2. blunt or penetrating trauma to head, face or neck; and
3. loss of airway control.

Cricothyrotomy is preferable to emergency tracheostomy in adults because the cricothyroid membrane is closer to the skin surface, and less dissection is required to reach the airway. Emergent tracheostomy is indicated however, in patients with intralaryngeal or subglottic airway obstruction.

Contraindications

Children: Cricothyrotomy should not be performed in children due to the small size of the intralaryngeal airway. Emergency tracheostomy is preferred.

Preop

Cricothyrotomy is a procedure resulting in a laryngostomy and is utilized to obtain emergent airway access only and cannot be regarded as an elective alternative to tracheostomy. In the elective setting, tracheostomy is preferred because long-term intubation through the cricothyroid membrane carries an unacceptably high risk of serious and possibly irreversible laryngeal damage. As the cricoid cartilage provides the only circumferential support of the subglottic airway, ischemic necrosis or erosion due to an indwelling prosthesis can result in irreparable subglottic stenosis. Although often thought of as a quick and reliable method of obtaining control of the airway, patients in whom it is indicated may already be in extremis, agitated or apneic, or may have severe anatomic distortion making the procedure far from trivial. Because cricothyrotomy remains a rare procedure, the most challenging aspect is to obtain and maintain a satisfactory level of skill, while restricting its use to the infrequent patient in whom it is truly the airway procedure of choice.

Procedure

Step 1. The patient is positioned supine with the neck in a neutral position. An adequate sterile field is prepared, and local anesthetic is infiltrated if the situation allows.

Step 2. The following landmarks are palpated: thyroid cartilage, cricothyroid membrane and cricoid cartilage.

Step 3. The thyroid cartilage is retracted cephalad and stabilized. For this, it is helpful to stand on the patient's right side (for a right-handed surgeon) and stabilize the cartilaginous framework by holding the sides of the thyroid cartilage with the left thumb and index fingers.

Step 4. A horizontal skin incision is made just below the thyroid with a No. 11 blade. Alternatively, a vertical incision is performed to avoid subcutaneous vessels and the anterior jugular veins.

Step 5. The cricothyroid membrane is palpated and incised horizontally with a short, stabbing motion. It is important to avoid pushing the blade too far to avoid injury to the posterior wall of the larynx or esophagus. Electrocautery is not required.

Step 6. A blunt scalpel handle, a Trousseau dilator, a tracheal spreader or a hemostat clamp is inserted into the incision and spread vertically to open the airway.

Step 7. The thyroid cartilage is elevated with a tracheal hook and a small cuffed endotracheal tube (5-6) or tracheostomy tube (4-6) is inserted and secured. In most adults, a No. 6 airway is the largest that can be inserted but a No 4 is adequate for initial placement. If an endotracheal tube is used, care is taken to avoid inserting the tube more than 5 cm to prevent intubation of a main bronchus.

Step 8. Correct placement is confirmed by colorimetric capnography or mass spectrometry and bilateral lung auscultation. The tube is connected to the ventilator circuit and pulse oximetry is monitored.

Postop

An arterial blood gas and a chest X-ray should be performed to insure the proper functioning and position of the cricothyrotomy tube. Additional care is similar to a tracheostomy tube with regard to suctioning to control secretions and maintain airway patency.

Complications

1. *Incorrect placement.* This occurs more commonly through the thyrohyoid membrane.
2. *Hemorrhage.* Usually it is secondary to injury to the anterior jugular veins and can be avoided by limiting the size of a horizontal incision or by performing a vertical midline incision.
3. *Intralaryngeal stenosis.*

Follow-Up

An emergency cricothyrotomy should always be converted to a standard tracheostomy soon after the patient is stabilized, usually during the first two to four days. If left in place, the tube will erode the cricoid and often the thyroid cartilage, leading to stenosis. Following elective creation of a tracheostomy, the cricothyrotomy stoma is closed with sutures because a stricture in this location is far more difficult to treat than a similar injury in the trachea.

Chapter 95

Median Sternotomy and Cardiopulmonary Bypass

S. Chris Malaisrie, Richard Lee, Edwin McGee and Patrick M. McCarthy

Indications

Cardiopulmonary bypass (CPB) involves an extracorporeal circuit consisting of a membrane oxygenator and blood pump. CPB is indicated when manipulation of the heart compromises hemodynamic stability or when intracardiac surgery is required.

Preop

After a thorough history and physical examination, the chest radiograph and coronary angiogram should be inspected for signs of aortic calcification that would preclude safe cannulation of the aorta. Computer tomography should be performed to evaluate a calcified aorta. Alternative sites of arterial cannulation include the axillary artery and femoral artery.

Procedure—Median Sternotomy

Step 1. The procedure is performed under general anesthesia with a single-lumen endotracheal tube.

Step 2. The skin incision extends from the midpoint between the sternal notch and the Angle of Louis to below the xyphoid process (Fig. 95.1). The subcutaneous tissue is divided with electrocautery.

Step 3. The midline is identified by palpating the sternal borders at the interspaces, and the periosteum is scored with electrocautery. The interclavicular ligament at the sternal notch is divided allowing finger palpation of the posterior sternum. The fascia overlying the xyphoid process is divided in the midline.

Step 4. An oscillating sternal saw is used to divide the sternum and can be performed either from the top-down or from the bottom-up depending on surgeon preference. Bleeding points in the periosteum are selectively cauterized.

Step 5. A sternal retractor is placed with the blades on the lower two-thirds of the sternum to minimize the risk of sternal fracture.

Step 6. The two halves of the thymus gland are divided up to the innominate vein and the thymic veins draining into the innominate vein are ligated or clipped.

Step 7. The pericardium is opened vertically with care not to injure the underlying structures. The inferior pericardial attachment to the diaphragm is divided transversely to make a T-shaped incision.

Step 8. 0 silk sutures are used to suspend the heart in a pericardial cradle (Fig. 95.2).

Northwestern Handbook of Surgical Procedures, 2nd Edition, edited by Nathaniel J. Soper and Dixon B. Kaufman. ©2011 Landes Bioscience.

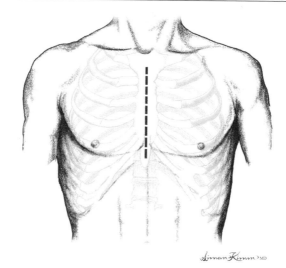

Figure 95.1. Median Sternotomy.

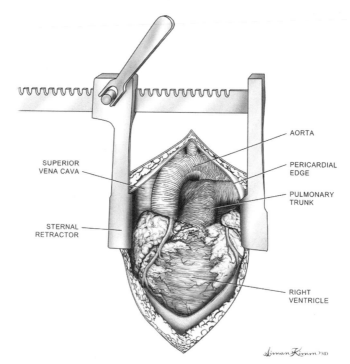

Figure 95.2. Exposure of the heart.

Procedure—Initiating Cardiopulmonary Bypass

Step 9. The patient is systemically heparinized at a dose of 300 units/kilogram to achieve an ACT (Activated Clotting Time) >480 seconds. Occasionally, fresh frozen plasma must be transfused along with additional heparin to achieve an adequate ACT.

Step 10. A suitable area on the distal ascending aorta that is free of athersclerotic plaques is identified for cannulation using manual palpation, transesophageal echocardiography, or epi-aortic ultrasonography.

Step 11. Two partial-thickness, concentric, diamond-shaped pursestring sutures are placed with 3-0 polypropylene suture (Fig. 95.3). The ends of the sutures are placed through rubber tourniquets. The adventitia is divided, and an 11 blade is used to make a transverse aortotomy within the pursestrings. The aortic cannula is inserted through the aortotomy with the tip pointing into the aortic arch. The tourniquets are tightened, the cannula is tied to the tourniquets with 0 silk suture. Both the cannula and the tourniquet is secured to the skin to prevent inadvertent dislodgment.

Step 12. The aortic cannula is de-aired and connected to the arterial end of the CPB circuit. Correct positioning of the cannula is confirmed by a pulsatile waveform with the perfusionist.

Step 13. A 3-0 polypropylene stitch is placed around the right atrial appendage as a pursestring and placed through a rubber tourniquet. A two-staged venous cannula (side holes at the cannula tip for IVC drainage and side holes in the body of the cannula for right atrial drainage) is passed through the amputated tip of the right atrial appendage. Small bidging trabeculae within the right atrium may be divided to facilitate passage of the cannula. After confirmation that the tip of the cannula is within the IVC, the tourniquet is tightened and secured to the cannula. The cannula is then connected to the venous end of the CPB circuit.

Step 14. After confirming that an adequate ACT is achieved, cardiopulmonary bypass is initiated, and lung ventilation can be ceased. Complete venous drainage and subsequent decompression of the heart signifies successful initiation of CPB.

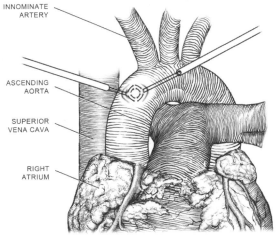

INNOMINATE ARTERY

ASCENDING AORTA

SUPERIOR VENA CAVA

RIGHT ATRIUM

95

Figure 95.3. Aortic cannulation.

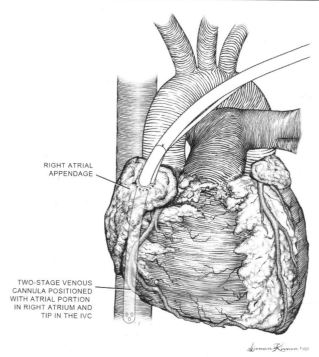

RIGHT ATRIAL
APPENDAGE

TWO-STAGE VENOUS
CANNULA POSITIONED
WITH ATRIAL PORTION
IN RIGHT ATRIUM AND
TIP IN THE IVC

Figure 95.4. Venous cannulation.

Procedure—Terminating Cardiopulmonary Bypass

Step 15. The patient is warmed to normothermia if necessary. Lung ventilation is re-initiated. The heart is allowed to regain normal sinus rhythm with adequate contractility. Rate, rhythm, preload, afterload, and inotropy are adjusted accordingly. Metabolic derangements such as anemia, hyperkalemia, and acidosis are resolved prior to discontinuing cardiopulmonary bypass.

Step 16. As CPB is weaned, volume is returned to the heart from the venous reservoir. When hemodynamic stability is achieved, protamine is administered at a dose of 1 mg per 100 units of heparin. The patient is monitored for a protamine reaction. The heart is decannulated and cannulation sites are checked for hemostasis.

Procedure—Sternal Closure

Step 17. One or two chest tubes are left in the mediastinal cavity and brought through stab incisions below the sternal incision with care to avoid entry into the peritoneal cavity. Temporary epicardial pacing wires may be placed on the right ventricle and right atrium as indicated.

Step 18. Interrupted #6 or #8 steel wires are used to re-approximate the sternal edges. As a rule-of-thumb, if an interrupted configuration is used, 1 wire per 10 kg of body weight should be placed. A figure-of-eight configuration can also be used. The wires are crossed and twisted ensuring complete apposition of the sternal edges with no laxity in the wires.

95

Step 19. The fascia is closed over the sternum and the upper abdominal fascia is closed over the xyphoid process with an 0 braided polypropylene suture. The subcutaneous tissue is reapproximated with 2-0 dissolvable suture and the skin closed with a 4-0 dissolvable monofilament suture. Skin staples are not recommended in case emergent reopening of the chest is required postoperatively.

Postop

Patients are extubated in the intensive care unit after hemodynamic stability has been established and no signs of bleeding are present. The chest tubes should be kept on suction and removed when the tubes are draining less than 200 ml per 24 hours.

Complications

1. Sternal wound infection
2. Sternal dehiscence
3. Myocardial ischemia-reperfusion injury
4. Low-cardiac output syndrome
5. Mediastinal hemorrhage
6. Pericardial tamponade
7. Stroke
8. Renal failure
9. Respiratory failure
10. Mesenteric and peripheral ischemia
11. Death

Follow-Up

Follow-up is dependent on the nature of the procedure performed.

SECTION 6: TRANSPLANTATION

Section Editor: Dixon B. Kaufman

Chapter 96

Arteriovenous Graft (AVG)

Anton I. Skaro

Indications

1. Long-term hemodialysis access;
2. inability to perform continuous ambulatory peritoneal dialysis (CAPD); and
3. inability to place arteriovenous fistula.

Preop

Routine preoperative screening with special emphasis on cardiovascular system. Use of the nondominant upper extremity is preferred. Preoperative vascular assessment should include arterial and venous evaluation: arterial (Allen's test, sonography), venous (venography [iodinated contrast and/or carbon dioxide in patients not yet on dialysis], sonography). Choices of anesthesia includes: (1) regional, (2) local, (3) monitored anesthesia care (MAC), and (4) general.

Procedure

Step 1. Supine position with target extremity abducted on armboard.

Step 2. Circumferential sterile prep/surgical drape from finger tips to lateral border of pectoralis major.

Step 3. Antecubital fossa: Mark skin over the palpated brachial artery, and, if visible, antecubital veins.

Step 4. Incision: 2 cm distal to the antecubital fold, hemostasis using electrocautery and absorbable suture material.

Step 5. Venous dissection: Target vein (cephalic or median antecubital vein) isolated and division of vein branches avoided (maintain venous drainage of extremity and preserve veins for later use in secondary AVG or revisions of primary AVG). Obtain vascular control for about 3-4 cm.

Step 6. Arterial dissection: Divide the subcutaneous tissue down to the biceps aponeurosis. Divide the aponeurosis sharply in a cruciate formation for optimal exposure to the artery and concomitant veins below. Palpate the brachial artery, separate but preserve concomitant paired deep brachial veins (located on each side of artery). Arterial dissection is greatly simplified by dissecting in the plane of Leriche (periadventitial plane) for about 3 cm to facilitate vascular control. Vessel loops are placed around vascular structures to assist in control prior to anastomosis.

Step 7. Graft selection: Thin-walled 4-7 mm tapered polytetrafluoroethylene (PTFE) grafts are preferred as the small (4 mm) end is anastomosed to the arterial inflow to avoid postoperative "steal" syndrome. Use graft length ensuring redundancy so that kinking and tension are avoided.

Northwestern Handbook of Surgical Procedures, 2nd Edition, edited by Nathaniel J. Soper and Dixon B. Kaufman. ©2011 Landes Bioscience.

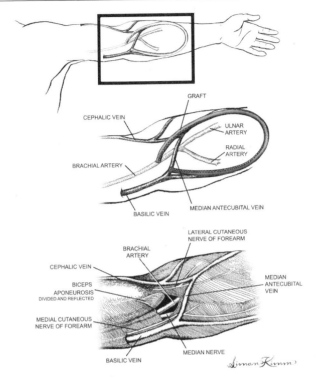

Figure 96.1. Arteriovenous graft (AVG).

Step 8. Subcutaneous tunnel: Select a semicircular tunnel device (Noon, Kel-ly-Weck, or sheath) of a diameter slightly greater than that of the selected PTFE graft. Advance the tunnel device from the antecubital incision distally along the medial/volar aspect of the forearm. Create a small cutdown incision distally and then pass the PTFE graft (retrograde) through the tunnel. Repeat these steps on the lateral/volar aspects of the forearm to create an oval graft tunnel where the tapered (4 mm) end of the graft lies next to the brachial artery and the redundant end of the graft lies next to the target vein. Attention should be given to avoid twisting or kinking of the graft as it is passed through the tunnel. The graft is marked to aid in maintaining proper orientation during the tunneling procedure. Irrigation of the graft with heparinized-saline will ensure its patency prior to anastomosis.

Step 9. Venous anastomosis: Occlude the vein (and any tributaries) with atraumatic vascular clamps (e.g., bulldog or Heifets clamps) so that sutures cannot be caught in the clamps. Incise the vein using an 11 blade at the intended site of anastomosis; irrigate with heparinized saline. Extend the venotomy with Pott's scissors to a length to match the graft, estimating the angle at which the graft will be cut. Tailor the graft to match venotomy and anastomose (end-to-side) using fine (5-0 or 6-0) nonabsorb-able monofilament suture (polypropylene). Vascular control is relieved and blood is allowed to flow back through the venous anastomosis; the graft is flushed with heparinized-saline and occluded with a soft vascular clamp.

96

Step 10. Arterial anastomosis: Obtain vascular control (atraumatic clamps) proximal and distal to the site of anastomosis. Create a small arteriotomy (11 blade) with care to avoid injuring the posterior arterial wall. Flush with heparinized saline. Extend the arteriotomy to approximately 6 mm (4 mm graft end will be cut at a slight angle, matching the arteriotomy site). Perform the anastomosis using 6-0 or 7-0 nonabsorbable monofilament suture (polypropylene) as done with the vein. Attention should be paid to avoid intimal injury during tissue handling while also ensuring that all layers of the arterial wall are incorporated to prevent intimal dissection/anastomotic bleeding.

Step 11. Graft perfusion: Remove the graft clamp (outflow) and remove the distal arterial clamp (back bleed). Then remove the proximal arterial clamp. There should be brisk flow, often felt as a thrill through the graft and the vascular anastomosis sites. However, flow can be difficult to discern from the graft, whereas a palpable thrill is usually apparent in the outflow vein immediately adjacent to the venous anastomosis.

Step 12. Hemostasis: Avoid unnecessary manipulation of the anastomosis. Surgical gauze with gentle pressure will stop needle-hole bleeding. Alternatively, judicious use of topical hemostatic agents can be a helpful adjunct.

Step 13. Closure: The wound is closed with 3-0 or 4-0 absorbable subcutaneous sutures; skin is closed with 4-0 or 5-0 subcuticular sutures.

Step 14. Dressing: Loose gauze. Avoid circular or semicircular dressings or tapes (patients risk postoperative swelling and dangerous tourniquet effect).

Postop

Use arm freely; elevate arm while at rest. Oral analgesics.

Complications

Early complications: AVG thrombosis (urgent thrombectomy/revision), postoperative swelling (surveillance for infection), bleeding (reoperation, hemostasis), skin infection (antibiotics), graft infection (removal), arterial steal (graft revision).

Follow-Up

Avoid puncturing the AVG for 8 weeks to allow time for tissue incorporation of PTFE. Doing so will prevent needle-hole perigraft hematoma, graft thrombosis, and reduce the risk of graft infection due to repeat access.

Acknowledgment

The editors and author wish to acknowledge Jon S. Matsumura for contributing to the previous version of this chapter.

Chapter 97

Primary Radial Artery-Cephalic Vein Fistula for Hemodialysis Access

Joseph R. Leventhal

Indications

The indication for primary radial artery-cephalic vein fistula is end stage renal disease requiring long-term hemodialysis access. Ideally, fistula creation should precede the need for hemodialysis by several months in order to allow for adequate maturation before use. Careful physical examination should be performed to rule out arterial insufficiency in the upper extremity intended for use. Examination of the cephalic vein under tourniquet should be performed to ensure the absence of stenosis or thrombosis in the forearm. Patients with a history of previous central lines, dialysis catheters, and neck/chest trauma should be evaluated to rule out central venous obstruction.

Preop

Administer prophylactic antibiotics. Monitored anesthesia care with supplemental local analgesia should be used. Do not use local with epinephrine since it causes vasospasm. Position the patient supine, with the outstretched upper extremity placed on an armboard. Prep the arm from finger tips to chest wall. Place a stockinette over the hand and drape the patient to keep the axilla exposed in case more proximal dissection of a fistula or graft placement is required.

Procedure

Step 1. Mark the course of the radial artery and cephalic vein in the distal half of the forearm and mark the proposed longitudinal skin incision midway between the two. If possible, identify the common dorsal branch of the cephalic vein. The use of this branch provides a "patch" anastomosis which is optimal from a hemodynamic standpoint and also technically easier to perform.

Step 2. Infiltrate the proposed incision with 1% lidocaine solution. Using a 15 blade create a 5-6 cm incision midway between the artery and vein. The distal limit of the incision should be proximal to the radial styloid process.

Step 3. Dissect through subcutaneous tissue using blunt (mosquito hemostat) techniques to expose the cephalic vein. Carefully divide tissues using scissors or electrocautery. Avoid grasping the vein proper; isolate the vein and place a vessel loop around it to aid in retraction exposure. Isolate the vein for a length of 5-6 cm. Carefully ligate and divide small branches of the cephalic vein.

Northwestern Handbook of Surgical Procedures, 2nd Edition, edited by Nathaniel J. Soper and Dixon B. Kaufman. ©2011 Landes Bioscience.

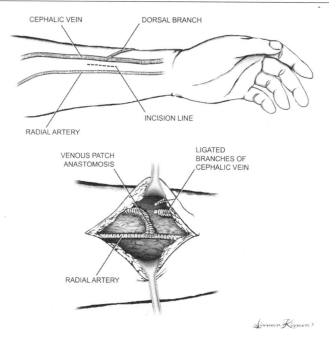

Figure 97.1. Radial artery-cephalic vein fistula.

Step 4. Dissect out and isolate the radial artery using techniques described for venous dissection. Take care to separate the concomitant veins which run on either side of the artery. Ligate or cauterize small arterial branches of the main radial artery to allow for complete mobilization. Use vessel loops to retract and move the artery. Avoid grasping the radial artery proper.

Step 5. Ligate the cephalic vein distally. Open the cephalic vein just proximal to the ligature and flush the vein with heparinized saline (5 ml of 1000 units/ml in 500 ml of saline). Apply a small bulldog clamp to the vein to prevent backflow. Using the dorsal branch of the vein, create a vein patch for anastomosis to the artery. Alternatively, spatulate open the main cephalic vein to allow for a generous anastomosis.

Step 6. Using vessel loops or bulldog clamps, occlude the radial artery proximal and distal to the proposed arteriotomy site.

Step 7. Make a small (1-2 mm) arteriotomy in the radial artery. Irrigate the artery with heparinized saline. Extend the arteriotomy to a length of 8-10 mm using Potts or Dietrich scissors.

Step 8. Sew the cephalic vein to the radial artery in an end-to-side fashion using 7-0 polypropylene suture.

Step 9. Remove the bulldog clamp from the vein. Release the distal artery, then the proximal artery vessel loop/vascular clamp.

Step 10. Establish hemostasis using gentle pressure at the anastomosis. Inspect the anastomosis and course of the cephalic vein. Be sure the vein is not kinked or twisted as it courses proximally. Feel for a thrill in the vein proximal to the anastomisis.

Step 11. Close the incision with absorbable suture.

Postop

Elevate the arm to prevent hand/arm swelling. Remove the dressing 24 hours postoperatively. Begin exercises to mature fistula (squeezing a ball several times a day for 10 minutes).

Complications

Complications include bleeding, thrombosis, infection, hand ischemia ("steal"), paresthesias from peripheral nerve injury during surgery.

Follow-Up

The patient should be examined at regular intervals to ensure fistula maturation before use in dialysis unit.

97

Chapter 98

Laparoscopic Donor Nephrectomy

Joseph R. Leventhal

Indications

The indication for laparoscopic donor nephrectomy is living donor renal trans-plant-ation. Donor evaluation should ensure that the renal donor has adequate functional reserve in the remaining kidney to allow donation to proceed. CT angiography or MR imaging should be performed to define renal anatomy before proceeding with donor nephrectomy.

Preop

Bowel prep with magnesium citrate is performed the night before surgery. Endotracheal intubation—avoid use of nitrous oxide to prevent bowel distension. Orogastric suction, bladder catheter drainage, prophylactic antibiotics, and antithrombotic sequential leg compression devices are routinely used. The patient is placed in the right decubitus position for left donor nephrectomy, and vice versa. Flex the operating table at a point midway between patient's iliac crest and rib cage and elevate the kidney rest in order to maximize exposure during the procedure. Prep and drape the patient to allow for—if necessary—open conversion to extended subcostal or standard flank approach if required. The operating surgeon stands facing the patient's abdomen, with the camera operator caudad. An assistant and scrub nurse are positioned opposite the surgeon. Standard laparoscopic instrumentation, along with a 30° laparoscope and ultrasonic scalpel are used. More than 95% of donor nephrectomies remove the left kidney, in order to obtain longer renal vein length. The operative procedure is described for left donor nephrectomy.

Procedure

Step 1. Insert a Veress needle in the subcostal location. Create a pneumoperitoneum of no more than 15 mm Hg.

Step 2. Introduce the laparoscope into abdomen using a 10 mm Visiport™.

Step 3. Place two 12 mm operating ports underneath the rib cage; a 5 mm port may be placed subcostally in the posterior axillary line to assist in retraction of the kidney. Port placement will vary slightly from patient to patient depending upon patient girth and length of the torso.

Step 4. Using an ultrasonic scalpel, mobilize the left colon and spleen medially, away from the kidney posteriorly.

Step 5. Open Gerota's fascia and dissect out the renal vein. Clip and divide gonadal, adrenal, and lumbar vein branches to obtain maximal length.

Step 6. Identify renal artery/arteries. Avoid overdissection of vessels to prevent vasospasm.

Step 7. Dissect adrenal gland off the upper pole of the kidney.

Northwestern Handbook of Surgical Procedures, 2nd Edition, edited by Nathaniel J. Soper and Dixon B. Kaufman. ©2011 Landes Bioscience.

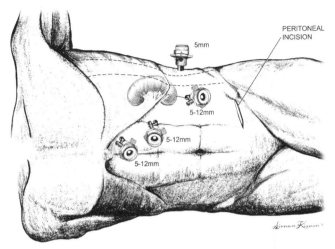

Figure 98.1. Laparoscopic donor nephrectomy. Trocar placement.

Step 8. Identify the ureter inferior to the lower pole of the kidney. Dissect adequate length of ureter allowing for easy transplantation of kidney in recipient, usually to the point where the ureter courses over iliac vessels. Avoid stripping the ureteral blood supply.

Step 9. Completely mobilize the kidney. Ensure that the kidney is completely free except for blood vessels and ureter. Avoid torsion of kidney.

Step 10. Create a 6-7 cm extraction incision in the left lower quadrant, without violation of the peritoneum.

Step 11. Anticoagulate the patient with 5,000 units of heparin sodium. Distally clip and divide the ureter.

Step 12. Divide the renal artery, followed by the renal vein, with a linear vascular laparoscopic stapler. Reverse the heparin with protamine sulfate. The kidney is now completely free.

Step 13. Open the peritoneum at the extraction site and deliver the kidney through this wound by hand into an iced saline solution.

Step 14. Remove the staple lines from the donor kidney blood vessels and flush out the kidney with a chilled preservation solution (Collins, Viaspan, etc.).

Step 15. Inspect the abdomen through the extraction incision. Confirm secure placement of clips on the distal ureter. Palpate the iliac artery to confirm good pulsations.

Step 16. Close the extraction incision with #1 or 0 absorbable suture. Reestablish pneumoperitoneum and inspect the operative field laparoscopically. Confirm adequate hemostasis, remove ports under direct visualization, and desufflate the abdomen. Close all incisions.

Postop

Foley catheter drainage until postoperative day 1. Intravenous fluids at 50 ml/hour until midnight, then heparin-lock IV. Parenteral analgesia until postoperative day 1, then switch to oral. Clear liquid diet the night after surgery, ambulate in the evening. Hemogram and renal chemistry panel on postoperative day 1. Anticipate discharge on postoperative day 1 or 2.

Complications

Complications include subcutaneous emphysema, atelectasis, infection, bleeding from vascular staple lines, splenic injury, and vascular injury.

Follow-Up

Follow-up in clinic first week postoperatively. Check renal function. No other office follow-up needed.

Chapter 99

Kidney Transplantation

Dixon B. Kaufman

Indications

Indications for kidney transplantation are irreversible chronic or end-stage renal disease. Contraindications include: ABO incompatibility without preparatory desensitization, active or chronic infection, active or recent (<3 years) malignancy, active glomerulonephritis, life expectancy <1 year, sensitization to donor HLA antigens without preparatory desensitization, serious (untreatable) preexisting comorbidities, medical noncompliance, active substance abuse, uncontrolled psychiatric disorders.

Preop

General orotracheal anesthesia, Foley bladder catheter, central venous catheter (optional, usually for recipients of deceased donor grafts). Upper extremity positioning to prevent occlusion of arteriovenous graft/fistula, perioperative antibiotics, confirmation of ABO compatibility, DVT prophylaxis .

Procedure

Step 1. Prone position. Right (or left) lower quadrant "hockey-stick" incision positioned two fingerbreadths medial to anterior superior iliac spine and one fingerbreadth superior to pubis.

Step 2. Incision is carried through the skin, subcutaneous tissue, and external obique aponeurosis. The medial edge of external/internal oblique musculature lateral to the edge of rectus sheath are opened down to (but not through) the peritoneum. The inferior epigastric vessels are suture-ligated and divided (optional). The round ligament (in females) is suture-ligated and divided. The spermatic cord (in males) is encircled with a Penrose drain and retracted medially and protected. The peritoneum is mobilized medially and cephalad to expose the retroperitoneal iliac fossa. Mechanical retractors are often used to aid in exposure.

Step 3. Expose and mobilize external iliac vein and artery by suture ligation and division of overlying lymphatics. It is sometimes necessary to expose and mobilize the hypogastric artery and common iliac artery. If necessary to gain maximal exposure of iliac vein, suture ligate (stick-ties) and divide hypogastric vein(s). This allows the external/common iliac vein to be elevated and mobilized lateral to the iliac artery.

Step 4. The location of the vascular anastomoses depends on the geometric relationships of the length of the renal vessels (short with living donor organs), size of the renal allograft, and anterior/inferior ascent of the psoas muscle. Position the renal allograft proximal enough to lie flat in the iliac fossa without the lower pole of the kidney being excessively tipped up by the psoas muscle. Also, the transplant position should be planned so that the renal transplant artery and vein will be properly aligned without undue tension. Often intravenous diurectics are slowly administered (over 30 minutes) during the vascular anastomoses.

Northwestern Handbook of Surgical Procedures, 2nd Edition, edited by Nathaniel J. Soper and Dixon B. Kaufman. ©2011 Landes Bioscience.

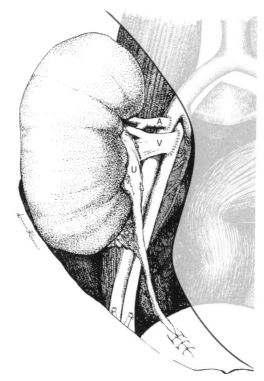

Figure 99.1. Kidney transplantation. Completion of anastomoses.

Step 5. If the patient is preuremic, administer parenteral heparin. Uremic patients do not require heparin unless a hypercoagulable state is known. Place a vascular clamp on the iliac vein. Bring the renal allograft into the operative field and keep as cold as possible. Perform a venotomy and irrigate with heparinized saline. Anastomose the renal vein end-to-side to the iliac vein using 5-0 monofilament nonabsorbable suture.

Step 6. Place vascular clamps on the iliac artery proximally relative to the venous anastomosis. The site of the arteriotomy may be the external or common iliac artery. Alternatively, the hypogastric artery may be used by suture ligation of the distal aspect with proximal placement of a bulldog clamp (maintains iliac arterial bloodflow; end-to-end anastomosis performed). Perform an arteriotomy and irrigate with heparinized saline. Perform an end-to-side anastomosis with 6-0 monofilament nonabsorbable suture.

Step 7. Slowly release the venous clamp and then release arterial clamps to allow full perfusion of the renal allograft. Control bleeding points with suture ligation or electocautery.

Step 8. Begin ureterovesical construction by infusing sterile saline into the bladder via side-port in cysto-tubing. Clamp Foley tubing to keep the bladder distended. Properly align the ureter, measure to the appropriate length and suture ligate and divide the distal ureteral artery. Cut and spatulate the distal ureter. In males, slip the ureter under the spermatic cord.

Step 9. (optional) Place a 6 F double-J silastic ureteral stent.

Step 10. For anterior ureteroneocystotomy, incise the bladder musculature down to (not through) the bladder mucosa. Create the cystotomy (approximately 2 cm). Lay the spatualted tip of the ureter on the open bladder mucosa (with distal stent in bladder) and perform mucosa-to-mucosa anastomosis with 5-0 monofilament absorbable suture. Create an antireflux tunnel over the distal aspect of the ureter by approximating the muscular layer with 4-0 monofilament absorbable suture as interrupted stitches.

Step 11. Irrigate the surgical field with antibiotic solution and position renal allograft to lie flat, avoiding tension on the vascular anastomosis and ureter.

Step 12. Close the incision by approximating the internal/external oblique musculature and external oblique aponeurosis in a single layer with #1 monofilament nonabsorbable suture. Approximate the skin edges with a stapler. Dress the incision.

Postop

Continue immunosuppressive medical therapy according to institutional protocols. Manage fluid replacement according to the rate of urine output and intravascular fluid status (clinical assessment ± central venous pressure measurement). Administer diuretics according to urine output and intravascular fluid balance. Routine assessment of urine output, CBC, and chemistries.

Complications

Complications include hemorrhage, vascular thrombosis, ureteral leak, lymphocele. Metabolic complications include delayed graft function (low urine output secondary to acute tubular necrosis), electrolyte disorders (hyperkalemia, hypocalcemia, hypomagnesemia).

Follow-Up

Immuosuppression to prevent rejection with periodic trough concentration monitoring to guide dosing. Routine measurement of serum blood urea nitrogen and creatinine to assess renal allograft function. Antibacterial and antiviral prophylaxis are typically prescribed.

Chapter 100

Distal Splenorenal (Warren) Shunt

Michael Abecassis

Indications

The indication for distal splenorenal (Warren) shunt is portal hypertensive bleeding refractory to medical and endoscopic therapy.

Preop

Hemodynamic stabilization; mesenteric venous examination (Doppler ultrasound, magnetic resonance venography).

Procedure

Step 1. The patient is positioned in the supine position under general anesthesia and endotracheal intubation. A Foley catheter is inserted for monitoring urine output, and a central venous catheter is utilized for monitoring central venous pressure.

Step 2. Antiseptic prep is used from the nipple line to the groin. The body is draped in sterile fashion. Appropriate antibiotics are given prior to skin incision.

Step 3. This operation consists of two essential components. First, the portosystemic shunt itself between the splenic vein and the left renal vein and, second, the disconnection between the portal vascular bed and the shunted portion of the portal bed, i.e., the splenic venous drainage.

Step 4. A transverse upper abdominal incision is used which is parallel to the long axis of the pancreas and which extends from the right to the left costal margins. Alternatively, in patients with narrow costal margins and a long epigastrium, a long upper midline incision can also be used.

Step 5. Once the peritoneal cavity is entered and appropriate retractors are placed, the lesser sac is entered through the gastrocolic omentum. This dissection is carried from the duodenum laterally to, but not including, the short gastric vessels. In the process of entering the lesser sac, the gastroepiploic vessels can be multiply ligated in order to disconnect the portal system from the venous system draining the spleen. At the point of the short gastric vessels, the interruption of these collaterals can be carried out more laterally inferior to the spleen by dividing the lienal-colic ligaments as well as possible. There are often very large collaterals in this area which can act as a sump for the splenic bed. Mobilization of the splenic flexure of the colon may be helpful and necessary in order to divide and disconnect these collaterals.

Northwestern Handbook of Surgical Procedures, 2nd Edition, edited by Nathaniel J. Soper and Dixon B. Kaufman. ©2011 Landes Bioscience.

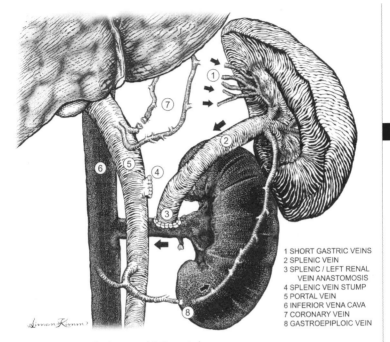

1 SHORT GASTRIC VEINS
2 SPLENIC VEIN
3 SPLENIC / LEFT RENAL
   VEIN ANASTOMOSIS
4 SPLENIC VEIN STUMP
5 PORTAL VEIN
6 INFERIOR VENA CAVA
7 CORONARY VEIN
8 GASTROEPIPLOIC VEIN

Figure 100.1. Distal splenorenal (Warren) shunt.

Step 6. Once the lesser sac is exposed, the inferior border of the pancreas is identified and the peritoneum overlying this inferior border is incised from the area of the superior mesenteric vein (SMV) laterally towards the tail of the pancreas. An avascular plane should be developed which allows extensive superior mobilization of the body and tail of the pancreas. When necessary, lymphatic vessels should be carefully ligated in order to prevent the potential complication of chylous/lymphatic ascites postoperatively. The purpose of the pancreatic mobilization is to identify and mobilize the splenic vein. The position of the splenic vein with respect to the body and tail of the pancreas can be quite variable, and if the splenic vein is not immediately obvious, the inferior mesenteric vein (IMV) can be used as a road map to find the SMV. In some cases, the SMV may be quite superiorly placed behind the pancreas making this dissection more difficult.

Step 7. Once the splenic vein is identified, it is circumferentially dissected and mobilized. This often requires ligation and division of numerous pancreatic branches which drain the pancreas into the splenic vein. The limits of the dissection medially include the confluence of the SMV and the portal vein and laterally, division of the splenic vein into splenic branches. The IMV often enters the SMV directly or at the confluence of the SMV with the splenic vein. Regardless, it must be divided in order to afford better mobilization of the splenic vein.

Step 8. The splenic vein is mobilized along its entirety. In doing so, one may encounter the coronary vein in the medial aspect of the dissection. If it can safely be done, the coronary vein should be ligated and divided at its origin from the splenic vein. Alternatively, the left gastric vein must be divided in the lesser gastric omentum.

Step 9. Once the splenic vein is completely mobilized, our attention is turned to dissection of the left renal vein. This is done by incising the retroperitoneum posterior to the inferior aspect of the pancreas. Again, this tissue can be thick and rich in lymphatics, and great care must be taken to ligate all lymphatic tissue. The left renal vein is identified. The left adrenal and left gonadal veins are divided in order to afford better mobilization of the left renal vein. Occasionally, the IVC can be mobilized and, depending on the distance between the splenic vein and the left renal vein or IVC, either an end-to-side splenic to left renal vein anastomosis can be constructed, or the end of the splenic vein can be anastomosed directly to the side of the IVC.

Step 10. Prior to constructing the anastomosis, a vascular clamp is applied distally on the splenic vein and appropriate proximal and distal control of either the renal vein or the IVC is then achieved. The left renal vein should be used close to the IVC since it may have a functional narrowing as it crosses over the aorta.

Step 11. With appropriate vascular control, the splenic vein is cut flush almost with the SMV and the confluence with the portal vein, and this is oversewn with fine polypropylene.

Step 12. The splenic vein is then swung inferolaterally making sure that there are no kinks and no redundancy, and an end-to-side anastomosis is carried out with running fine polypropylene.

Step 13. Once the anastomosis is completed, the clamps are removed in sequence. There should be a thrill palpable over the splenic vein.

Step 14. The retractors are released and the pancreas is allowed to rest gently on the shunt. For optimal results, the splenic vein should have a large diameter and have excellent flow through it. In the case where the splenic vein is small, spontaneous splenorenal shunts may be present and, in this situation, the splenorenal shunt may not remain patent. Because portal pressure is essentially unchanged, there is no need to measure either portal pressure or portosystemic gradients following these selective shunts.

Step 15. Once the shunt is completed, a thorough examination for collaterals is made and every attempt is made to disconnect the collaterals between the portal system and the venous drainage system of the spleen. This disconnection will avoid collateralization between the portal venous and systemic venous systems.

Step 16. At this point, the operative field is checked for hemostasis.

Step 17. A generous wedge-shaped liver biopsy is taken from the left lateral segment using hemostatic chromic stitches.

Step 18. The abdominal cavity is irrigated copiously with saline, and the suture lines are inspected for hemostasis.

Step 19. The retractors are removed, and the position of the NG tube is checked. The abdomen is then closed in layers using monofilament closure in two layers for the fascia closing first the posterior sheath and, subsequently, the anterior rectus sheath.

Step 20. A watertight skin closure is then applied in expectation of ascites formation. No intraperitoneal drains are placed.

Postop

The patient is then sent to the intensive care unit for careful monitoring of hemodynamics and of liver function.

Complications

Recurrent bleeding, ascites, encephalopathy.

Follow-Up

Doppler ultrasound, encephalopathy.

Chapter 101

H-Interposition Mesocaval Shunt

Michael Abecassis

Indications

The indication for an H-interposition mesocaval shunt is portal hypertensive bleeding refractory to medical and endoscopic therapy.

Preop

Hemodynamic stabilization; mesenteric venous examination (Doppler ultrasound, magnetic resonance venography).

Procedure

Step 1. The patient is positioned in the supine position under general anesthesia and endotracheal intubation. A Foley catheter is inserted for monitoring urine output and a central venous catheter is placed.

Step 2. An antiseptic prep is used from the nipple line to the groin. The body is draped in sterile fashion. Appropriate antibiotics are given prior to skin incision.

Step 3. Depending on the patient's body habitus, either a long midline or a bilateral subcostal incision can be used. In patients with narrow costal margins and a long epigastrium, the midline incision is preferred.

Step 4. The superior mesenteric vein (SMV) is exposed by lifting the transverse mesocolon and identifying the root of the small bowel mesentery. The peritoneum overlying the superior mesenteric vein is incised.

Step 5. The SMV is identified. The superior border of dissection consists of the origin of the middle colic vein. The SMV is then exposed anteriorly until it bifurcates caudally into numerous mesenteric branches. There is almost invariably a branch to the right of the SMV which needs to be ligated and divided in order to afford better mobilization of the SMV. The SMV is then mobilized circumferentially for a total distance of 2-3 cm.

Step 6. Next we turn our attention to the inferior vena cava (IVC). Exposure of the IVC requires mobilization of the third and fourth portions of the duodenum in order to sweep the duodenum upwards and create a path for the graft. Both during the dissection of the SMV and the IVC, all lymphatics should be ligated carefully in order to avoid the development of chylous/lymphatic ascites postoperatively.

Step 7. Extensive Kocherization of the second and third portions of the duodenum may be necessary in order to gain full mobilization, and this may require taking down the hepatic flexure of the colon with medial mobilization of the hepatic flexure. Once the anterior aspect of the infrahepatic IVC is exposed, minimal mobilization should ensue so that a side-biting vascular clamp can be applied for an anterior anastomosis between the IVC and the graft.

Northwestern Handbook of Surgical Procedures, 2nd Edition, edited by Nathaniel J. Soper and Dixon B. Kaufman. ©2011 Landes Bioscience.

101

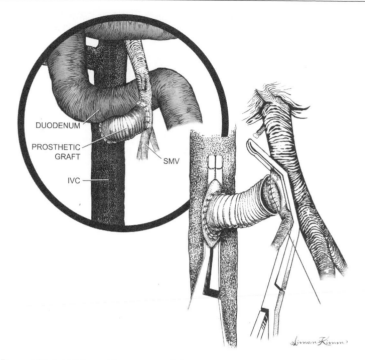

Figure 101.1. H-interposition mesocaval shunt.

Step 8. The next step is direct measurement of portal pressure. This is achieved by encircling a segment of the gastroepiploic vein easily found in the gastrocolic ligament. The proximal side is ligated and a transverse venotomy is made in the epiploic vein. A catheter is inserted in the direction of the main portal vein, and the catheter is attached to a manometer so that the portal pressure can be measured. Typically, a 5 F feeding catheter is used for this purpose, and a sterile IV tubing which is flushed with saline can be attached to the hub of the catheter and the other end passed to the anesthesiologist who can then monitor the pressure in mm Hg. This is important because a gradient is then determined between the central venous pressure and the direct portal pressure. This gradient is recorded. The catheter is secured and remains in place until completion of the shunt.

Step 9. Next, a 14 or 16 mm woven Dacron graft is used for the interposition. Alternatively, a ringed Gore-Tex® of the same size can also be used. The IVC anastomosis which will be the most dependent anastomosis is performed first. This is done using fine polypropylene running suture having removed a small ellipse of anterior IVC.

Step 10. The graft is gently curved around the fourth portion of the duodenum anteromedially and is brought to the area of the previously exposed SMV. A clamp can be placed on the graft itself, removing the side-biting clamp on the IVC in order to test that suture line.

Step 11. Next, the SMV is controlled either with a side-biting vascular clamp or, alternatively, with proximal and distal control.

Step 12. A lateral venotomy is performed on the SMV, and an end-to-side anastomosis is performed using fine running polypropylene suture. The SMV may be thin and one should avoid tension on this anastomosis in order to prevent tearing the SMV. If necessary, the venotomy can be carried down caudally into one of the mesenteric branches of the SMV. Again, the superior aspect of the venotomy is dictated by the origin of the middle colic vein.

Step 13. Once the SMV to graft anastomosis is completed, the clamps are removed. The SMV should be immediately noted to be nicely decompressed and there should be a thrill palpable over the graft. There should be a gentle curve on the graft, especially as it lies inferior and slightly posterior to the fourth portion of the duodenum.

Step 14. Once hemostasis of the suture line is secured, the portal pressure is again measured directly and a second reading of the central venous pressure is recorded so that the portosystemic gradient can be calculated. A comparison of the gradient is made to the earlier recording.

Step 15. At this point, the operative field is checked for hemostasis.

Step 16. The catheter is removed from the gastroepiploic vein, and the vein is ligated.

Step 17. A generous wedge-shaped liver biopsy is taken from the left lateral segment using hemostatic chromic stitches.

Step 18. The abdominal cavity is irrigated copiously with saline, and the suture lines are inspected for hemostasis.

Step 19. The retractors are removed, and the position of the NG tube is checked. The abdomen is then closed in layers using monofilament for the fascia.

Step 20. A watertight skin closure is then applied in expectation of ascites formation. No intraperitoneal drains are placed.

Postop

The patient is sent to the intensive care unit for careful monitoring of hemodynamics and of liver function.

Complications

Liver failure, thrombosis, encephalopathy.

Follow-Up

Doppler ultrasound, encephalopathy.

Chapter 102

Portacaval Shunts

Michael Abecassis

Indications

The indication for portacaval shunt is portal hypertensive bleeding refractory to medical and endoscopic therapy.

Preop

Hemodynamic stabilization; mesenteric venous examination (Doppler ultrasound, magnetic resonance venography).

Procedure

Step 1. The patient is positioned in the supine position under general anesthesia and endotracheal intubation. A Foley catheter is inserted for monitoring urine output, and a central venous catheter is utilized for monitoring central venous pressure.

Step 2. An antiseptic prep is used from the nipple line to the groin. The body is draped in sterile fashion. Appropriate antibiotics are given prior to skin incision.

Step 3. A right subcostal incision is made. Because of the presence of portal hypertension and frequent thrombocytopenia, liberal use of cautery is advised. The incision is extended through the abdominal muscular layers, and the peritoneal cavity is entered. Ascites may be encountered which needs to be submitted for cell count and differential analysis in order to rule out the possibility of spontaneous bacterial peritonitis.

Step 4. The ligamentum teres is divided in order to facilitate access to the right upper quadrant, as is the falciform ligament. A large patent umbilical vein is often encountered within the ligamentum teres, and suture ligature of this vein will prevent slippage of the ties and consequent bleeding. If necessary for exposure, the incision is extended across the midline into a left subcostal incision. Appropriate packs are placed above and behind the liver in order to bring the porta into view. A mechanical retractor is used in order to retract the costal margins proximally. A quick inspection and palpation of the hilum is essential in order to confirm patency of the portal vein which can easily be ballotted posteriorly.

Step 5. The next step is direct measurement of portal pressure. This is achieved by encircling a segment of the gastroepiploic vein easily found in the gastrocolic ligament. The proximal side is ligated, and a transverse venotomy is made in the epiploic vein. A catheter is inserted in the direction of the main portal vein, and the catheter is attached to a manometer so that the portal pressure can be measured. Typically, a 5 F feeding catheter is used for this purpose, and a sterile IV tubing which is flushed with saline can be attached to the hub of the catheter and the other end passed to the anesthesiologist who can then monitor the pressure in mm Hg. This is important because a gradient is then determined between the central venous pressure and the direct portal pressure. This gradient is recorded. The catheter is secured and remains in place after completion of the shunt. Therefore, it is important that the catheter not be placed past the location of portal vein transection.

Northwestern Handbook of Surgical Procedures, 2nd Edition, edited by Nathaniel J. Soper and Dixon B. Kaufman. ©2011 Landes Bioscience.

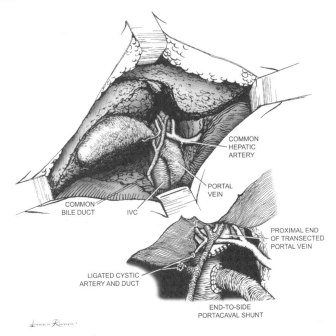

Figure 102.1. Portacaval shunt.

Step 6. The next step involves mobilizing the portal vein. It may be necessary to perform a cholecystectomy in order to achieve complete exposure of the posterior hilum. The cystic structures are transected and the gallbladder is dissected away from the fossa taking great care not to violate Glisson's capsule.

Step 7. The hepatic arterial anatomy is determined at this point, and aberrant right hepatic arterial anomalies are identified. An aberrant right hepatic artery emanating either from the superior mesenteric artery or directly from the aorta can be found posterior to the bile duct and lateral to the portal vein and may require extensive mobilization prior to mobilization of the portal vein. The portal vein can be approached anteriorly by encircling the common bile duct, placing a vessel loop around it and retracting laterally on the vessel loop while identifying the main hepatic artery, encircling the artery with a second vessel loop, and retracting medially on the artery. The plane between the hepatic artery and the bile duct will lead directly to the anterior aspect of the portal vein and is typically devoid of important vascular or biliary structures. Typically, the superior dissection of this anterior approach is limited by the right hepatic artery, and the inferior limit of the dissection is encompassed by the gastroduodenal artery.

Step 8. Division of the gastroduodenal artery may be necessary to better expose the portal vein. Great care should be taken not to devascularize the bile duct, and all lymphatic tissue in the area should be carefully ligated in order to prevent the development of postoperative chylous ascites. Once the bile duct is mobilized, a vein retractor can be very useful in retracting the bile duct anteriorly so that the portal vein can be then be approached posterolaterally.

Step 9. Using blunt dissection with a peanut, the portal vein is mobilized postero-laterally from the rest of the porta. This dissection is done under direct vision, and occasionally small branches of the portal vein may need to be divided. This dissection is carried out until the bifurcation of the left and right portal veins is visualized. It is important to appreciate that most of this dissection is performed from a posterior and lateral approach to the porta hepatis. Once the portal vein can be encircled with a vessel loop, the dissection proceeds more easily with retraction of the portal vein posterolaterally away from the hilum.

Step 10. The next step consists of identifying the anterior aspect of the infrahepatic inferior vena cava (IVC). The peritoneum between the caudate lobe and the IVC is incised in order to expose the IVC anteriorly. Caudate venous branches may need to be divided to improve exposure. The cava is dissected free of the peritoneum caudally and needs to be mobilized enough for a side-biting vascular clamp to be applied. The direction of the shunt needs to be envisioned in order to plan the end-to-side shunt so that no kinking of the portal vein occurs. This may require dissection posterior to the portal vein, typically of lymphatic tissue and, again, these should be ligated carefully in order to avoid postoperative chylous/lymphatic ascites. In contrast to a side-to-side portacaval shunt, minimal dissection of the cava should be necessary.

Step 11. At this point, a vascular clamp is applied to the portal vein proximally, and the liver side of the portal vein should be controlled. Preferably, the left and right portal veins are tied individually with heavy silk, and the portal vein is transected short of the bifurcation in order to prevent slippage of the portal venous ties. Alternatively, the liver side of the portal vein can be clamped with a crushing clamp and oversewn with fine polypropylene. Finally, vascular staples may be applied to the hepatic side of the portal vein. Regardless, the vein is transected and swung posterolaterally to the IVC.

Step 12. A side-biting vascular clamp is then applied to the IVC, and optimally an ellipse of vein is removed from the anterior aspect of the IVC, leaving a defect that matches the orifice of the portal vein.

Step 13. An end-to-side anastomosis is constructed, performing the back wall anastomosis from the inside and completing the anterior wall anastomosis. There should not be much redundancy of the portal vein in order to prevent kinking, and great care should be taken so that the portal vein is not angled acutely possibly resulting in obstruction of the shunt.

Step 14. The side-biting clamp on the cava is released first so that any bleeders in the anastomotic line can be identified, and the portal vein clamp is finally released. There should be a thrill palpable over the shunt. The portal vein should be soft and noticeably decompressed.

Step 15. Once hemostasis of the suture line is secured, the portal pressure is again measured directly and a second reading of the central venous pressure is recorded so that the portosystemic gradient can be calculated. A comparison of the gradient is made to the earlier recording.

Step 16. At this point, the operative field is checked for hemostasis. The bile duct and hepatic artery are inspected. The catheter is removed from the gastroepiploic vein and the vein is ligated.

Step 17. A generous wedge-shaped liver biopsy is taken from the left lateral segment using hemostatic chromic stitches.

Step 18. The abdominal cavity is irrigated copiously with saline, and the suture line on the liver side of the portal vein is inspected for hemostasis.

102

Step 19. The retractors are removed, and the position of the NG tube is checked. The abdomen is then closed in layers using monofilament closure in two layers for the fascia closing first the posterior sheath and, subsequently, the anterior rectus sheath.

Step 20. A watertight skin closure is then applied in expectation of ascites formation. No intraperitoneal drains are placed.

Postop

The patient is sent to the intensive care unit for careful monitoring of hemodynamics and of liver function.

Complications

Ascites, liver failure, thrombosis, encephalopathy.

Follow-Up

Doppler ultrasound, encephalopathy.

102

Chapter 103

Liver Transplantation

Jonathan Fryer

Indications

Liver transplantation is indicated in circumstances where a life-threatening pathological process that involves the liver cannot be overcome, without replacing the entire liver. Most commonly, this occurs when end-stage liver disease (ESLD) has developed in a cirrhotic liver. Other situations where liver transplantation may be needed when cirrhosis is not present include acute liver failure, nonmetastatic tumors that are otherwise unresectable, polycystic liver disease, severe hepatic trauma, or acute failure of a transplanted liver. Disease entities that commonly cause cirrhosis in adults include hepatitis B and C, alcohol abuse, nonalcoholic steatohepatitis (NASH), primary biliary cirrhosis (PBC), and primary sclerosing cholangitis (PSC). In children common causes of cirrhosis include biliary atresia and metabolic liver diseases like Wilson's or alpha 1 antitrypsin deficiency. Other conditions, like acute liver failure and inborn errors of metabolism, may also necessitate transplantation in children. Not all cirrhotic patients need liver transplants. Some cirrhotic patients who have not developed significant complications attributable to their liver disease (i.e., ESLD) may not require liver transplant. The complications of cirrhosis that typically warrant consideration of transplantation include poorly controlled portal hypertension, bleeding, ascites or peripheral edema, hepatic encephalopathy, hydrothorax, hepatorenal syndrome, hepatopulmonary syndrome, hepatocellular carcinoma, severe fatigue, and severe pruritus. If these conditions cannot be adequately controlled with more conservative therapies, liver transplantation should be considered.

Preop

Since one cannot always predict when cadaveric donor organs will become available, liver transplant candidates on the waiting list require ongoing monitoring to ensure they are kept in optimal condition for surgery at all times. They must be followed closely for infection, bleeding, malnutrition, or other problems that could compromise their transplant eligibility when a liver donor becomes available. If they have hepatocellular carcinoma, they must be frequently reevaluated to rule out tumor progression. Conversely, if a segment of liver from a living donor is going to be used for transplantation, a single preoperative evaluation of the donor and recipient is sufficient since the transplant can be scheduled electively. When a suitable cadaveric donor liver has been offered, the donor procurement team is sent to the donor hospital where they perform a final inspection of the liver to ensure that it is a suitable size match for the recipient and does not have significant traumatic injuries or fatty changes. A liver biopsy is sometimes necessary to rule out a fatty liver or other pathologic entities. Meanwhile, the potential recipient is brought in to the hospital and prepared for surgery. If final evaluations of the donor liver and the liver transplant candidate reveal no surprises, they are cross-matched for packed red blood cells, plasma, and platelets and transferred to the operating room.

Northwestern Handbook of Surgical Procedures, 2nd Edition, edited by Nathaniel J. Soper and Dixon B. Kaufman. ©2011 Landes Bioscience.

After intubation and induction of general anesthesia, several large-bore venous lines are placed to provide hemodynamic monitoring and to allow rapid infusion of blood products, medications, and fluids. Arterial lines are placed both for blood pressure monitoring and for obtaining blood specimens to monitor arterial blood gases, electrolytes, CBC, and coagulation. A Foley catheter is placed to monitor urine output. To provide much-needed gastric decompression, an orogastric or nasogastric tube is placed at the beginning of the case, with great care given not to initiate bleeding from esophagogastric varices. In addition to the chest and abdomen, the left groin and left upper arm, including axilla, are shaved and prepped to provide sterile access to the sapheno-femoral junction and axillary veins respectively, should cannulation for veno-veno bypass be necessary.

Procedure

Step 1. The patient is positioned supine with both arms partially extended (45°) to optimize access to the axillae for veno-veno bypass without risking hyperextension injuries. The legs are partially abducted and elevated to optimize exposure to the groins for veno-veno bypass and to promote venous return.

Step 2. The entire abdomen, chest, left axilla, and left groin should be surgically prepped and draped in continuity. Access to the chest may be necessary intraoperatively to optimize surgical exposure, insert chest tubes, or to administer manual cardiac compression in the event of cardiac arrest. Access to the left groin and axilla will be needed if veno-veno bypass becomes necessary. The right groin and/or axilla can also be used if necessary.

Step 3. A bilateral subcostal incision is used often with a cephalad midline extension to the xiphoid process. To provide exposure to the right lobe the right extension of the subcostal incision usually exceeds the left.

Step 4. The umbilical vein, encountered in the midline, is encircled, ligated, and divided. The falciform ligament is divided using cautery as far cephalad as exposure allows.

Step 5. Using a fixed mechanical retractor, the rib cage is retracted cephalad and lateral to optimize exposure to suprahepatic structures. Similarly the infrahepatic viscera are gently retracted caudad to optimize exposure to the porta hepatis and infrahepatic structures. Adhesions should be carefully divided prior to placing retractors to avoid tearing adjacent structures.

Step 6. With retractors in place, the falciform ligament is divided cephalad to where it splits into left and right coronary ligaments. Taking care to protect the stomach and spleen, the left triangular and coronary ligaments are divided from lateral to medial using cautery.

Step 7. Retracting the left lobe of the liver to the right, the gastrohepatic ligament (lesser omentum) is exposed. If a replaced or accessory left hepatic artery is identified, it should be ligated and divided. The gastrohepatic ligament is divided from the cut edge of the left coronary ligament to the porta hepatis using cautery, thereby exposing the caudate lobe.

Step 8. The porta hepatis is dissected in a plane close to the liver. All venous and lymphatic structures should be ligated and divided. The hepatic arteries and common hepatic duct are ligated and divided close to the liver. The cystic duct and artery are ligated and divided close to the gallbladder. The portal vein is skeletonized along an adequate length to allow placement of vascular clamps and a bypass cannula if portal decompression is necessary using veno-veno bypass.

103

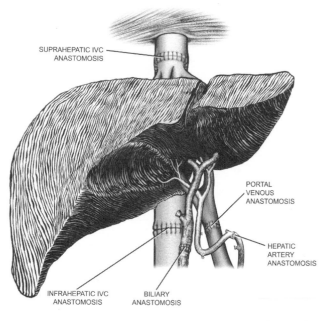

Figure 103.1. Liver transplantation. Caval anastomoses.

Step 9. The right lobe of the liver is gently retracted to the left exposing the right triangular ligament which is divided using cautery. The peritoneal attachments to the right lobe are divided as gradually progressive leftward retraction is applied until the retrohepatic vena cava is exposed.

Step 10. The left and caudate lobes are retracted to the right exposing the posterior peritoneal reflections of the caudate and retrohepatic vena cava, which are carefully divided using cautery. Care should be taken not to injure venous branches of the cava.

Step 11. There are two techniques commonly utilized for performing the caval anastomoses: the standard technique (Fig. 103.1) and the cavaplasty technique (Fig. 103.3). With both techniques the suprahepatic cava is encircled. With cavaplasty, no further dissection is performed. With the standard technique, the infrahepatic vena cava is also encircled. The right adrenal vein should be identified to avoid injury and divided if necessary. The posterior surface of the retrohepatic vena cava and the caudate lobe are freed from their retroperitoneal attachments.

Step 12. In coordination with anesthesia, the portal vein, infrahepatic cava, and suprahepatic cava are sequentially cross-clamped. If the patient tolerates this hemodynamically, the liver is excised. The portal vein is divided close to the liver. With the standard technique the entire retrohepatic cava is removed with the liver while with the cavaplasty technique the liver is separated from the cava which is kept intact. The explanted liver is sent to pathology.

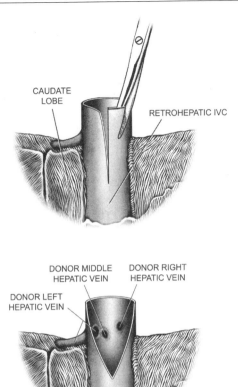

CAUDATE
LOBE

RETROHEPATIC IVC

103

DONOR MIDDLE   DONOR RIGHT
HEPATIC VEIN   HEPATIC VEIN

DONOR LEFT
HEPATIC VEIN

Figure 103.2. Liver transplantation. Caval anastomoses by the cavaplasty technique.

Step 13. The donor liver is brought up to the surgical field. The recipient is placed in reverse Trendelenburg to optimize exposure. With the standard technique, the suprahepatic cavae of the donor and recipient are anastomosed end-to-end with 4-0 polypropylene in a running fashion. With the cavaplasty technique, a posterior slit is made from the cephalad end of the donor retrohepatic cava (Fig. 103.2) and an anterior slit is made in the recipient cava extending caudad from the joined hepatic vein orifices. A triangular anastomosis is then performed between the posterior aspect of the donor cava and the anterior aspect of the recipient cava using 4-0 polypropylene in a running fashion. (Fig. 103.3).

Step 14. The patient is placed in slight Trendelenburg. To eliminate residual preservation solution, the liver graft is flushed with 1L of Ringer's Lactate which is infused via the portal vein and vented via the infrahepatic vena cava. With the cavaplasty technique, the infrahepatic cava is then stapled closed. With the standard technique, the infrahepatic cavae of the donor and recipient are anastomosed end-to-end using 4-0 polypropylene (Fig. 103.1).

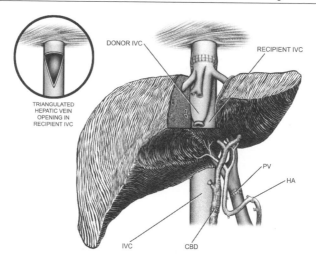

Figure 103.3. Liver transplantation. Caval anastomoses with a triangulated hepatic vein opening in the recipient IVC.

Step 15. The donor and recipient portal veins are cut to appropriate lengths to allow end-to-end anastomosis without tension or redundancy. The anastomosis is performed using 5-0 polypropylene and a running technique. A growth stitch (i.e., "air-knot") is left to allow expansion at the anastomotic site. The portal, infrahepatic, and suprahepatic caval clamps are sequentially removed, thereby reperfusing the liver.

Step 16. After surgical hemostasis has been achieved, the arterial anastomosis is performed. After being flushed with heparinized saline, small atraumatic vascular clamps are applied to the donor and recipient arteries. The sites for anastomosis with the donor and recipient arteries are selected based on their relative calibers, lengths, and quality. A tensionless anastomosis without excessive redundancy between two vessels of comparable quality and luminal diameter is the goal. A common practice is to fashion a Carrel patch with the donor artery using the origin of the celiac trunk, or "branch patches" in donor and recipient vessels using any suitable arterial bifurcation to help prevent narrowing at the anastomotic site. The anastomosis is performed using 6-0 polypropylene and a running technique. When complete the clamps are removed. If the recipient artery is thought to be unsuitable, an arterial conduit, fashioned from donor arteries, can be anastomosed to the infrarenal aorta and passed through a retrocolic or retropancreatic tunnel where it is anastomosed to the donor artery.

Step 17. The biliary anastomosis is then performed. Most commonly, this is performed end-to-end between donor and recipient ducts using an 5-0 absorbable monofilament suture and an interrupted technique. Great care must be taken to preserve the blood supply to both the donor and recipient ducts. T-tubes or stents are unneccesary. If a duct-to-duct anastomosis cannot be achieved safely due to recipient disease (i.e., PSC, biliary atresia) or donor issues (i.e., too small), a biliary enteric anastomosis is performed using a loop of jejunum fashioned as a Roux-en-Y. This anastomosis is similarly performed using interrupted 5-0 absorbable monofilament sutures.

Step 18. Prior to closure a vigorous attempt should be made to eliminate all surgical bleeding and reverse coagulopathy. When this is achieved the retractors are removed. Consideration should be given to placing Jackson-Pratt drains to monitor bleeding and detect bile leaks. Closure should be performed in two layers using 0-polypropylene sutures.

Postop

Because bleeding, sepsis, low systemic vascular resistance, and massive fluid shifts are common in the immediate posttransplant period, close monitoring in an ICU setting is essential until hemodynamic stability is achieved. However, when stable, patients should be extubated and transferred to the regular surgical floor. With most straightforward transplants, the recipients can be transferred out of ICU within 24-48 hours. As soon as feasible, patients should be mobilized and nasogastric (NG) tubes, Foley catheters, and invasive vascular lines removed and oral intake initiated. In patients with Roux-en-Y anastomoses, the NG tube should be retained until intestinal function resumes. In optimal circumstances these patients can be discharged in 2-3 days. In patients that experience complications or that were severely debilitated pretransplant, a longer hospital stay is sometimes necessary.

Complications

Potential postoperative complications include bleeding, infection, bile duct problems, hepatic arterial thrombosis, and fluid retention.

Follow-Up

In the first several postoperative days, close monitoring of liver enzymes (ALT, AST, alkaline phosphatase) and parameters of liver function (INR, bilirubin, acidosis) is necessary to evaluate for evidence of hepatic artery thrombosis (HAT) or primary nonfunction (PNF) of the liver graft. If hepatic artery patency is in question, liver ultrasound with Doppler evaluation of the hepatic vasculature and/or hepatic angiography are required. If HAT or PNF are diagnosed within the first 7 days posttransplant, a recipient is eligible for relisting for another liver transplant as a status 1. Once stability is achieved, liver transplant recipients can be discharged home as soon as they are mobile and capable of taking care of their own bodily needs. In the first several months posttransplant, recipients need to be closely monitored for rejection, infectious complications, biliary complications, recurrent disease, and complications related to immunosuppressive drugs (nephrotoxicity, neurotoxicity, neutropenia, hyperglycemia, gastrointestinal toxicity) or to generalized immunosuppression such as posttransplant lymphoproliferative disorders. In the long-term, the most significant concerns are recurrent disease (hepatitis C, hepatitis B, alcohol abuse), chronic rejection, and the consequences of long-term immunosuppressive therapy such as renal failure, hypertension, diabetes, and bone disease. Hepatitis C is the most common etiology leading to liver transplantation and these patients are especially difficult to follow posttransplant. Hepatitis C recurs in essentially all liver transplant recipients. While hepatitis C recurrence is usually indolent and slowly progressive, approximately 10% of hepatitis C recipients experience an aggressive recurrence rapidly progressing to end-stage liver disease within months. Further, hepatitis C is difficult to differentiate histologically from other hepatic inflammatory processes, including acute rejection.

103

Chapter 104

Pancreas Transplantation

Dixon B. Kaufman

Indications

Pancreas transplantation is indicated for Type I diabetes complicated by difficulty with conventional exogenous insulin therapy. Often patients with chronic or end-stage renal failure with diabetes are candidates for combined pancreas and kidney transplantation. Contraindications include: ABO incompatibility without preparatory desensitization, active or chronic infection, active or recent (<3 years) malignancy, life expectancy <1 year, sensitization to donor HLA antigens without preparatory desensitization, serious (untreatable) preexisting comorbidities, medical noncompliance, active substance abuse, uncontrolled psychiatric disorders.

Preop

General orotracheal anesthesia, central venous catheter, Foley bladder catheter, and arterial line (optional), perioperative antibiotics, confirmation of ABO compatibility, DVT prophylaxis.

Procedure

Step 1. Prone position. Midline abdominal incision starting approximately 6-8 cm below xiphoid and extending to the pubis.

Step 2. Mobilize the ascending colon medially by taking down the avascular plane at the white line of Toldt to the hepatic flexure to expose the common iliac vein and artery. Alternatively, open the retroperitoneal tissue medial to the ascending colon to expose the common iliac artery, distal vena cava, and common iliac vein.

Step 3. Partially mobilize and expose the distal vena cava and common iliac vein by suture ligation and division of overlying lymphatics.

Step 4. If the patient is preuremic, administer parenteral heparin. Uremic patients do not require heparin unless a hypercoagulable state is known. Place vascular clamps on the proximal common iliac vein.

Step 5. Anastomose the pancreas allograft portal vein (with or without short extension graft from donor external iliac vein) using 5-0 monofilament suture.

Step 6. Place vascular clamps (e.g., Fogarty clamps with inserts) on the common iliac artery (or external iliac artery) distally relative to the position of venous anastomosis.

Step 7. Anastomose the common iliac artery portion of the pancreas allograft extension Y-graft that was previously anastomosed to the donor splenic and superior mesenteric arteries (from donor internal/external/common iliac artery complex) to the recipient common iliac artery slightly distal to the position of the venous anastomosis using 6-0 monofilament suture.

Step 8. Slowly remove venous clamps and then release the arterial clamps to allow full perfusion of the pancreas allograft.

Northwestern Handbook of Surgical Procedures, 2nd Edition, edited by Nathaniel J. Soper and Dixon B. Kaufman. ©2011 Landes Bioscience.

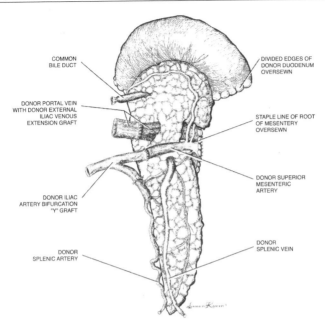

COMMON
BILE DUCT

DIVIDED EDGES OF
DONOR DUODENUM
OVERSEWN

DONOR PORTAL VEIN
WITH DONOR EXTERNAL
ILIAC VENOUS
EXTENSION GRAFT

STAPLE LINE OF ROOT
OF MESENTERY
OVERSEWN

DONOR SUPERIOR
MESENTERIC
ARTERY

DONOR ILIAC
ARTERY BIFURCATION
"Y" GRAFT

DONOR
SPLENIC VEIN

DONOR
SPLENIC ARTERY

104

Figure 104.1. Pancreas transplantation. Pancreas allograft.

Step 9. Control bleeding points with suture ligation or electrocautery.

Step 10. Perform donor-to-recipient duodenoenterostomy to midjejunum approximately 20-30 cm distal to ligament of Treitz.

Step 11. Handsewn two-layer anastomosis performed with bowel closed by first placing back wall of 3-0 silk sutures as interrupted Lembert stitches approximating the antimesenteric borders of the donor duodenum to the recipient jejunum. Length of anastomosis approximately 3-4 inches.

Step 12. Open donor duodenum, aspirating contents, and open adjacent recipient jejunum.

Step 13. Place inner layer mucosa-to-mucosa anastomotic suture line using 3-0 absorbable monofilament suture as a running hemostatic stitch.

Step 14. Place outer second anastomotic layer of interrupted Lembert stitches using 3-0 silk suture.

Step 15. Position the pancreas allograft with tail tucked into the retrocystic pouch of Douglas and the pancreaticoduodenal head cephalad.

Step 16. Irrigate abdomen with 2 L of antibiotic solution and close midline fascia in a single layer with running monofilament suture.

Postop

Continue immunosuppressive medical therapy according to institutional protocols. Administer mild anticoagulation and/or antiplatelet agents to minimize graft thrombosis risk. Administer parenteral antibiotics for 3-7 days to reduce infectious risk. Follow hemoglobin levels for bleeding. Follow serial blood glucose measurements to assess pancreatic function. Antibacterial and antiviral prophylaxis are typically prescribed.

104

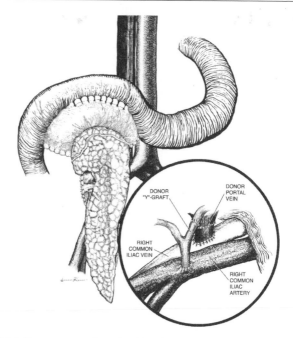

Figure 104.2. Pancreas transplantation. Vascular anastomoses.

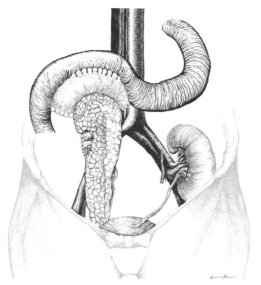

Figure 104.3. Pancreas transplantation. Complete with tail tucked into the retrocystic pouch of Douglas and the pancreaticoduodenal head cephalad.

Complications

Rejection, vascular thrombosis, transplant pancreatitis, enteric leak and intra-abdominal abscess, GI bleeding from duodenoenteric anastomosis.

Follow-Up

Immunosuppression to prevent rejection. Monitoring of chemistries to assess pancreas transplant function and to ensure adequate exposure to immunosuppression.

104

SECTION 7: VASCULAR SURGERY
Section Editor: Mark K. Eskandari

Chapter 105

Carotid Endarterectomy

Mark D. Morasch

Indications

Carotid endarterectomy is indicated for prevention of stroke in patients with symptomatic carotid bifurcation stenosis (greater than 50%) and asymptomatic carotid bifurcation stenosis (greater than 60%) contingent upon the patient's estimated perioperative stroke and death rate.

Preop

General oral endotracheal anesthesia or local regional anesthesia. Continuous radial arterial blood pressure monitoring.

Procedure

Step 1. Position the patient supine with the torso and head slightly elevated and turned to expose the neck on the surgical side.

Step 2. Longitudinal incision along the sternocleidomastoid muscle in the upper neck or transverse incision in a skin crease with the creation of subplatysmal flaps.

Step 3. Mobilize the sternocleidomastoid muscle posteriorly to expose the jugular vein, taking care not to injure the greater auricular nerve.

Step 4. Lateral mobilization of the jugular vein and jugulodigastric lymph node group. Ligate and divide all venous branches from the anterior aspect of the jugular vein. This includes ligation and division of the facial vein in most cases. The anterior aspect of the jugular vein should be completely mobilized from the level of the strap muscles to a site above the angle of the mandible.

Step 5. Gentle circumferential dissection of the common carotid artery at the base of the neck. Care must be taken to avoid vigorous manipulation of the common carotid artery. Similarly, the vagus nerve should be retracted laterally with minimal dissection or manipulation.

Step 6. Circumferential dissection of both the distal internal carotid artery and the external carotid artery, again, taking care to avoid excessive manipulation of the vessels. The carotid bulb should be left undissected until after clamping. Both the vagus and hypoglossal nerves require gentle retraction away from the operative field.

Step 7. Administration of systemic heparin for anticoagulation through an intravenous catheter; measurement of carotid back stump pressures (optional); and clamping of internal, external, and common carotid arteries, in that order.

Step 8. Mobilization of the area of the carotid bifurcation (after clamps are applied) to allow rotation of the carotid bifurcation. This allows for creation of a longitudinal arteriotomy along the lateral aspect of the common and internal carotid arteries through the area of bifurcation stenosis. Eversion endarterectomy can be performed as an alternative technique. Eversion endarterectomy is completed after transversely amputating the internal carotid artery just distal to the bifurcation.

Northwestern Handbook of Surgical Procedures, 2nd Edition, edited by Nathaniel J. Soper and Dixon B. Kaufman. ©2011 Landes Bioscience.

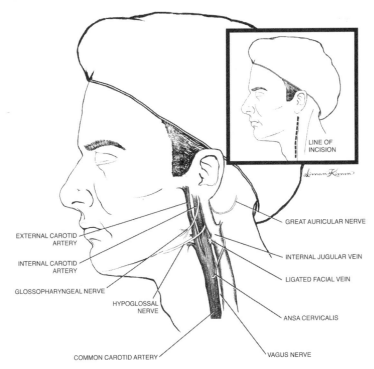

Figure 105.1. Carotid endarterectomy.

Step 9. The endarterectomy is started by identifying the appropriate plane between the inner one-third and the outer two-thirds of the media using an endarterectomy spatula. The plaque specimen is then transected in the common carotid artery proximal to the bulk of the disease. Eversion technique is utilized to clear the external carotid artery of plaque. Great care must be taken to feather the distal end of the plaque as it ends in the internal carotid artery. Gentle downward traction on the plaque by the operator or by the assistant will facilitate the creation of a smooth endpoint.

Step 10. Irrigate the remaining vessel wall to identify and remove remaining small fronds of media and to create a smooth new vessel lumen.

Step 11. Close the longitudinal arteriotomy with a single small monofilament suture. Alternatively, a patch closure can be utilized to further widen the bulb if necessary. Polytetrafluoroethylene (PTFE), Dacron, or thigh saphenous vein are the usual choices for patch closure.

Step 12. After backbleeding and forebleeding all three vessels and vigorously flushing the site with heparinized saline, the arteriotomy can be closed completely and flow reestablished. The clamp on the internal carotid should be released last.

Step 13. Completion imaging of the endarterectomy site utilizing angiography, intraoperative duplex ultrasound, or angioscopy.

Step 14. Hemostasis and wound closure by reapproximating the platysma and then the skin with a plastics-type closure.

Step 15. A carotid shunt can be placed routinely or selectively in patients considered high risk for perioperative stroke.

Step 16. A small, closed-suction drain can be left in place at the discretion of the surgeon.

Postop

The patient should be monitored in the recovery room for at least 2 hours and, if neurologically and hemodynamically stable, may be transferred postoperatively to a surgical ward. If complications arise or if the patient requires intravenous infusions to control hypertension, hypotension, or arrhythmias, the patient should be transferred from the recovery room to an intensive care unit for further monitoring.

Complications

Permanent or transient central neurologic deficits (stroke or TIA); bleeding complications requiring neck reexploration and hematoma evacuation; peripheral nerve injuries including injuries to the hypoglossal, vagus, recurrent laryngeal, marginal mandibular, posterior auricular, and, more rarely, the glossopharyngeal and spinal accessory nerves.

Follow-Up

Patients are usually discharged on the day following the operation and should be seen back in the outpatient setting in 3 weeks to assure normal wound healing. The patient should be monitored with duplex ultrasound after 6 months and yearly thereafter.

Chapter 106

Repair Infrarenal Aortic Aneurysm: Elective

Mark K. Eskandari

Indications

Elective repair of an infrarenal aortic aneurysm is indicated in fit patients for an abdominal aortic aneurysm that is 5.5 cm or larger, enlarging >5 mm/6 months, symptomatic, ruptured, or infected (mycotic).

Operative Principles

Secure vascular control and visualization of the aorta and iliac arteries usually requires generous exposure and lighting. Specific attention is given to avoiding inadvertent venous injuries. Gentle handling and retraction of the bowel and keeping the small intestine inside the abdomen can minimize postoperative ileus and shorten recovery. Both mental preplanning and manual dexterity are contributory elements for rapid vascular reconstruction which minimizes cross-clamp time, bleeding, and occlusion risk. Meticulous review of the preoperative imaging assists in formulation of this individualized cognitive plan. Image review includes an assessment of mesenteric and renal artery patency, iliac artery occlusive disease, and position of the left renal vein which is typically anterior to the aorta but on occasion can run posterior and inferior to the aorta.

Preop

The patient is given a mild, nondehydrating bowel laxative the day before. Two units of packed red blood cells are crossmatched. One gram of cefazolin is commonly given intravenously. A cell saver collection is attached to the suction but is only processed if there is a need for transfusion. The abdomen and groins are shaved immediately prior to the procedure. Active warming is used after the patient gowns, through the operative period, and until normal body temperature is achieved after operation.

Procedure

Step 1. Supine position, bladder catheter, prep from nipples to the knees, midline incision from the xiphoid to lower abdomen.

Step 2. Avoid bowel injury when entering the peritoneum; anterior adhesions are taken down carefully. Any enterotomy would necessitate aborting the aortic reconstruction.

Step 3. Retract the transverse colon cephalad, small bowel to the right, and sigmoid colon to the left using padded self-retaining retractor system (i.e., Omni retractor).

Step 4. Proximal aortic neck exposure: mobilize the duodenum to the right side of the abdomen, identify and loop the left renal vein, and finger dissect tissue lateral to infrarenal neck. Test vascular clamp application.

Northwestern Handbook of Surgical Procedures, 2nd Edition, edited by Nathaniel J. Soper and Dixon B. Kaufman. ©2011 Landes Bioscience.

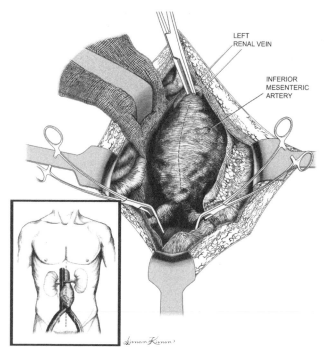

LEFT
RENAL VEIN

INFERIOR
MESENTERIC
ARTERY

106

Figure 106.1. Repair of infrarenal aortic aneurysm. Elective.

Step 5. Distal exposure of iliac arteries: preoperative imaging should predict the extent of dissection but reassess after direct observation and palpation of the arteries, as a different graft configuration may be indicated. If necessary, expose femoral arteries for anastomosis. When these dissections are mandatory, they can be performed prior to opening the abdomen to decrease the laparotomy time. Create retroureteral/retrocolic tunnels by tunneling just anterior to the external and common iliac arteries.

Step 6. Systemic heparin is given prior to applying iliac clamps and aortic clamp. The distal clamps should be placed first to minimize the risk of distal embolization.

Step 7. Open aneurysm sac and oversew lumbar arteries with 2-0 nonabsorbable suture. The inferior mesenteric artery may be controlled prior to opening the sac or searched for on the underside of the left edge of the sac.

Step 8. Select a graft, trim the aortic portion if a bifurcated prosthesis, and suture to the aorta with 3-0 polypropylene. Liberal use of felt strips or pledgets can help support a tenuous anastomosis. After moving the clamp onto the graft, carefully inspect for hemostasis (particularly posterior half) prior to working on distal anastomosis. Flush the graft antegrade and retrograde just prior to completing the distal anastomosis.

Step 9. Verify hemostasis of all suture lines and recheck lumbar vessels (after distal flow is restored, back pressure may increase.) Observe backbleeding through inferior mesenteric artery and reimplant into the prosthetic graft if bleeding is sluggish or both hypogastric arteries are occluded. Palpate for femoral pulses.

Step 10. Close the sac and approximate retroperitoneal tissues with running 2-0 absorbable suture so that the aortic graft is covered and not in contact with bowel.

Step 11. Close the abdomen with running #2 nylon with the knots buried. Avoid electrocautery near this suture line.

Step 12. Check distal pulses/Doppler indices prior to moving patient off the operating room table.

Postop

Rapid transport to intensive/postanesthetic care unit is necessary to minimize the gap in monitoring, resuscitation, and rewarming. Monitor organ function, particularly myocardial and renal function. Correct significant coagulation and electrolyte abnormalities. Maintain treatment of chronic medical comorbidities. Prescribe appropriate pain control. Check a chest radiograph if central line attempted.

Complications

Clinical outcome after technically successful aortic operations is dependent on anticipation, attempted prevention, early identification, and appropriate management of the many potential complications. Patients with aortic aneurysm often have serious pathology of critical organs that remains uncorrected after the aortic operation. Bleeding, renal failure, myocardial infarction, and limb ischemia are the main concerns in the first 24 postoperative hours. In the first few days, infections (pulmonary, urinary, and line), colon ischemia, fluid overload, and electrolyte/acid-base imbalance are additional concerns. After several days, wound complications, gastrointestinal hemorrhage, nutritional depletion, deconditioning, mental status changes, venous thromboembolism, and pressure ulcers come into play. While many of these complications become clinically manifest later, earlier attention can arrest the problems at a subclinical level. Long-term complications include ventral hernia, prosthetic graft infection, aortoenteric fistula, graft limb occlusion, and subsequent development of aneurysm.

106

Follow-Up

Patients are seen frequently postoperatively to monitor for the above complications. After recovery, they are seen annually for clinical exam and to reinforce lifestyle modification. A CT scan is done every 5 years to detect subsequent aneurysms or anastomotic pseudoaneurysms.

Acknowledgment

The editors and author wish to acknowledge Jon S. Matsumura for contributing to the previous version of this chapter.

Chapter 107

Repair Infrarenal Aortic Aneurysm: Emergent for Rupture

Mark K. Eskandari

Indications

The indication for emergent repair of infrarenal aortic aneurysm is rupture.

Operative Principles

Rapid vascular control of the aorta requires timely diagnosis and decisive tranportation to the operating room. Delayed resuscitation until bleeding is controlled decreases dilution-related coagulopathy. Therefore, after the diagnosis is made either clinically or on cross-sectional imaging, expedient transportation to the operating room is mandatory. Prelaparotomy transbrachial or transfemoral aortic occlusion balloon under local anesthetic may be feasible in some centers with hybrid operating room capabilities.

Preop

The patient is not sedated or paralyzed until just prior to laparotomy to avoid complete vascular collapse. Alternatively, percutaneous access for placement of an aortic balloon occlusion can be achieved under local anesthesia. Type and cross 6 units packed red blood cells and order cell saver. One gram of cefazolin may be given intravenously. Shave abdomen and groins.

Procedure

Step 1. Supine position, bladder catheter, prep from nipples to the knees, midline incision, avoid bowel injury.

Step 2. Supraceliac aortic control: The left lobe of the liver is retracted to the patient's right and the stomach to the left. Incise the gastrohepatic ligament to enter the lesser sac. Use the nasogastric (NG) tube to palpate the esophagus and move it to the left to expose the right crus of the diaphragm. Either use the aortic compressor to pinch aorta against the spine or divide the crus, finger dissect the lateral sides of the aorta and then place the vascular clamp over the index and middle fingers onto the aorta.

Step 3. Retract the transverse colon cranially, small bowel to the right and sigmoid colon to the left using self-retaining retractor system or two assistants.

Step 4. The proximal neck is often dissected by hematoma, otherwise mobilize the duodenum, identify and loop the left renal vein, and finger dissect tissue lateral to the infrarenal neck. Once a vascular clamp is applied here, remove the supraceliac clamp. A rapid alternative to supraceliac control is to reach into the retroperitoneal hematoma and feel for the aneurysm. Move the hand cephalad until the proximal neck is palpated and pinch the neck between the thumb and fingers until the area is exposed and clamped.

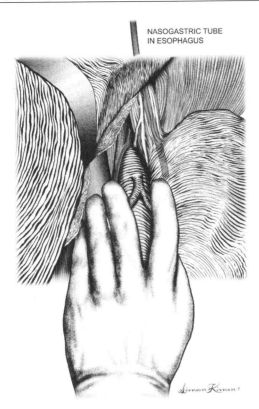

NASOGASTRIC TUBE
IN ESOPHAGUS

107

Figure 107.1. Repair of infrarenal aortic aneurysm. Emergent. Use the NG tube to palpate the esophagus and move it to the left to expose the right crus of the diaphragm.

Step 5. Distal control of the iliac arteries is obtained. Small (<2 cm) iliac aneurysms may not need to be addressed in emergent operation for aortic aneurysm rupture, thus avoiding the extra time for a bifurcated graft or femoral dissections. If necessary, expose the femoral arteries—beware of unsuspected femoral aneurysms—and create retroureteral/retrocolic tunnels.

Step 6. Heparin may be given if the patient has minimal bleeding and blood loss has been limited, but in some cases the procedure is done without systemic anticoagulation to avoid the risk of DIC from hemorrhagic shock.

Step 7. Open the sac and oversew the lumbar arteries with 2-0 nonabsorbable suture. The inferior mesenteric artery may backbleed from the underside of the left edge of the sac and needs to be looked for specifically.

Step 8. Select a graft, trim the aortic portion if a bifurcated prosthesis, and suture to the aorta with 3-0 polypropylene. Flush the anastomosis of air and debris, move clamp onto graft, and check hemostasis (particularly posterior half) prior to working on the distal end. Flush the graft antegrade and retrograde just prior to completing the distal anastomosis.

Step 9. Verify hemostasis of all suture lines and recheck the lumbar vessels. After distal flow is restored, back pressure may increase. Observe backbleeding through the inferior mesenteric artery and reimplant into the prosthetic graft if bleeding is sluggish or both hypogastric arteries are occluded. Palpate for femoral pulses.

Step 10. Close the sac and approximate retroperitoneal tissues with running 2-0 absorbable suture so that the aortic graft is covered and not in contact with the bowel.

Step 11. Close the abdomen with running #2 nylon. Consider retention sutures if fascia is attenuated or leave the abdomen open if abdominal compartment syndrome is likely.

Step 12. Check distal pulses/Doppler indices prior to moving the patient off the operating room table.

Postop

Rapid transport to the intensive/postanesthetic care unit is necessary to minimize the gap in monitoring, resuscitation, and rewarming. Third space losses, ongoing bleeding, and reperfusion syndromes generate significant intravascular fluid deficits. Monitor organ function, particularly myocardial and renal function. Correct coagulation and electrolyte abnormalities. Maintain treatment of chronic medical comorbidities. Often complete ventilator support and continuous sedative drips are necessary in the early postoperative period. Check all tube placements and consider relocating sites, as initial insertion is often done in extreme circumstances.

Complications

Clinical outcome after technically successful aortic operations is dependent on anticipation, attempted prevention, early identification, and appropriate management of the many potential complications. Patients with aortic aneurysm often have serious pathology of critical organs that remains uncorrected after the aortic operation. Bleeding, renal failure, myocardial infarction, and limb ischemia are the main concerns in the first 24 postoperative hours. In the first few days, pneumonia, urinary tract infection, catheter sepsis, colon ischemia, fluid overload, and electrolyte/acid-base imbalance are frequent additional concerns. After several days, wound complications, gastrointestinal hemorrhage, nutritional depletion, deconditioning, mental status changes, venous thromboembolism, and pressure ulcers come into play. While many of these complications become clinically manifest later, earlier attention can arrest the problems at a subclinical level. Long-term complications include ventral hernia, retained aneurysm growth, prosthetic graft infection, aorto-enteric fistula, graft limb occlusions, and subsequent development of an aneurysm.

Follow-Up

Patients are seen frequently postoperatively to monitor for the above complications. After recovery, they are seen annually for clinical exam and to reinforce lifestyle modification. A CT scan is done after full recovery and every 5 years to detect retained or subsequent aneurysms or anastomotic pseudoaneurysms.

Acknowledgment

The editors and author wish to acknowledge Jon S. Matsumura for contributing to the previous version of this chapter.

Chapter 108

Endovascular Repair of Infrarenal Aortic Aneurysm

Heron E. Rodriguez

Indications

Endovascular repair of infrarenal aortic aneurysm (EVAR) is indicated in patients with abdominal aortic aneurysms of 5.5 cm or larger and with suitable anatomy for the available endovascular devices. Occasionally, EVAR is indicated in aneurysms smaller than 5.5 cm when there is documented enlargement of 5 mm/6 months, presence of symptoms (pain, embolism) or if associated with other risk factors for rupture. Major contraindications for open aneurysm repair (other major comorbidities limiting the patient's life expectancy) are almost always also a contraindication for EVAR.

Operative Principles

EVAR requires extensive preoperative imaging and judgement to assess for anatomic qualifications, anticipate potential intraoperative problems, and select appropriate endovascular devices and sizes. Additional counseling and long-term radiographic surveillance are necessary to insure long-term clinical success. Fluoroscopic guidance is used during the procedure and may be supplemented by intravascular ultrasound. There are several device-specific directions for use and potential pitfalls that are learned through extensive training and experience, which are not covered in this brief description.

108

Preop

EVAR is usually done under general anesthesia but can also be performed under regional and local anesthesia. Two units of packed red blood cells are cross-matched. One gram of cefazolin is given intravenously. The abdomen and groins are shaved.

Procedure

Step 1. Supine position, bladder catheter, check fluoroscopy field is not obstructed, and prep from nipples to the knees.

Step 2. Percutaneous access is commonly used in selected institutions. Alternatively, the femoral artery is exposed through a transverse incision above the inguinal skin crease. The external iliac artery is double looped and the common femoral artery and branches controlled. A separate stab wound for sheath placement may be useful to minimize sheath kinking.

Step 3. Retrograde cannulation of the aorta is performed with an atraumatic guidewire on each side. A short angled guide catheter may be useful if the iliac arteries are tortuous or diseased and to exchange for a stiff guidewire.

Northwestern Handbook of Surgical Procedures, 2nd Edition, edited by Nathaniel J. Soper and Dixon B. Kaufman. ©2011 Landes Bioscience.

108

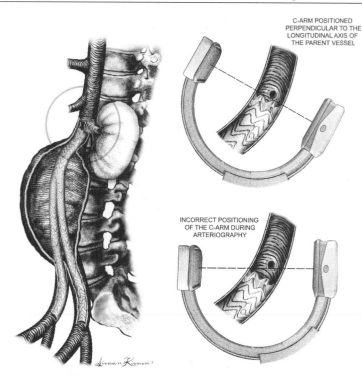

C-ARM POSITIONED
PERPENDICULAR TO THE
LONGITUDINAL AXIS OF
THE PARENT VESSEL

INCORRECT POSITIONING
OF THE C-ARM DURING
ARTERIOGRAPHY

Figure 108.1. Endovascular repair of infrarenal aoric aneurysm. Anatomy.

Step 4. Arteriography is performed to identify and assess the position of the branch arteries in relation to the aneurysm and to confirm the intraluminal distances obtained during preoperative planning. Particularly important are the distances between the lowest renal artery and the beginning of the aneurysm and the most distal portion of the aneurysm and the internal iliac arteries. A complete reassessment of the anatomic suitability for endovascular repair is performed. The procedure may be continued, aborted, or converted to open repair as necessary. Confirmatory arteriography—with the C-arm positioned perpendicular to the longitudinal axis of the parent vessel and the origin of the important branch vessel—may be performed just prior to each stage of deployment.

Step 5. Systemic heparin (5000 units) may be given and hemostatic sheaths of appropriate size are inserted under fluoroscopic guidance. Iliac injury should be suspected if there is difficulty inserting the larger sheaths, and occasionally direct exposure is necessary for repair and access.

Step 6. Main trunk device deployment. The device is inserted and rotated for optimal orientation—usually in a position that facilitates contralateral gate junction cannulation. The proximal end is deployed immediately after radiographic location of the important branch vessels. The contralateral limb and the ipsilateral limb are similarly placed. Care is taken to select balloon size, position and inflate properly. The deployment system is removed.

Step 7. Extenders are used as needed to establish longer overlap and seal zones.

Step 8. After completion arteriography shows suitable endograft deployment and configuration, the sheaths are removed. A retrograde sheath injection may be performed during removal to rule out iliac injury.

Step 9. Verify hemostasis of all suture lines or closure devices. Palpate for femoral pulses.

Step 10. Close incisions with absorbable suture in layers and skin with the knots buried.

Step 11. Check distal pulses/Doppler indices prior to moving patient off the operating room table.

Postop

Monitor organ function, particularly renal function and lower extremity perfusion. Maintain treatment of chronic medical comorbidities. Prescribe appropriate pain control.

Complications

Bleeding, renal failure, cardiac complications and distal embolization are the main concerns in the first 24 postoperative hours.

Follow-Up

108

Clinical outcomes after technically successful endovascular repair are dependent on long-term clinical and radiographic follow-up with selective reintervention. Specific protocols for imaging follow-up exist and should be carefully observed. Long-term abnormalities and complications include endoleak (continued blood flow in the sac), aneurysm growth and rupture, device migration, device material failure, prosthetic graft infection, graft limb occlusion, and subsequent development of a new aneurysm.

Acknowledgment

The editors and author wish to acknowledge Jon S. Matsumura for contributing to the previous version of this chapter.

Chapter 109

Aortofemoral Bypass for Obstructive Disease

Mark K. Eskandari

Indications

Indications include extensive bilateral iliac disease and the presence of buttock or thigh claudication.

Preop

The patient is assessed using a standard angiogram or a high-quality magnetic resonance angiogram to delineate the level of the infrarenal aorta and rule out a potential aneurysm and also to delineate the distal targets, typically the common femoral artery at its bifurcation. The patient is treated preoperatively with intravenous antibiotics.

Procedure

Step 1. The patient is placed in a supine position.

Step 2. The first two incisions are made obliquely in each of the groins, isolating the common femoral artery, superficial femoral artery, and profunda femoris artery bilaterally.

Step 3. Through a midline laparotomy incision, the infrarenal aorta is approached and carefully dissected at the level just below the renal arteries.

Step 4. The aorta-bifemoral bypass graft may be sewn in either an end-to-end or end-to-side fashion. End-to-end is used in cases where the extent of external iliac disease is minimal, allowing retrograde flow into the pelvis. An end-to-side anastomosis is utilized in cases where there is an extremely small aorta, typically in middle-aged women, or in cases where there is extensive external iliac disease, not allowing adequate retrograde perfusion to the pelvis. In both situations, the anastomosis is performed below the renal arteries.

Step 5. The tunnels are performed for the limbs of the graft from the groin up into the pelvis, along the native vessels underneath the ureter.

Step 6. The patient receives heparin systemically, typically 5000 units.

Step 7. The aorta is controlled proximally and distally. Typically complete aorta cross-clamp is necessary because a side clamp is very difficult to place due to the small size of the aorta. A longitudinal arteriotomy is fashioned and an end-to-side anastomosis with a bifurcated graft is performed with a 2-0 polypropylene suture. Alternatively, if an end-to-end anastomosis is performed, it is done in a standard fashion and the distal stump is oversewn with a running mattress and running over-and-over suture of 2-0 polypropylene.

Step 8. The limbs of the graft are tunneled down to the level of the groins. After proximal and distal control of the femoral arteries is attained, the anastomosis is sewn to the bifurcation, not above it.

Step 9. The clamps are removed and flow is restored. Hemostasis is obtained.

Northwestern Handbook of Surgical Procedures, 2nd Edition, edited by Nathaniel J. Soper and Dixon B. Kaufman. ©2011 Landes Bioscience.

The retroperitoneum is closed in layers, the abdominal cavity is closed, and the groins are closed in the standard fashion.

Postop

Postoperative considerations include monitoring of cardiac function, fluid requirements, and assessment of renal functions.

Complications

Complications include bleeding from the anastomosis or from the retroperitoneum requiring reoperation and pelvic ischemia from inadequate collateral flow resulting in colonic ischemia. Other potential long-term complications are an aortic-duodenal fistula as well as limb occlusion of one of the limbs of the prosthetic graft.

Follow-Up

Patients are usually discharged 1 week following the operation and should be seen back in the outpatient setting in 3 weeks to assure normal wound healing. The patient should be monitored with duplex ultrasound after 6 months and yearly thereafter.

Chapter 110

Axillofemoral Bypass

Melina R. Kibbe

Indications

Axillofemoral bypasses are performed for several different indications; however, the predominate indication remains bilateral severe aortoiliac occlusive disease. While an axillofemoral bypass is not the first-line therapy offered to most patients, it is considered ideal for patients with bilateral aortoiliac inflow disease and severe co-morbidities that make open abdominal aortic surgery prohibitive with respect to perioperative risk. Axillofemoral bypass is a good option for patients who have undergone extensive prior intra-abdominal aortic surgery where redo open aortic surgery would pose significant anatomic risk. Axillofemoral bypass is best for patients who present with acute ischemia of both lower extremities when open aortic surgery is less ideal due to time or anatomic reason, as the dissection required to perform an axillofemoral bypass is well-tolerated and can be performed quickly. Another indication to perform an axillofemoral bypass is for patients who present with infected aortic or iliac prosthetic devices. An axillofemoral bypass can be performed in an extra-anatomic manner (i.e., to the lateral profunda femoral arteries or the popliteal arteries) followed by removal of the infected prosthetic material in a staged manner.

Preop

Preoperatively, it is imperative that the patient be optimized with respect to cardiopulmonary risk factors and be well beta-blocked, unless contraindicated. This is not always possible for urgent and emergent cases; however, every precaution should be taken when operating on a patient emergently. The patient should also be receiving aspirin since prosthetic material will be used and has a higher thrombotic potential. Before offering this procedure, the systolic blood pressures (SBP) must be obtained in each arm to ensure that there is no disease on the side chosen for inflow. If a difference in SBP is detected between the arms, the arm with the higher SBP should be used for the axillary anastomosis. If time permits, we prefer to obtain a physiologic study in the vascular lab in which segmental SPB and Doppler waveforms of the upper extremities are obtained. In addition, imaging modality can be helpful to assess the proximal subclavian and axillary artery for disease. This includes a magnetic resonance angiogram (MRA), computed tomography angiogram (CTA), or conventional contrast angiography. The last is rarely obtained at our institution given the invasive nature of the procedure. If proximal inflow disease is suspected based on SBP or Doppler waveforms, further imaging studies are mandatory to assess inflow. With respect to outflow, imaging studies such as MRA, CTA, or contrast angiography will identify suitable outflow vessels and are of particular importance when performing axillofemoral bypass grafts for infected prosthetic material where extra-anatomic approaches will be utilized for the profunda femoral artery or popliteal artery. Lastly,

Northwestern Handbook of Surgical Procedures, 2nd Edition, edited by Nathaniel J. Soper and Dixon B. Kaufman. ©2011 Landes Bioscience.

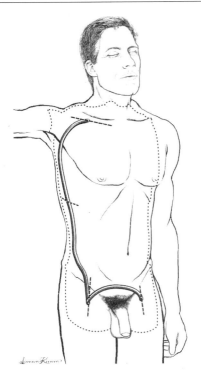

Figure 110.1. Axillofemoral bypass with femorofemoral anastomosis.

if the indication for intervention is an infected prosthesis, systemic control of the infection is optimal with the use of broad spectrum antibiotics unless speciation and sensitivities have been obtained. If a local source of infection exists, such as a septic foot, this should be addressed prior to implanting the prosthetic material.

Procedure

Step 1. *Positioning.* The patient should be positioned on the operating room table with the arm abducted to 90° on the side of the axillary inflow. The contralateral arm should be tucked for easy access for the assistant surgeon to all of the operative fields. Any arterial lines, SBP cuffs, or intravenous catheters should be placed in the contralateral, nonused, arm. A Foley catheter is advisable. Anesthesia is typically general anesthesia. If severe cardiopulmonary risk factors exist, regional anesthesia can be used; however, this is extremely rare.

Step 2. *Prep and Draping.* A wide sterile field is required for this operation given the breadth of the body the graft spans. Landmarks included in the sterile operative field include the sternal notch superiorly, the mid thighs inferiorly, the posterior axillary line on the side of the axillary inflow, and the anterior axillary line contralaterally (Fig. 110.1). Since prosthetic material will be placed, we routinely use a 3M Ioban drape so that no skin will come in contact with the prosthetic material. If two teams will be working simultaneously, it is beneficial to include the use of two Bovie cauteries.

Step 3. *Incisions.* An 8 cm long transverse incision for the axillary anastomosis is made approximately 2 fingerbreadths below the clavical in the midclavicular line. The femoral incisions are made directly overlying the femoral artery. Since there will most likely be no pulse, a vertical incision is made approximately 2 fingerbreadths lateral to the pubic tubercle. It should extend from approximately 1 cm above the inguinal ligament (which spans from the pubic tubercle to the anterior superior iliac spine) to 6-7 cm below the inguinal ligament. This length allows for exposure of the common femoral artery at the inguinal ligament, the superior most aspect of the deep dissection, as well as the profunda femoral and superficial femoral arteries distally.

Step 4. *Axillary Dissection.* For the axillary dissection, after the incision is made, the dissection is carried through the subcutaneous tissue and fascia to the level of the pectoralis major muscle. The muscle fibers of the pectoralis major are split bluntly in the direction of the incision and a self-retaining retractor, such as a Weitlaner retractor, is used to maintain this exposure. The clavipectoral fascia is then sharply incised to expose the underlying axillary sheath which encases the corresponding neurovascular bundle. Laterally, the pectoralis minor is identified and retracted laterally to provide additional exposure. Once the axillary sheath is opened, the axillary vein is the first structure identified. The axillary artery lies superior and deep to the vein. By retracting the vein caudally, the artery can be easily identified. During the course of this dissection, several venous tributaries will require ligation and division to aid with the exposure of the axillary artery. Care should be taken to avoid injury to the lateral pectoral nerve which crosses the artery and vein anteriorly. Once the axillary artery is identified, it should be mobilized as far proximally and distally as possible within the confines of the incision and controlled with vessel loops.

Step 5. *Femoral Dissection.* The common femoral, superficial femoral, and profunda femoral arteries are dissected and controlled with vessel loops in standard fashion. If the indication for the axillofemoral bypass is for an infected aortic or iliac prosthesis, then the approach to the femoral arteries may be through a lateral incision in order to stay in an extra-anatomic plane. If the distal profunda femoral artery is the target for outflow, the incision is typically lateral to the border of the Sartorius muscle. Through this incision, the distal profunda femoral artery can be dissected and controlled with vessel loops.

Step 6. *Graft Selection.* Studies have shown that an 8 mm diameter graft is the smallest diameter that should be utilized for an axillary inflow procedure. Graft diameters of smaller caliber become a rate-limiting factor with respect to flow. No consensus exists regarding use of Dacron versus ePTFE, supported versus unsupported grafts, standard wall ePTFE versus thin wall ePTFE grafts, or preformed versus nonpreformed grafts. Our practice is to use 8 mm ePTFE ringed grafts of either standard wall or thin wall. This author prefers to use a thin walled graft as they theoretically should develop less neointimal hyperplasia. Furthermore, this author prefers to use preformed axillofemoral grafts, as it minimizes the need to create one more anastomosis. Some surgeons prefer to perform their own graft-to-graft anastomosis to theoretically decrease the risk of thrombosis of the graft material distal to the graft-to-graft anastomosis. Traditional teachings have taught this; however, no studies have supported this theory. For those that support this, a "C" configuration can be created with the graft-to-graft anastomosis being performed in a retrograde fashion (Fig. 110.2). Or, to preserve antegrade flow to the contralateral limb, a "lazy S" shape can be fashioned.

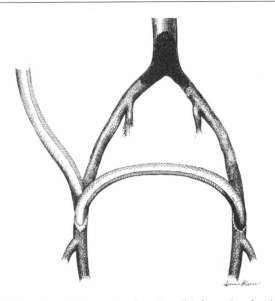

Figure 110.2. View of the inferior portion of a right axillobifemoral graft with the aorta, iliac, and femoral arteries behind it. This is the most common axillobifemoral bypass configuration with the axillofemoral limb to proximal (in this case right) femorofemoral limb anastomosis placed as distally as possible to maintain maximum flow through the entire axillofemoral limb.

110

Step 7. *Tunneling.* Several important concepts exist with respect to tunneling the graft. First, it is imperative to have sufficient redundancy, or slack, of the graft allowing for the patient to abduct his/her arm, but not so much redundancy as to cause graft kinking. The graft should also be tunneled behind the pectoralis major muscle, leaving sufficient slack for arm abduction. It is also recommended to have the graft travel laterally in parallel with the axillary artery for as much distance as possible before taking a 90° turn inferiorly down the chest wall. This will minimize the chance for anastomotic disruption should there not be enough slack in the graft during arm abduction. Second, it is imperative to tunnel the graft in the midaxillary line in order to minimize the effect of torso flexion, both anterior-posteriorly and medial-laterally, on the graft. If tunneled in the midaxillary line, the effect of such flexion is minimized. Often, it is helpful to create a counter incision for tunneling one-half to two-thirds of the way down the chest. If using the long 65 cm Gore® Tunneler, a counter incision may not be necessary, as this tunneling device will easily reach from the groin incision to the axillary incision, the preferred direction of passing the Tunneler to avoid injury to the brachial plexus. If using a preformed graft, a counter incision is necessary at the location of the bifurcation and the Tunneler must be carefully passed the opposite direction. Tunneling of the short limbs can be performed with a large aortic clamp.

Step 8. *Anticoagulation.* After the graft has been tunneled, the patient should be systemically anticoagulated with heparin. We reserve anticoagulation until after tunneling to minimize hematoma formation.

Step 9. *Axillary anastomosis.* In order to minimize the effect of arm abduction on the anastomosis, it is crucial to create the anastomosis on the axillary artery as medial as possible, where the anatomy of the artery is most stationary. By pulling up on the vessel loops, the artery is brought into the field, anterior to the vein and far from the brachial plexus. When placing the vascular clamps, care should be taken to avoid injury to the brachial plexus, which lies deep to the axillary artery. Therefore, blind placement of the vascular clamps should be avoided. The anastomosis is created in standard spatulated end-to-side fashion to the axillary artery with a running suture line, noting that the axillary artery is a much softer artery to sew to. Too much tension on the suture line can cause the sutures to pull through the artery. The angle of the graft to artery should also be kept to a minimum, preferably less than 75°. After creation of the anastomosis and restoration of flow to the arm, the graft is clamped just distal to the anastomosis with noncrushing clamps, such as Fogarty hydrogrips. The graft is clamped as close to the anastomosis as possible to avoid having a stagnant column of blood in the graft while the distal anastomoses are created.

Step 10. *Femoral anastomoses.* The femoral anastomoses are created in standard fashion. If the surgeon prefers to create her own graft-to-graft anastomosis, this anastomosis is typically created as close to the femoral anastomosis as possible. Some surgeons prefer to perform a femorofemoral bypass first, and then anastomose the axillofemoral anastomosis onto the femorofemoral graft. This "inverted C" loop configuration is thought to minimize the chance of thrombosis of the graft distal to the graft-to-graft anastomosis due to low flow (Fig. 110.2); however, no studies have definitively shown this. Prior to completion of the femoral anastomoses, it is imperative to flush not only the native femoral arteries, but also the axillary portion of the newly created graft.

Step 11. *Hemostasis and closure.* Once the graft is completed, hemostasis is achieved in standard fashion. The femoral incisions are closed in standard fashion, taking care to close at least three layers of tissue above the prosthetic graft material. The axillary incision is closed in several layers, taking care not to close the fascia in the event a hematoma develops. Drain placement is at the discretion of the surgeon; however, we try to avoid placement of drains except in instances of reoperative incisions where lymph leakage might be higher.

Postop

Postoperatively, the patient should be monitored in the recovery room for at least 2 hours. If hemodynamically and neurologically stable, the patient may be transferred to a surgical ward. The SBP should be measured immediately postoperatively to ensure no steal. A chest X-ray should be obtained in the recovery room to ensure there was no injury to the phrenic nerve (i.e., elevated hemidiaphragm) and no entry into the chest cavity (i.e., pneumothorax). The patients may be given a clear liquid diet that evening and a regular diet on postoperative day (POD) 1. Intravenous fluids can be discontinued on POD 1 if the patient tolerates a diet. The patient may resume regular activity on POD 1. Perioperative antibiotics are given once preoperatively and once postoperatively, unless there is evidence of an active infection for which prolonged antibiotics will be administered. If a drain was placed, the drain output should be monitored. The drain may be removed when the output is less than 30 ml per 24 hours. Unless contraindicated, the patient should receive aspirin and warfarin for the life of the graft. We start the warfarin on POD 1. The patient should also be instructed not to sleep in a lateral decubitus position on the side of the graft. As with any patient who has graft material placed, they should be instructed to obtain prophylactic antibiotics for any subsequent medical or dental procedure associated with transient bacteremia (i.e., dental work, colonoscopy, etc.).

Complications

Any of the operative incisions are at risk for hematoma formation. If a hematoma develops in the axillary wound, close monitoring for nerve compression is mandatory. We have a low threshold for re-exploration for axillary wound hematomas. Additional complications can include brachial plexus nerve injury, venous injuries, anastomotic disruption of the axillary anastomosis, graft thrombosis, distal emboli, wound infection, and graft infection. While this procedure is typically well-tolerated, patients undergoing this operation typically have significant cardiopulmonary risk factors. Thus, close monitoring in the postoperative period for myocardial infarction, congestive heart failure, and pneumonia is prudent.

Follow-Up

Postoperatively, the ankle-brachial index and brachial SBP are monitored and serve as a baseline for future follow-up appointments. After discharge, the patient is evaluated at 2 weeks, 6 months, then yearly thereafter. An arterial noninvasive physiologic study with segmental SBP and Doppler waveforms may assist with surveillance; however, no studies have demonstrated the utility of these studies to predict impending failure of the graft.

Acknowledgment

The editors and author wish to acknowledge Joseph R. Schneider for contributing to the previous version of this chapter.

110

Chapter 111

Femorofemoral Bypass

Melina R. Kibbe

Indications

Femorofemoral bypass is most commonly indicated for severe unilateral iliac disease that is not amenable to percutaneous options. This operation is also ideal for completion of unilateral aortoiliac endografts, for salvage of a thrombosed limb of an aortobifemoral bypass graft, or for other unilateral iliac pathology such as isolated iliac artery aneurysms or localized infections of one limb of an aortobifemoral bypass graft.

Preop

Assessment of iliac artery inflow is mandatory before performing a femorofemoral bypass. Assessment should consist of a pulse examination combined with an arterial noninvasive study with Doppler waveform analysis. If the pulse is normal and the Doppler waveform reveals a normal triphasic signal, adequate inflow can be assumed. If the Doppler waveform is not normal, further assessment with imaging modalities is necessary. These imaging modalities can consist of a magnetic resonance angiogram (MRA), computed tomography angiogram (CTA), or conventional digital subtraction contrast angiogram. The imaging modality will be best determined based on the quality of the image that can be provided by the equipment at the individual hospital, and taking into account that contrast angiography is an invasive procedure. However, a benefit of performing contrast angiography is that if inflow disease is detected, it can be treated percutaneously with balloon angioplasty and stenting in the same setting. Assessment of outflow is also necessary before performing a femorofemoral bypass in order to determine where the recipient anastomosis will be placed and to determine if additional outflow procedures, such as an endarterectomy, are necessary. Again, this assessment can be performed with any of the imaging modalities described above. Ankle-brachial indexes (ABI) are performed preoperatively as a baseline for comparison after the procedure. The patient should receive aspirin preoperatively to minimize the risk of thrombosis of the prosthetic material.

Procedure

Step 1. *Positioning and Anesthesia.* The patient should be positioned supine on the operating room table with both arms tucked, or one arm to the side to provide access for anesthesia. Anesthesia can be performed using general, regional, or local anesthesia. Local anesthesia is reserved for patients with severe cardiopulmonary risk factors. The patient should receive preoperative antibiotics. If an infectious source exists, this should be addressed prior to implantation of a prosthetic material. If the bypass is being performed due to an infectious process, this should be controlled with broad spectrum antibiotics until speciation and sensitivities are obtained. The patient should have an arterial catheter, intravenous access, and a Foley catheter in place.

Northwestern Handbook of Surgical Procedures, 2nd Edition, edited by Nathaniel J. Soper and Dixon B. Kaufman. ©2011 Landes Bioscience.

Step 2. *Prep and Draping.* If performing a femorofemoral bypass and no additional outflow procedure, the abdomen, both groins, and proximal half of the thighs should be prepped and draped in the sterile field. If an outflow procedure is to be performed in addition to the femorofemoral bypass, that entire leg should be prepped and draped into the field.

Step 3. *Incisions.* Routine vertical groin incisions should be created. Typically this is approximately 2 fingerbreadths lateral to the pubic tubercle. The incision should extend from approximately 1 cm above the inguinal ligament (which spans from the pubic tubercle to the anterior superior iliac spine) to 6-7 cm below the inguinal ligament. This length allows for exposure of the common femoral artery at the inguinal ligament, the superior most aspect of the deep dissection, as well as the profunda femoral and superficial femoral arteries distally.

Step 4. *Graft Selection.* For femorofemoral bypasses, an 8 mm graft is utilized in order to minimize effects on flow due to inadequate conduit diameter. No consensus exists as to the type of graft material (ePTFE versus Dacron), or if using ePTFE, standard versus thin wall. However, most surgeons do agree that if using ePTFE, supported external rings provide a theoretical advantage to resist kinking. This author prefers thin wall, externally supported, 8 mm ePTFE grafts. Thin walled ePTFE has the theoretical advantage of developing less neointimal hyperplasia.

Step 5. *Tunneling.* After the common femoral, profunda femoral, and superficial femoral arteries are dissected and controlled with vessel loops, the subcutaneous suprapubic tunnel is created bluntly. This is best started by creating the tunnel in the superior-medial aspect of each incision with hemostats, deep enough in the subcutaneous tissue so that closure of sufficient tissue above the graft is possible. The tunnel is then brought more superficial as it crosses superior to the pubis using blunt finger dissection. Once fingers from each side meet in the middle and the tunnel is created, umbilical tape can be passed through the tunnel using an aortic clamp. Alternatively, a Goretex Tunneler can be used. However, this author prefers the blunt dissection with fingers to avoid injury to the bladder or rectum.

Step 6. *Anticoagulation.* After the tunnel is created, the patient is systemically anticoagulated with heparin. The graft is passed through the tunnel.

Step 7. *Anastomoses.* The graft is anastomosed to the common femoral artery if outflow is normal. If the superficial femoral artery is occluded, the anastomosis is typically carried down to the proximal portion of the profunda femoral artery. If disease is noted in the profunda femoral artery, an endarterectomy may be necessary. The lay of the graft is typically in an inverted C position (Fig. 111.1). However, some surgeons prefer to create the inflow anastomosis using a lazy S shape so that flow in the graft is antegrade, not retrograde. No data has shown superiority of one method over the other. The actual anastomosis is performed in a standard spatulated end-to-side manner with running suture, typically 5-0 polypropylene suture. Prior to completion of the anastomosis, all native vessels are bled and flushed with heparinized saline. After completion of the first anastomosis, the graft is clamped with a noncrushing graft clamp, such as the Fogarty hydrogrip clamp, as close to the anastomosis as possible to avoid creation of a stagnant column of blood in the proximal graft. Flow is restored to the leg. The second anastomosis is created. Prior to completion of the second anastomosis, it is imperative to bleed and flush the native arterial system as well as the graft to avoid embolization of clot. After completion, flow is restored to the leg via the newly created femorofemoral bypass graft. Intraoperative Doppler can be used to confirm enhancement of signal in the recipient leg with flow through the graft as well as no change in the Doppler signal on the donor limb with flow through the graft.

111

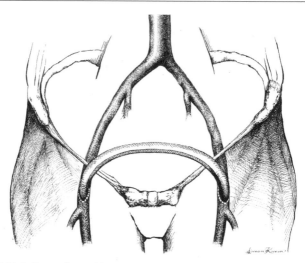

Figure 111.1. Femorofemoral bypass.

Step 8. *Hemostasis and Closure.* After hemostasis is achieved, the groin wounds are closed in at least three layers using either interrupted or running absorbable suture. We use 2-0 Vicryl closure for closure, using running suture lines for virgin groins and interrupted suture for reoperative groins. Drains are only placed if a lymphatic leak is suspected or the incision is a reoperative wound. Skin is closed with staples.

Postop

ABI are measured in both limbs postoperatively to insure improvement with the bypass graft and no steal in the donor limb. These ABI will also serve as a baseline for subsequent follow-up. The night of the procedure the patient may have a liquid diet. Postoperative day (POD) 1 the patient may have a regular diet. If the patient is able to tolerate liquids, the intravenous fluids are discontinued. Normal activity may resume on POD1. If drains were placed, output should be monitored every 8 hours. The drains may be removed once the output is less than 30 ml per 24 hours. In addition to the preoperative antibiotic dose, one postoperative antibiotic dose should be administered. The patient should receive aspirin for the life of the graft. As with any patient that has graft material placed, they should be instructed to obtain prophylactic antibiotics for any subsequent procedure associated with transient bacteremia (i.e., dental work, colonoscopy, etc.).

Complications

Femorofemoral bypass grafts are generally well-tolerated. However, often these procedures are performed on patients with significant cardiopulmonary risk factors. As such, monitoring in the postoperative period for myocardial infarctions, congestive heart failure, and pneumonia are imperative as death due to cardiac morbidity is the most common cause of death following these procedures. Other complications can include hematoma formation, graft thrombosis, distal emboli, wound infection, graft infection, femoral neuropathy, and arterial steal from the donor limb. Any infectious process detected in the postoperative period should be treated promptly with antibiotics and long-term administration should be considered.

Follow-Up

Postoperatively, the ABI are monitored and serve as a baseline for future follow-up appointments. After discharge, the patient is evaluated at 2 weeks, 6 months, then yearly thereafter. An arterial noninvasive physiologic study with segmental SBP and Doppler waveforms should be obtained prior to discharge and at each follow-up appointment. However unlike femoral-distal bypass grafts, no studies have demonstrated the utility of surveillance studies to predict impending femorofemoral bypass graft failure.

Acknowledgment

The editors and author wish to acknowledge Joseph R. Schneider for contributing to the previous version of this chapter.

111

Chapter 112

Femoral-Popliteal Bypass with a Vein or Prosthetic Graft

Heron E. Rodriguez

Indications

The indications for intervention on patients suffering from lower extremity occlusive disease are three: life-style limiting claudication not responsive to medical management, ischemic rest pain and tissue loss due to ischemia (nonhealing ulceration or gangrene). Current invasive therapies for disease localized to the superficial femoral artery include endovascular intervention (percutaneous transluminal angioplasty and stenting, remote endarterectomy) and femoro-popliteal bypass. The choice between these modalities is a matter of heated controversy. In general, endovascular approaches are best suited for short segment occlusion or stenoses, for patients at high risk for complications following general anesthesia and those with inadequate saphenous vein. On the other hand, a bypass is favored in patients with long segment occlusion and those with previous failed endovascular treatments. The patency of femoro-popliteal bypasses with autogenous conduit (i.e. saphenous vein) is better than when a prosthetic graft is used.

Preop

Accurate delineation of the arterial anatomy of the lower extremities is mandatory. At present, MRA and CTA have virtually replaced conventional angiography. Duplex venous mapping is used to determine whether or not the great saphenous vein is an adequate conduit. Preoperative cardiac risk stratification and baseline ABI's are mandatory. Prophylactic antibiotics are ordered and blood is typed and crossed.

Procedure

Step 1. *Positioning.* The procedure is usually performed under general anesthesia, occasionally under epidural anesthesia and extremely rarely under local anesthesia. The patient is placed in the supine position with the lower abdomen, both groins and the entire affected leg prepped. A bump under the leg is helpful to expose the popliteal vessels. The bump is placed below the knee for explorations of the above the knee popliteal and under the thigh when exploring the popliteal artery below the knee.

Step 2. *Exposure.* The femoral artery is exposed using a longuitudinal incision in the groin. The inguinal ligament is identified first and the artery is dissected distal to this landmark. The popliteal artery is exposed above or below the knee depending on the distal target identified on the angiogram. A medial incision is done and deepened carefully to prevent injury to the greater saphenous vein. The fascia is incised to open the popliteal space. The neurovascular bundle is identified, the artery is separated from the vein and nerve, and a 3 cm segment of vessel beyond the occlusive disease is isolated and surrounded by vessel loops proximally and distally.

Step 3. *Conduit harvesting.* If the great saphenous vein is used as a conduit, it is harvested. This can be done in several ways. A single, uninterrupted incision can be made to expose the vein. Alternatively, "skip" separate incisions with the creation of small subcutaneous tunnels can be used. Also, a minimally invasive technique with the use of specially designed, endoscopic instruments can be chosen and avoids the creation of large wounds. Once the vein is exposed by any of the described methods, all venous tributaries are double ligated and transected, the vein is transected proximally at the sapheno-femoral junction and distally at the lowest end of the incision. The stump of the sapheno-femoral junction and the distal vein are oversewn.

Step 4. *Conduit configuration.* Once the saphenous vein is harvested, two decisions are made: first, whether or not to "reverse" the vein and second, whether to leave the vein "in situ" or to reposition it in an anatomic fashion. If the caudal end of the harvested vein is of large enough caliber, the vein can be "reversed" using the caudal end to create the proximal anastomosis. By doing this, the direction of flow will be the same as it was originally in the saphenous circulation and its valves will not interfere with the flow in the bypass. When the caudal end of the vein is inadequate for creation of the proximal anastomosis, the vein is not reversed and the cephalad end is used to create the proximal anastomosis. This requires the use of a valvulotome to lyse the valves, since now the flow will be in the opposite direction of the original saphenous circulation and any functional valve would compromise it. In an "in situ" bypass, the saphenous vein is left on its bed, whereas in an "anatomic" bypass, the vein is removed and "tunneled" underneath the sartorious.

Step 5. *Tunnel creation.* If a prosthetic graft is used or if an "in situ" technique is not used, a subsartorial tunnel adjacent to the course of the superficial femoral artery is created through which the graft is passed. This is usually performed with a tunneling instrument which has a removable tip and a hollow core through which the reversed saphenous vein or prosthetic graft is passed without twisting or kinking.

112

Step 6. *Proximal anastomosis.* After completing the tunnel, intravenous heparin is given. Clamps are applied proximal and distal to the site of anastomosis and the anterior wall of the femoral artery is opened with an 11 blade. The arteriotomy is extended with an angled Potts arteriotomy scissors. The proximal end of the vein or prosthetic graft is beveled and sewn end-to-side to the common femoral artery with a 5-0 polypropylene suture and under loupe magnification. Once the anastomosis is completed, a vascular clamp is placed across the origin of the graft, and the clamps on the femoral artery are removed.

Step 7. *Distal anastomosis.* The popliteal artery is then clamped proximally and distally to isolate a 3 cm segment. The medial surface is opened with a scalpel, and a Potts scissors is used to extend the arteriotomy. Extreme care is taken to avoid an intimal dissection at the distal end of the arteriotomy. The distal end of the graft is shortened and beveled to the appropriate length and shape and sewn end-to-side to the artery with a 6-0 polypropylene suture. Just prior to the completion of the anastomosis, the graft and the artery are flushed of air and debris. The anastomosis is completed and secured. The clamps are removed from the origin of the graft and the proximal artery. After several seconds, the distal clamp is removed to initiate flow to the distal leg.

Step 8. *Completion arteriography.* The distal anastomosis and outflow vessels are evaluated using digital substraction techniques after injecting contrast through a 20-gauge angiocatheter inserted directly into the proximal portion of the graft. Any radiographic abnormality at the distal anastomosis requires correction. If a nonreversed configuration was used, the entire graft should be imaged to confirm that all of the valves have been completely lysed. Alternatively, intraoperative duplex US can be used instead of angiography. The presence of a palpable pedal pulse is generally expected after most femoro-popliteal bypasses.

Step 9. *Closure.* Hemostasis is verified, and the subcutaneous tissues are reapproximated with a synthetic absorbable suture. The skin is closed with either staples or a synthetic absorbable suture.

Postop

Patients require intensive or intermediate monitoring during the first 24 hours after surgery. Clearly determine and document pulse examination, Doppler signals, ABIs and neurologic function. Any change requires immediate attention. Pay extreme attention to local wound care. The use of postoperative antibiotics and anticoagulation is determined on an individual basis. Perioperative beta-blockade has been proven to decrease coronary events.

Complications

The most common complications after leg bypass surgery are cardiopulmonary in nature. Wound infection, postoperative bleeding and thrombosis are complications specific to this procedure.

Follow-Up

Pulse examination, Doppler signals and/or ABIs are performed in the recovery room, every shift for the first day and every day after. The wounds are examined in the surgeon's office a week or two after surgery. The use of duplex ultrasound to evaluate bypass grafts is an effective strategy to prolong patency and is routinely used in most vascular centers. The first one of these "surveillance" studies is done within the first postoperative month.

Acknowledgment

The editors and author wish to acknowledge John V. White and Christopher Bulger for contributing to the previous version of this chapter.

112

Chapter 113

Composite Sequential Bypass

William H. Pearce

Indications

A composite sequential bypass is indicated in patients with limb-threatening ischemia in whom there is not a sufficient length of vein to span the femoral tibial segment. Limb-threatening ischemia includes patients with nonhealing ulcers, ischemic rest pain, and gangrene. For infrainguinal bypass, an all-autogenous reconstruction (the greater saphenous vein or splicing veins from upper and lower extremities) functions better than prosthetic grafts. When these options are not available, the composite sequential offers a greater long-term patency than an all-prosthetic bypass.

Operative Principles

This operation is a combination of a prosthetic femoral-popliteal bypass and a distal venous graft to a tibial vessel. This hybrid graft consists of both a prosthetic and a venous segment. The venous segment is most commonly derived from the lesser saphenous vein of the ipsilateral leg. The operative exposures are those that are described previously for femoral popliteal bypasses and tibial bypasses. The operative procedure begins with the harvesting of the lesser saphenous vein. This is done with the patient supine and the leg elevated. This is the most difficult part of the procedure. Once the lesser saphenous vein is harvested, the femoral, popliteal, and distal tibial arteries are exposed. The intermediate anastomosis, that is the anastomosis between the vein and the prosthetic, is unique to this procedure.

This anastomosis is created with an end vein to side of prosthetic graft. This large anastomosis is perhaps the reason for success of this procedure. The large diameter of this anastomosis allows for intimal hyperplasia to occur without impeding the flow of the graft.

113

Preop

As with all patients with lower extremity arterial disease, a preoperative blood flow study and an arteriogram are required. Additionally, these patients also require complete vein mapping for all usable vein. Usable vein is marked preoperatively.

Procedure

Step 1. The patient is placed in the supine position on the operating table and, after general anesthesia, the leg is prepped and draped in a standard sterile fashion.

Step 2. The leg is elevated. The operation begins with harvesting the lesser saphenous vein at the previously marked site; the posterior skin overlying the lesser saphenous is opened. The vein is harvested from ankle to popliteal fossa. The wound is left open. If the vein is less than 3-4 mm in diameter it is not usable. Alternative sources of vein are then necessary (upper extremity).

Northwestern Handbook of Surgical Procedures, 2nd Edition, edited by Nathaniel J. Soper and Dixon B. Kaufman. ©2011 Landes Bioscience.

Step 3. Exposure of the femoral artery, popliteal artery either above or below the knee, and the distal tibial vessel is accomplished using the techniques previously described.

Step 4. Next, the femoral popliteal bypass using prosthetic graft is completed initially. This procedure uses an 8 mm polytetrafluoroethylene (PTFE) graft following systemic heparinization with 5000 U of heparin. The anastomosis is created end-to-side in both locations.

Step 5. The venous segment can be implanted either in an orthograde or retrograde direction. The orthograde technique requires a proximal anastomosis to be completed first, with stripping of the valves using a Mills valvulotome. Using the reverse technique, this step is omitted and a proximal anastomosis is completed.

Step 6. The distal anastomosis is created to the tibial vessel again, using similar techniques as described above. Upon completion, the entire system is flushed of debris.

Step 7. An intraoperative completion arteriogram is performed to demonstrate technical adequacy of all anastomoses.

Step 8. Following completion of the arteriogram, hemostasis is achieved and the wounds are closed in multiple layers. It is important not to forget to close the posterior lesser saphenous vein harvest site.

Step 9. Occasionally, these patients require postoperative heparinization and as a result a #10 Jackson Pratt drain is left in calf and mid-thigh incisions.

Postop

These patients are treated like those with any distal bypass. They are observed in the intensive care unit overnight for bleeding or early graft occlusion. Postoperative Doppler tones or ankle-brachial indices are performed every 2-4 hours. In general, most patients are discharged on aspirin. However, based on surgeon's preference, the patients may be discharged on coumadin or other anticoagulants or antiplatelet agents.

Complications

Early graft failure may occur in approximately 1-10% of patients. In addition, postoperative hematomas are common, particularly in patients with perioperative anticoagulation. Postoperative graft infections are uncommon. Skin and wound necrosis is, however, a common problem occurring in 2-20% of patients. The posterior incision for the lesser saphenous vein is rarely a significant problem. Similar to all lower extremity revascularization, the preoperative mortality is 2-6%.

Follow-Up

Graft surveillance should be performed at 3 month intervals for the first year and at 6-month intervals thereafter.

Chapter 114

Infrapopliteal Bypass: Vein or Prosthetic

Heron E. Rodriguez

Indications

The indication to perform a distal (tibial) bypass is critical limb ischemia. This includes patients with ischemic rest pain and tissue loss due to ischemia (nonhealing ulceration or gangrene). Advances in technology have made endovascular interventions (percutaneous transluminal angioplasty and stenting, remote endarterectomy) a feasible option in patients in whom performing a bypass is not optimal (poor risk for general anesthesia, inadequate spahenous vein).

Preop

Preoperative cardiac risk stratification and baseline ABI's are mandatory. Three variables need to be determined prior to the operation: inflow vessel, conduit and outflow target. To determine the inflow and outflow vessels, accurate delineation of the arterial anatomy of the lower extremities is mandatory. At present, MRA and CTA have virtually replaced conventional angiography. Duplex venous mapping is used to determine whether or not the great saphenous vein is an adequate conduit. If the ipsilateral great saphenous is inadequate, the contralateral great saphenous vein can be used. Alternatives include the small saphenous vein, cephalic vein, basilic vein, or the use of prosthetic graft (polytetrafluoroethylene [PTFE], Dacron). Prophylactic antibiotics are ordered and blood is typed and crossed.

Procedure

Step 1. *Positioning.* The procedure is usually performed under general anesthesia, occasionally under epidural anesthesia and extremely rarely under local anesthesia. The patient is placed in the supine position with the lower abdomen, both groins and the entire affected leg prepped.

Step 2. *Exposure.* Almost always, the femoral artery is the inflow vessel but ocasioanlly a bypass can be based off the profunda femoris, distal SFA and popliteal arteries. If the common femoral is chosen, the inguinal ligament is identified first and the artery is dissected distal to this landmark. The target vessel is then exposed. The anterior tibialis is usually exposed via a lateral incision in the lower leg, whereas the posterior tibialis and the proximal and mid peroneal are exposed through medial leg incisions. The distal peroneal artery can be exposed by removing the distal third of the fibula. Some bypasses are created using the pedal arteries as distal target (dorsalis pedis, posterior tibialis, tarsal or pedal branches).

114

Step 3. *Conduit harvesting.* If the great saphenous vein is used as a conduit, harvesting is done. This can be done in several ways. A single, uninterrupted incision can be made to expose the vein. Alternatively, "skip" separate incisions with the creation of small subcutaneous tunnels can be used. Also, a minimally invasive technique with the use of specially designed, endoscopic instruments can be chosen and avoids the creation of large wounds. Once the vein is exposed by any of the described methods, all venous tributaries are double ligated and transected, the vein is transected proximally at the sapheno-femoral junction and distally at the lowest end of the incision. The stump of the sapheno-femoral junction and the distal vein are oversewn.

Step 4. *Conduit configuration.* Once the saphenous vein is harvested, two decisions are made: first, whether or not to "reverse" the vein and second, whether to leave the vein "in situ" or to reposition it in an anatomic fashion. If the caudal end of the harvested vein is of large enough caliber, the vein can be "reversed" using this caudal end to create the proximal anastomosis. By doing this, the direction of flow will be the same as it was originally in the saphenous circulation and its valves will not interfere with the flow in the bypass. When the caudal end of the vein is inadequate for creation of the proximal anastomosis, the vein is not reversed and the cephalad end is used to create the proximal anastomosis. This requires the use of a valvulotome to lyse the valves, since now the flow will be in the opposite direction of the original saphenous circulation and any functional valve would compromise it. In an "in situ" bypass, the saphenous vein is left on its bed, whereas in an "anatomic" bypass, the vein is removed and "tunneled" underneath the sartorious.

Step 5. *Tunnel creation.* If a prosthetic graft is used or if an "in situ" technique is not used, a subsartorial tunnel adjacent to the course of the superficial femoral artery is created through which the graft is passed. This is usually performed with a tunneling instrument which has a removable tip and a hollow core through which the reversed saphenous vein or prosthetic graft is passed without twisting or kinking. If the graft needs to reach the lateral lower leg, a tunnel is created through the interosseus membrane.

Step 6. *Proximal anastomosis.* After completeting the tunnel, intravenous heparin is given. Clamps are applied proximal and distal to the site of anastomosis, and the anterior wall of the femoral artery is opened with an 11 blade. The arteriotomy is extended with an angled Potts arteriotomy scissors. The proximal end of the vein or prosthetic graft is beveled and sewn end-to-side to the common femoral artery with a 5-0 polypropylene suture and under loupe magnification. Once the anastomosis is completed, a vascular clamp is placed across the origin of the graft and the clamps on the femoral artery are removed.

Step 7. *Distal anastomosis.* The target vessel is clamped proximally and distally to isolate a 3 cm segment. Sometimes, when extreme calcification is present the use of a sterile tourniquete or the application of intraluminal occlusion devices is needed. An arteriotomy is performed with a scalpel, and a Potts scissors is used to extend the arteriotomy. Extreme care is taken to avoid an intimal dissection at the distal end of the arteriotomy. The distal end of the graft is shortened and beveled to the appropriate length and shape and sewn end-to-side to the artery with a 6-0 polypropylene suture. Just prior to the completion of the anastomosis, the graft and the artery are flushed of air and debris. The anastomosis is completed and secured. The clamps are removed from the origin of the graft and the proximal artery. After several seconds, the distal clamp is removed to initiate flow to the distal leg. If a prosthetic graft is used, several techniques to create "venous patches" at the site of the distal anastomosis have been described.

Step 8. *Completion arteriography.* The distal anastomosis and outflow vessels are evaluated using digital substraction techniques after injecting contrast through a 20-gauge angiocatheter inserted directly into the proximal portion of the graft. Any radiographic abnormality at the distal anastomosis requires correction. If a nonreversed configuration was used, the entire graft should be imaged to confirm that all of the valves have been completely lysed. Alternatively, intraoperative duplex US can be used instead of angiography. The presence of a palpable pedal pulse is generally expected after most femoro-popliteal bypasses.

Step 9. *Closure.* Hemostasis is verified, the subcutaneous tissues are reapproximated with a synthetic absorbable suture. The skin is closed with either staples or a synthetic absorbable suture.

Postop

Patients require intensive or intermediate monitoring during the first 24 hours after surgery. Clearly determine and document pulse examination, Doppler signals, ABIs and neurologic function. Any change requires immediate attention. Pay extreme attention to local wound care. The use of postoperative antibiotics and anticoagulation is determined on an individual basis. Perioperative beta-blockade has been proven to decrease coronary events.

Complications

The most common complications after leg bypass surgery are cardiopulmonary in nature. Wound infection, postoperative bleeding and thrombosis are complications specific to this procedure.

Follow-Up

Pulse examination, Doppler signals and/or ABIs are performed in the recovery room, every shift for the first day and every day after. The wounds are examined in the surgeon's office a week or two after surgery. The use of duplex ultrasound to evaluate bypass grafts is an effective strategy to prolong patency and is routinely used in most vascular centers. The first one of these "surveillance" studies is done within the first postoperative month.

114

Acknowledgment

The editors and author wish to acknowledge Thomas W. Kornmesser for contributing to the previous version of this chapter.

Chapter 115

Lower Extremity Thrombectomy/Embolectomy

Mark K. Eskandari

Indications

Indications for a lower extremity embolectomy or thrombectomy are acute occlusions of the lower extremity necessitating immediate revascularization.

Preop

Most patients who require this procedure have an acute occlusive event; therefore, they require systemic heparinization prior to their procedure and adequate resuscitation. Additional preoperative workup includes an assessment of their basic serum chemistry profile, including CBC, platelets, coagulation, and a chemistry panel. The patient should also be consented for a thrombectomy/embolectomy and possible revascularization. These procedures are typically done under general anesthesia. However, they also may be done under spinal or epidural anesthesia or local anesthesia.

Procedure

Step 1. The patient is on intravenous heparin at a continuous infusion and given a dose of intravenous antibiotics.

Step 2. With the patient in a supine position, the lower extremity is prepped from the umbilicus down to the toes bilaterally.

Step 3. Incision is made in the groin over the common femoral artery of the ischemic limb. The common femoral artery as well as the superficial femoral artery and profunda femoris artery are carefully dissected and encircled with vessel loops, taking care not to injure the crossing lateral circumflex vein, which is found at the crux of the femoral artery bifurcation.

Step 4. With the patient on heparin, proximal and distal control is obtained by clamping the common femoral artery, the superficial femoral, and the profunda femoris artery.

Step 5. A transverse arteriotomy is made in the common femoral artery near the femoral bifurcation.

Step 6. A Fogarty embolectomy catheter is passed first distally. Usually for the lower extremity, a 4 Fogarty embolectomy catheter is used. The catheter is passed down the superficial femoral artery as far as it can go, and the balloon is carefully inflated with gentle back-tension retrieving the clot.

Step 7. A similar maneuver is performed after cannulating the profunda femoris artery. These steps are repeated until there is no retrieval of clot after two passages through both arteries.

Step 8. The thromboembolectomy is performed through the proximal portion of the common femoral artery using the same techniques.

Northwestern Handbook of Surgical Procedures, 2nd Edition, edited by Nathaniel J. Soper and Dixon B. Kaufman. ©2011 Landes Bioscience.

Step 9. At this point, the transverse arteriotomy is closed with interrupted 6-0 polypropylene sutures. Prior to the completion of the anastomosis, it is flushed proximally and distally, the anastomosis completed, and then the clamps removed sequentially.

Step 10. The clamps are first removed from the common femoral artery, the profunda femoris artery, and then the superficial femoral artery, restoring flow to the lower extremity.

Step 11. A completion angiogram is performed to delineate the adequacy of the thromboembolectomy.

Step 12. If this is inadequate or if there is residual clot in the tibial vessels, the next approach is to cut down on the below-knee popliteal artery through a medial incision.

Step 13. The incision is made just below the knee in the soft part of the medial aspect of the leg approximately 2 cm posterior to the posterior edge of the tibia. In making the incision, it is important to be careful of the greater saphenous vein traveling along this course.

Step 14. The incision is carried down to the level of the superficial fascia, which is incised. The gastrocnemius muscle is retracted posteriorly, and the vascular fossa is identified.

Step 15. With careful dissection in this avascular plane, the below-knee popliteal artery is carefully identified as well as its paired popliteal veins. The artery is carefully dissected from the popliteal veins and encircled with vessel loops.

Step 16. A thromboembolectomy is performed using a 3 Fogarty catheter as described above in a similar manner through transverse arteriotomy and repaired in a similar manner.

Step 17. A completion angiogram is performed, confirming the adequacy of the thromboembolectomy.

Step 18. Any residual clot of significance should be retrieved through repeat of the above-outlined steps.

Postop

The patient is extubated, taken to the intensive care unit, resuscitated, and maintained on anticoagulation. The source of thromboembolic disease is identified with additional testing, an EKG to rule out atrial fibrillation, a transesophageal echocardiogram to rule out mural thrombus within the heart, CT scan of the entire aorta and iliac system to rule out an arterial source for embolic disease and plaque rupture, and a hypercoagulable screen as well. The patient is maintained on anticoagulation until the source of the acute ischemic event is delineated.

115

Complications

Complications associated with this procedure include trauma to the arteries resulting in a dissection necessitating a local bypass operation. Additional complications include rupture of either the balloon or the artery and arterial ruptures identified with the completion angiogram necessitating repair with an interposition vein graft; prolonged ischemic event prior to the thromboembolectomy, and the resultant development of acute compartment syndrome that may require fasciotomy at a later time.

Follow-Up

Patients are usually discharged on the day following the operation and should be seen back in the outpatient setting in 3 weeks to assure wound healing. Patients should be monitored with duplex ultrasound after 6 months and yearly thereafter.

Chapter 116

Repair Popliteal Aneurysm: Emergent (Thrombosed)

Mark K. Eskandari

Indications

Indications for repair of a popliteal aneurysm include thrombosis of the aneurysm or distal embolization resulting in thrombosis of the tibial vessels.

Preop

Diagnosis of a popliteal aneurysm may be made by physical exam, duplex ultrasonography, CT scan, or MRI. An angiogram or high-quality magnetic resonance angiogram is typically required to delineate the proximal extent of the aneurysm and distal targets for a potential bypass. If no distal target is appreciated on an angiogram, thrombolysis may be utilized for 24 to 48 hours to open an acutely occluded tibial vessel.

Procedure

Step 1. Anesthetic options include spinal or general anesthesia.

Step 2. Intravenous antibiotics are administered.

Step 3. Two incisions are made over the popliteal artery. One is above the knee medially and one is below the knee medially. Through these incisions, the greater saphenous vein is harvested.

Step 4. The popliteal artery above the knee and below the knee is carefully dissected and circled with vessel loops.

Step 5. After obtaining control, the patient receives systemic heparinization, typically 5000 U IV.

Step 6. Proximal control is obtained in the above-knee popliteal artery.

Step 7. The bypass is performed in an end-to-side fashion between the greater saphenous vein and the above-knee popliteal.

Step 8. The vein graft can be tunneled deep between the two heads of the gastrocnemius muscle or subcutaneously.

Step 9. Distal anastomosis is performed in a similar manner to an outflow vessel, typically the below-knee popliteal artery or one of the tibial arteries.

Step 10. The popliteal artery below the proximal anastomosis is ligated using an umbilical tape or is oversewn with a polypropylene suture.

Step 11. The distal below-knee popliteal artery proximal to the distal anastomosis also is ligated using an umbilical tape or is oversewn with a polypropylene suture. This excludes direct perfusion to the aneurysm.

Step 12. A completion angiogram is performed.

Northwestern Handbook of Surgical Procedures, 2nd Edition, edited by Nathaniel J. Soper and Dixon B. Kaufman. ©2011 Landes Bioscience.

Alternatives to this procedure include a posterior approach with the patient in a prone position. This is typically not necessary in cases of an emergent thrombosis of the popliteal artery.

Postop

Postoperative considerations include the adequacy of revascularization. No anticoagulation is necessary with a vein bypass graft.

Complications

Complications include an inability to do a distal bypass necessitating an amputation; wound infections; absence of a vein for bypass necessitating a prosthetic graft; or bypass ultimately requiring anticoagulation for the prosthetic bypass graft.

Follow-Up

Patients are usually discharged 4-7 days following the operation and should be seen back in the outpatient setting in 3 weeks to assure normal wound healing. The patient should be monitored with duplex ultrasound after 6 months and yearly thereafter.

116

Chapter 117

Exploration for Postoperative Thrombosis

Heron E. Rodriguez

Indications

The decision to explore a patient who recently underwent bypass surgery and is suspected to have thrombosis, needs to be supported by objective evidence. It is of utmost importance to clearly document (1) the pulse examination, (2) ankle brachial index (ABI) values, and (3) motor and sensory function prior to and immediately after the original operation. Only with accurate knowledge of the "baseline" values for these variables, can a significant change be detected and a prompt decision to re-explore can be made. In general, exploration for thrombosis is indicated when "hard signs" of ischemia are present (pallor, paralysis, paresthesias, etc.), when a previously palpable pulse disappears, when an audible Doppler signal is no longer present, or if an ABI value decreases by more than 0.15. In most cases, by carefully assessing these variables, the presence of graft thrombosis can be determined with certainty. Only in very rare instances, will a duplex US or other imaging study (CTA, MRA) be needed to determine whether or not thrombosis has occurred.

Preop

When graft thrombosis is suspected, several immediate actions should be taken: NPO status is ordered, the most senior member of the team is notified and anticoagulation—if not already being administered—is started. If the patient is already receiving anticoagulants, coagulation parameters (PT, PTT, INR) are evaluated to determine the effectiveness of therapy. Heparin-induced thrombocytopenia and other hypercoagulable conditions should be suspected and investigated. Blood is typed and crossed, venous and arterial access is secured. The urgency to re-explore is dictated by findings on physical exam. The presence of audible signals on hand-held Doppler along with a normal neurologic exam, are signs of adequate minimal perfusion and allow for time to prepare and to plan the re-exploration. If no Doppler tones are heard and/or in the presence of paresthesias or paralysis, immediate surgical intervention is needed due to the imminent risk for limb loss. Catheter-directed thrombolysis is generally not considered because of the proximity of the past surgery. A thorough but expeditious analysis of the patient's history and physical exam should identify systemic problems that may preclude re-exploration.

Procedure

Step 1. Exploration of the bypass either proximally or distally depending on where the problem is thought to be.

Step 2. A transverse graftotomy is usually made through which a series of Fogarty thrombectomy catheters are advanced proximally and distally until flow is re-established.

Northwestern Handbook of Surgical Procedures, 2nd Edition, edited by Nathaniel J. Soper and Dixon B. Kaufman. ©2011 Landes Bioscience.

Step 3. Intraoperative arteriography (be aware of kidney function) and/or intraoperative duplex examination are done to delineate the cause of the thrombosis. A systematic evaluation of (1) the inflow vessels, (2) the conduit and (3) the outflow vessels is performed.

Step 4. Repair of the cause of the thrombosis is attempted. I.e., patch angioplasty for a technically deficient anastomosis, angioplasty of an unsuspected lesion in the inflow vessels, lysis of a retained valve, ligation of arteriovenous fistula, repair of a graft kink or twist, etc.

Step 5. Completion arteriography and/or duplex scanning are done to be certain that the problem has been repaired and that no other abnormalities are present. For cases of distal embolism or diffuse thrombosis, arterial and venous access in the affected extremity are obtained, and intraoperative thrombolytic therapy can be used with an "isolated limb technique".

Step 6. Depending upon the duration and severity of ischemia, a fasciotomy should be considered.

Step 7. Anticoagulation is evaluated (PT, PTT, ACT) and continued.

Postop

ICU monitoring is routinely used, due to the high risk for re-thrombosis. Clearly determine and document pulse examination, Doppler signals, ABIs and neurologic function. These variables should be assessed every hour for the first 4 hrs after embolectomy and at least every 4 hours afterwards. Any change requires immediate attention. Maintain anticoagulation and/or antiplatelet therapy. If fasciotomies were not performed, close observation for possible compartment syndrome must be undertaken. Pay extreme attention to local wound care.

Complications

The morbidity and mortality associated with reoperative surgery in vascular patients is high. Cardiopulmonary complications and renal insufficiency are frequent. Recurrent thrombosis, bleeding, ischemic neuropathy and graft infection are common complications after redo bypass surgery.

Follow-Up

117

During the acute recovery, measurement of ABIs will detect new or recurrent thrombotic problems. A baseline duplex US is obtained soon around the time of discharge and this should be repeated 3 months after surgery. Grafts at high risk of failure—in which prophylactic intervention should be considered—are those with a velocity ratio of >3.5, a peak systolic velocity of >300 cm/sec a drop on ABI >0.15 or graft flow velocity lower than 45 cm/sec.

Acknowledgment

The editors and author wish to acknowledge Thomas W. Kornmesser for contributing to the previous version of this chapter.

Chapter 118

Fasciotomy: Lower Extremity

Mark K. Eskandari

Indications

Indications for a fasciotomy of the lower extremity include acute compartment syndrome after significant lower extremity trauma (i.e., crush injury), prolonged ischemia from acute occlusion, or excessive trauma related to strenuous exercise.

Preop

Most of these procedures are performed under general anesthesia. However, they may be performed under spinal or epidural anesthesia with the patient in a supine position. Additional preoperative assessment includes the use of compartment pressures. Anything above 40 mm Hg is diagnostic of acute compartment syndrome. Typically, the diagnosis is made on clinical grounds alone with evidence of swelling over the compartments, the anterior being the most predominantly affected, followed by the lateral, deep posterior, and superficial posterior. Clinically, the patient will have exquisite tenderness with palpation of the particular compartment and pain on passive plantar and dorsiflexion with normal vascular examination and normal overlying skin and sensory examination.

Procedure

Step 1. The patient is in a supine position. Most fasciotomies are performed through two incisions: one incision is medial and one incision lateral. The lateral incision is approximately 5-7 cm in length, about 2 cm posterior to the posterior aspect of the tibia carried down through the subcutaneous tissue.

Step 2. Both the anterior and the lateral fascial compartments and the septum separating these two are identified. The anterior compartment is then incised using electrocautery, and the whole length of the compartment is decompressed using scissors incising the fascial band, both proximally and distally.

Step 3. A similar technique is utilized for decompression of the lateral compartment, taking care to limit the amount of fascial incising proximally because of the location of the superficial peroneal nerve just beyond the head of the fibula on the lateral aspect of the leg.

Step 4. The posterior compartments, the deep and superficial, are approached through a medial incision. The incision is approximately 5-7 cm in length and approximately 2 cm beyond the posterior aspect of the tibia in the soft part of the leg. Through this longitudinal incision, frequently the saphenous vein will be identified, and care should be taken to preserve this if possible.

Step 5. The superficial fascia is incised, retracting the gastrocnemius muscle posteriorly.

Northwestern Handbook of Surgical Procedures, 2nd Edition, edited by Nathaniel J. Soper and Dixon B. Kaufman. ©2011 Landes Bioscience.

Step 6. The deep compartment is then decompressed by taking down the attachments of the soleus muscle to the posterior aspect of the tibia using electrocautery. This should be done carefully because of the proximity of the posterior tibial and peroneal vessels just below the fascial band. This is incised to the length of approximately 5 cm.

Step 7. Hemostasis is obtained and most fasciotomy sites are packed open with a moist gauze.

Step 8. After approximately 3 days, the patient may be returned to the operating room for primary closure, split-thickness skin graft, or these wounds may be allowed to close secondarily.

Postop

Additional postoperative care includes identifying postoperatively the integrity of the superficial peroneal nerve and the ability to dorsiflex the foot. Many of these patients with prolonged periods of ischemia and acute compartment syndrome can develop rhabdomyolysis so renal function and CPK levels should be assessed and treated appropriately.

Complications

Complications related to four-compartment fasciotomies include, first and foremost, injury to the superficial peroneal nerve during the decompression of the lateral compartment; wound infection; and bleeding from the site, particularly in patients who were anticoagulated, necessitating adequate hemostasis at the completion of the procedure.

118

Chapter 119

Toe Amputation

Mark K. Eskandari

Indications

Toe amputations are generally indicated for wet gangrene, dry gangrene, and/or osteomyelitis of the toe.

Preop

The procedure is usually performed in the operating room, with local anesthesia or an ankle block. While in a supine position, the foot to the knee is prepped and draped in a sterile fashion. Systemic intravenous antibiotics are administered immediately prior to the procedure.

Procedure

Step 1. The local anesthetic is given if desired using 1% lidocaine without epinephrine on either side of the web spaces of the desired toe amputation site. Or, if an ankle block or a spinal has been administered, this should be tested.

Step 2. If the toe amputation is to be either the second, third, or fourth toe, an elliptical incision is made with the apices of the incision on the dorsum and plantar aspect of the foot and the toe amputated either at the proximal phalanx site or proximal to the metatarsal head.

Step 3. The incision is carried down to the level of the bony tissue. Hemostasis is obtained at the digital vessels and the bone transected using a bone cutter. If the level of amputation requires resection of the metatarsal head, the joint space is entered and the metatarsal head resected back using rongeurs.

Step 4. The ligaments and tendons are transected back as far proximally as possible.

Step 5. The wound is irrigated with saline solution. If it is dry gangrene or a clean wound, it may be closed primarily with interrupted vertical mattress 3-0 nylon sutures. If it is infected or a wet gangrene amputation site, it is packed open with a moist saline-soaked gauze.

Step 6. If the toe amputation is of either the first or second toe, a racquet-shaped incision is utilized as opposed to an elliptical incision and the amputation performed as described above.

Postop

Postoperatively, patients should be evaluated to determine the causative factors of their gangrenous toes.

Northwestern Handbook of Surgical Procedures, 2nd Edition, edited by Nathaniel J. Soper and Dixon B. Kaufman. ©2011 Landes Bioscience.

Complications

Complications from this procedure are early toe amputation without revascularization leading to proximal skin necrosis. Therefore, an adequate assessment of perfusion to the foot should be delineated with arterial blood flows prior to any toe amputation unless the foot is septic.

Follow-Up

After wound healing is complete, patients should be referred to a prosthetist for shoe-wear evaluation. Patients should be given appropriate deep venous thrombosis prophylaxis until they are mobile.

119

Chapter 120

Transmetatarsal Amputation

Mark K. Eskandari

Indications

This operation is most commonly performed for gangrene or nonhealing ulcer of the toes. If amputation of more than two toes is required and gangrene extends proximal to the metatarsophalangeal joint, transmetatarsal amputation is indicated. Lesions may be due to emboli, atherosclerosis, or neuropathy.

Preop

Most important is the physical exam and noninvasive testing to determine if the patient has adequate blood flow for wound healing. A variety of tests are available to assess this, including ankle-brachial index, arterial Doppler, pulse volume recordings, transcutaneous oximetry, and laser Doppler. Diabetic patients need attention to glucose control perioperatively. Anesthesia may be regional (spinal or ankle block) or general.

Procedure

Step 1. Note: Gentle tissue handling throughout the operation is essential for successful wound healing. Areas of ulcer or gangrene should be covered with an occlusive dressing after prepping the patient to avoid contamination of the amputation wound. In cases of gross contamination, consideration should be given to a staged procedure—open guillotine amputation followed by a formal completion amputation 3-5 days later.

Step 2. An incision is made on the dorsal surface of the foot at the midmetatarsal level. This incision may be slightly curved to facilitate the creation of a plantar flap.

Step 3. Extend the incision to create a plantar flap, which should extend to about 1 cm proximal to the webspace while preserving only viable tissue. The plantar flap will include skin and subcutaneous tissue.

Step 4. The dorsal incision is carried down to the bone, dividing the extensor tendon as proximally as possible. It is important to remove all exposed tendons and ligaments to facilitate better wound healing.

Step 5. Metatarsal bones are divided with a sagittal saw just proximal to the plantar foot incision.

Step 6. Remaining exposed tendons are divided as proximally as possible with a scalpel.

Step 7. The wound is irrigated with antibiotic solution.

Step 8. The subcutaneous tissue is approximated with interrupted absorbable suture.

Step 9. The skin is closed with interrupted vertical mattress 3-0 or 4-0 monofilament sutures, being careful not to tie the sutures too tightly.

Step 10. A bulky protective dressing is applied.

Northwestern Handbook of Surgical Procedures, 2nd Edition, edited by Nathaniel J. Soper and Dixon B. Kaufman. ©2011 Landes Bioscience.

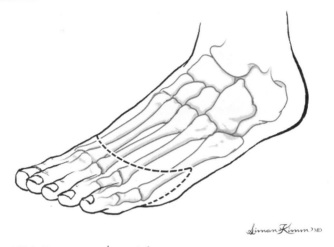

Figure 120.1. Transmetatarsal amputation.

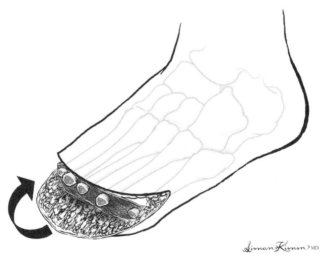

Figure 120.2. Transmetatarsal amputation.

120

Postop

Patients should keep the foot elevated and should not bear weight on the foot for 10-14 days. Patients should be given appropriate deep venous thrombosis (DVT) prophylaxis until they are mobile.

Complications

Failure to heal, wound infection, DVT, pulmonary embolism.

Follow-Up

After wound healing is complete, patients should be referred to a prosthetist for shoe-wear evaluation. Patients should be given appropriate DVT prophylaxis until they are mobile.

Acknowledgment

The editors and author wish to acknowledge Nancy Schindler for contributing to the previous version of this chapter.

Chapter 121

Below Knee Amputation (BKA)

Heron E. Rodriguez

Indications

The general indications for below knee amputation are infectious, ischemic, traumatic, neuropathic or diabetic wounds in the foot not amenable to revascularization/ reconstructive limb preserving surgery. Additionally, rest pain in patients who are not candidates for revascularization due to anatomy, conduit availability, or patient condition is a common indication for below knee amputation (BKA). Below knee amputation is contraindicated in debilitated, bedridden patients with knee contractures.

Preop

Most important is the physical exam and noninvasive testing to determine if the patient has adequate blood flow to heal the wound. A variety of tests are available to aid the surgeon in assessing the likelihood of healing: ankle-brachial index, arterial Doppler, pulse volume recordings, transcutaneous oximetry, and laser Doppler. Preoperative consultation with physical therapy and rehabilitation medicine is obtained in order to optimize prosthetic fitting and ambulation potential. Strict glycemic control and correction of underlying infection and nutritional deficits are critical. DVT prophylaxis and cardioprotection is initiated preoperatively. Anesthesia may be regional or general. A bladder catheter is inserted.

Procedure

Step 1. Note: Gentle tissue handling throughout the operation is essential to successful wound healing. Minimize the use of electrocautery and pay special attention to hemostasis. The procedure may be performed with a tourniquet if desired.

Step 2. The entire leg is prepped and draped. If possible any open or infected wounds should be covered with an occlusive dressing.

Step 3. The anterior incision is made approximately 10 cm below the tibial tuberosity and extends about two-thirds of the circumference of the leg.

Step 4. A long posterior flap incision is made. The posterior flap should be 20-30% longer than the measured distance from the tibia anteriorly to the base of the posterior flap.

Step 5. The anterior incision is carried down to the tibia. No anterior flap is created. The anterior tibial vessels are identified in the lateral wound. The artery and vein are identified, ligated, and divided. The nerve is placed under tension and divided.

Step 6. The musculature of the leg is divided and the fibula is exposed. It must be exposed about 3 cm above the skin incision.

Step 7. The musculature of the medial leg is divided. The posterior tibial and peroneal vessels are identified, ligated, and divided.

121

Northwestern Handbook of Surgical Procedures, 2nd Edition, edited by Nathaniel J. Soper and Dixon B. Kaufman. ©2011 Landes Bioscience.

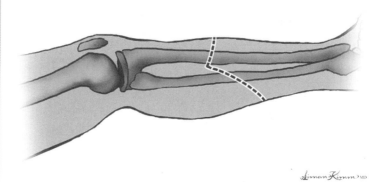

Figure 121.1. Below knee amputation.

Step 8. The tibia is then exposed and the periostium is elevated. The tibia is divided transversely 1-2 cm above the anterior skin incision. The fibula is then divided at the same level of transection of the tibia.

Step 9. The flail leg is then placed on mild traction. The tibial nerve is placed on mild tension, ligated, divided, and allowed to retract.

Step 10. An amputation knife is then used to create the posterior flap of gastrocnemius and soleus muscles. Long smooth strokes are used, and the flap is tapered.

Step 11. A rasp or saw is used to smooth the surface of the tibia.

Step 12. The subcutaneous tissue is closed with interrupted absorbable sutures. Interrupted monofilament suture or staples are used to close the skin.

Step 13. A dressing is applied that will keep the knee extended. A knee immobilizer or cast can be used for this purpose.

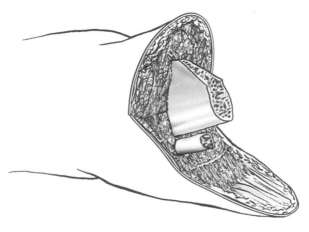

Figure 121.2. Below knee amputation.

Postop

Appropriate analgesia, cardioprotection and DVT prophylaxis are of critical importance in the immediate postoperative period. Physical and occupational therapy consultation are started as soon as possible. The leg should be kept in some type of dressing which maintains knee extension until most of the postoperative pain has resolved. Failure to do so may result in a contracture. Most patients will require an inpatient rehabilitative stay if they are candidates for ambulation. A prosthesis can be fitted when wound healing is complete.

Complications

Complications include failure to heal, wound infection, deep vein thrombosis, pulmonary embolism. Because these amputations are frequently performed in patients with severe vascular disease and diabetes, cardiac complications and death can also occur.

Follow-Up

Outpatient visits are scheduled until complete healing is observed. Analgesia and nutrition status are reassessed at every visit. Rehabilitation medicine should coordinate physical and occupational therapy and prepare the residual limb for prosthesis if appropriate.

Acknowledgment

The editors and author wish to acknowledge Nancy Schindler for contributing to the previous version of this chapter.

121

Chapter 122

Above Knee Amputation (AKA)

Heron E. Rodriguez

Indications

In general, above knee amputation is indicated in patients requiring a major lower extremity amputation (non reconstructable ischemia, advanced diabetic foot infection, trauma, etc.) who cannot undergo a below knee amputation. This includes patients in whom perfusion or infection make healing of a below knee wound unlikely, patients who are nonambulatory or with a knee contractures. Above knee amputation requires significantly higher energy expenditure for ambulation with a prosthesis and should be performed only when absolutely inevitable if a patient has the potential to ambulate.

Preop

A palpable femoral pulse and a normal triphasic signal at the femoral level are almost always indicative of adequate perfusion for healing of an above knee amputation. Diabetics need attention to glucose control in the perioperative period. Nutritional improvement and control of infection are of paramount importance. Preoperative consultation with physical therapy and rehabilitation services is obtained. Anesthesia may be regional or general. The leg is prepped and draped circumferentially including the groin. A bladder catheter is inserted. DVT prophylaxis and cardioprotection are initiated preoperatively.

Procedure

Step 1. Note: Gentle tissue handling throughout the operation is essential for successful wound healing. Minimize the use of electrocautery and pay attention to hemostasis.

Step 2. A wide fish-mouth or circumferential incision is made in the skin at the mid-thigh level and deepened to the fascia.

Step 3. The great saphenous vein is identified in the inner aspect of the thigh, ligated, and divided.

Step 4. Muscles are divided at the level of the skin and are allowed to retract upward.

Step 5. In the medial leg, the femoral artery and vein are identified. They are ligated and divided individually. The proximal end is suture ligated.

Step 6. The femur is exposed and a periosteal elevator is used to expose bone about 10 cm above the incision.

Step 7. In the posterior thigh, the sciatic nerve is identified. This is placed on traction and ligated as high as possible. It is then divided and allowed to retract high into the thigh.

Northwestern Handbook of Surgical Procedures, 2nd Edition, edited by Nathaniel J. Soper and Dixon B. Kaufman. ©2011 Landes Bioscience.

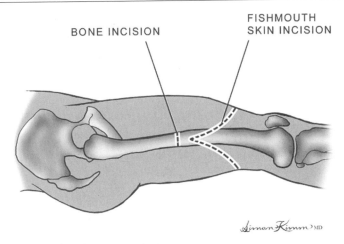

Figure 122.1. Above knee amputation.

Step 8. The femur is divided about 10 cm above the skin incision.
Step 9. A rasp or the saw is used to round off the edges of the bone.
Step 10. The fascia is closed with interrupted absorbable suture.
Step 11. The skin is closed with interrupted monofilament suture or staples.
Step 12. A soft bulky dressing is applied.

Postop

Adequate analgesia, cardioprotective measures and DVT prophylaxis are the most important aspect of the immediate postoperative care. Nutritional deficits and any remaining infection are treated aggressively. Physical and occupational therapy are started soon as possible. Most patients will require inpatient rehabilitative stay if they are candidates for ambulation. A prosthesis can be fitted when wound healing is complete.

Complications

They include perioperative myocardial infarction, failure to heal, wound infection, deep venous thrombosis, pulmonary embolism. Because these amputations are frequently performed in patients with severe vascular disease and diabetes, cardiac complications and death can also occur.

Follow-Up

Outpatient visits are scheduled until complete healing is obtained. Analgesia and nutrition status are reassessed at every visit. Rehabilitation medicine should coordinate physical and occupational therapy and prepare the residual limb for prosthesis if appropriate.

Acknowledgment

The editors and author wish to acknowledge Nancy Schindler for contributing to the previous version of this chapter.

Chapter 123

Varicose Veins

William H. Pearce

Indications

Varicose veins are a common clinical problem. Primary varicose veins occur as a result of saphenofemoral or saphenopopliteal valvular incompetence. Surgery is indicated for cosmesis, pain, swelling, stasis dermatitis, bleeding, and venous ulceration.

Operative Principles

Obliteration of significant sources of venous superficial reflux is the basis for this operation. It is important to identify the location of significant reflux. Reflux may arise at the saphenofemoral or saphenopopliteal or in intermediate locations where perforating branches join the superficial system. Once the reflux has been identified and ligated, enlarged tributaries are avulsed or stripped.

Preop

All patients undergoing excision of varicose veins should undergo duplex scanning. Duplex ultrasound scanning serves two functions. First it ensures that the deep residual veins are competent and normal. Patients with significant deep venous disease should not undergo stripping unless under special circumstances. Secondary duplex scans should be used to identify significant sources of reflux. The entire course of the greater saphenous and lesser saphenous veins is followed. Valsalva maneuvers are performed to identify areas of reflux; these areas may be marked preoperatively by the blood flow technician. When the patient arrives for the operative procedure, the veins are marked with a water-insoluble marker to identify all perforators and dilated venous tributaries.

Procedure

Step 1. For patients undergoing greater saphenous stripping, the patient is placed in the supine position. The patients undergoing lesser saphenous stripping are placed face down.

Step 2. The patients are prepped using a benzalkonium chloride 1:750 (Zephiran®) solution. This clear solution allows the previous markings to be identified.

Step 3. A small cutdown is performed on the source of reflux. In the saphenofemoral region, a small, transverse incision should be made just medial to the femoral artery just below the groin crease. The saphenofemoral junction is identified by tracing the saphenous vein to its confluence with the common femoral vein. All of the tributaries are divided (superficial inferior epigastric, superficial external pudendal, superficial circumflex iliac, and arteriolateral superficial veins).

Northwestern Handbook of Surgical Procedures, 2nd Edition, edited by Nathaniel J. Soper and Dixon B. Kaufman. ©2011 Landes Bioscience.

Step 4. Once the common femoral vein has been identified, it is safe to ligate the saphenous vein close to the common femoral vein. An external stripper may be placed either antegrade or retrograde through the saphenous system. The stripping system is tied to the saphenous vein, generally in the groin. If the catheter is placed retrograde, a separate stab incision is made just above the knee. The greater saphenous is stripped from groin to knee using an inversion technique.

Step 5. Occasionally, the greater saphenous may be completely stripped from groin to ankle. It is not often recommended to strip the greater below the knee since injury to the saphenous nerve is more common.

Step 6. Clusters of varicose veins are removed using the stab avulsion technique. Using an 11 blade a small incision less than 1/8 inch is made directly over the vein. Either using a vein hook, or a small hemostat, the vein is grasped and teased from the underlying tissue until the vein is avulsed. Generally, these small stab incisions need only to be closed with a steri-strip or a small 5-0 subcuticular suture. The leg is milked of any residual hematoma and the groin wound is closed in multiple layers with the final layer consisting of the 5-0 subcuticular absorbable suture.

Step 7. Should the patient require lesser saphenous stripping, the patient is placed prone on the operating room table. An incision is made directly over the saphenopopliteal junction. This junction should be identified preoperatively with duplex ultrasound. For the stripping of the lesser saphenous, it is important to realize that injury can occur to either the superficial peroneal or sural nerves. Therefore it is very important to clearly see the lesser saphenous vein before stripping. Stab avulsions are made in the lower segments of the lesser saphenous vein.

Step 8. Following completion of the procedure, the wounds are steri-stripped and a dressing of Kerlix and an external Ace wrap is placed from the metatarsal heads to the mid thigh.

Addsteps. To avoid injury to structures other than the vein to be stripped, meticulous dissection and clear identification of the structures is mandatory. Since inadvertent stripping of arterial and other venous structures has occurred, it is mandatory to identify the saphenofemoral junction and see the common femoral vein. A clear knowledge of the anatomy is important.

Postop

Postoperatively, the patient is discharged to home. The Ace wrap is kept in place for 24 hours. After 24 hours the Ace wrap is applied from the metatarsal heads to the tibial tuberosity. This wrap is used for approximately 2 weeks until the patient returns to the office.

Complications

The most common complication is bruising throughout the leg. The patient should be warned of this complication since it is so common. Injury to peripheral nerves including the saphenous, sural, and peroneal nerves may occur. The sensory neuropathy that occurs following these injuries will generally resolve over one year. However, motor injuries are usually permanent. Seromas occasionally occur along the course of a saphenous vein but are uncommon. Wound infections are also uncommon (<5%). Deep venous thrombosis is rare (0.6%). Fatal pulmonary embolism has been reported to occur in 1 in 30,000. In addition, older literature has reported inadvertent stripping of arterial structures or deep venous structures.

123

Follow-Up

The patient is seen 2 weeks following the surgical procedure. The sutures are removed; however, in most instances subcuticular stitches are placed so this is not required. No further follow-up is required. The patients are encouraged to wear support hose since they generally have a predisposition for venous disease.

SECTION 8: PEDIATRIC SURGERY

Section Editor: Marleta Reynolds

Chapter 124

Pediatric (Indirect or Congenital) Inguinal Hernia

Anthony C. Chin

Indications

Repair of an indirect inguinal hernia is indicated when the processus vaginalis, an outpouching of the peritoneal cavity that extends through the internal and external ring, fails to obliterate and allows intra-abdominal contents to herniate through. An inguinal hernia will not resolve spontaneously and should be repaired when diagnosed to prevent incarceration.

Procedure

Step 1. An incision is made along the skin fold overlying the external inguinal ring of the affected side.

Step 2. Subcutaneous tissues are dissected down to and through Scarpa's fascia with either blunt, sharp or cautery dissection.

Step 3. The lateral aspect of the external oblique is exposed, and blunt dissection is carried inferiorly and medially until the external inguinal ring is exposed.

Step 4. The cord structures will be appreciated emerging from the external inguinal ring.

Step 5. To facilitate exposure of the cord structures, the external inguinal ring may be divided, but this step is optional. Care is taken to preserve the ilioinguinal nerve.

Step 6. The cord is then carefully mobilized circumferentially with blunt dissection over and along the line of the spermatic cord.

Step 7. A hemostat is passed behind the cord to elevate the hernia and cord structures into the operating field.

Step 8. The cord structures are isolated and separated from the hernia by initially incising the cremasteric muscles that run along and overlie the cord.

Step 9. A fine layer of internal spermatic fascia is found underneath the cremasteric muscle and is divided. The spermatic vessels will be appreciated laterally and the vas deferens inferior-medially to the hernia sac.

Step 10. The thin hernia sac is held lightly anterior-medially, and the spermatic vessels and vas deferens are dissected gently off the hernia sac completely and reflected and protected laterally.

Step 11. The hernia may be gently grasped distally at the fundus in a blind ending sac or divided and grasped proximally with a pair of hemostats. The location of the vas deferens must be noted at all time prior to division of any tissue.

Step 12. If difficulty is encountered during the dissection, an alternative approach is to open the hernia sac and the bluntly dissect the sac off the cord structures below.

Northwestern Handbook of Surgical Procedures, 2nd Edition, edited by Nathaniel J. Soper and Dixon B. Kaufman. ©2011 Landes Bioscience.

Step 13. Gentle traction is placed on the hemostats holding the proximal sac, and the sac is dissected cephalad towards the internal ring until the extraperitoneal fat at the neck of the hernia sac is exposed. Tissue in the medial wall of the sac may be the bladder of a sliding hernia.

Step 14. The surgeon may twist the hernia sac to protect and reduce any herniated intra-abdominal contents.

Step 15. The sac is then highly ligated with a transfixing absorbable suture, twice, at the neck.

Step 16. The redundant hernia sac is excised distal to the sutures, and the proximal repair will retract back underneath the external inguinal ring.

Step 17. The attenuated or divided external inguinal ring is then repaired with interrupted sutures.

Step 18. The testicle is returned into the scrotal sac.

Step 19. Scarpa's fascia is closed with an interrupted stitch and the skin subsequently closed with a subcuticular stitch.

Postop

Term infants are generally discharged home with analgesics such as acetaminophen or ibuprofen. Narcotics are occasionally utilized for older children. Preterm infants less that 50 weeks postgestational age at the time of surgery are generally observed overnight because of the risk of apnea.

Complication

Complications following hernia repair are less than 2% and increase significantly following incarceration. They may include bleeding, wound infection and recurrence. Less commonly, injury to the vas deferens, a trapped testicle or testicular atrophy may occur.

Follow-Up

Patients follow-up 2 weeks following repair to inspect the incision, check for recurrence and palpate for a trapped testes. Long-term follow-up is not indicated unless a recurrence or other complications arise.

124

Chapter 125

Hypertrophic Pyloric Stenosis

Katherine A. Barsness

Indications

Pyloric stenosis is a gastric outlet obstruction secondary to obstructive hypertrophy of the pylorus. The patient presents in early infancy with a history of nonbilious and projectile emesis, dehydration and lethargy. Physical exam often finds a visible peristaltic wave in the epigastrium and a palpable mass ("olive") in the upper abdomen. Diagnosis is confirmed by ultrasonography with a pyloric channel length greater than 14 mm and a pyloric muscular width greater than 4 mm. Inconclusive ultrasonography results may warrant contrast upper gastrointestinal studies to further delineate the anatomy.

Preop

Infants with hypertrophic pyloric stenosis present with varying degrees of dehydration and electrolyte disturbances. The classic finding with gastric outlet obstruction is a hypokalemic, hypochloremic metabolic alkalosis with paradoxical aciduria. As the alkalosis worsens, these infants are prone to apneic episodes and must be on apnea/bradycardia monitors until normalization of all of the serum electrolytes. At presentation, the infants are given intravenous (IV) fluid boluses every hour with 20 ml/kg of normal saline until urine output resumes. The infants are then maintained on 1.5 times the rate of maintenance IV fluids until normalization of electrolytes (serum potassium greater than 3.4, CO_2 less than 30 and chloride greater than 100). Once the electrolytes are normal and the child has appropriate urine output, general anesthesia may be safely administered.

Procedure

The operative procedure may be performed via a right upper quadrant incision or a laparoscopic approach. Surgeon preference usually dictates the approach. The following procedure details the laparoscopic approach.

Step 1. The infant undergoes a rapid sequence general anesthetic after appropriate evacuation of residual gastric contents. The abdomen is then prepped and draped in the usual sterile fashion.

Step 2. The umbilicus is infiltrated with local anesthetic. A vertical incision is made in the base of the umbilicus. A 5 mm port is then placed into the abdomen. After confirmation of intra-abdominal position of the port, gas insufflation is begun with a flow of 8 L/minute and a pressure of 8 mm Hg. A 4 mm/30° telescope is advanced into the umbilical port. The thickened and enlarged pylorus is easily visible beneath the edge of the liver

Northwestern Handbook of Surgical Procedures, 2nd Edition, edited by Nathaniel J. Soper and Dixon B. Kaufman. ©2011 Landes Bioscience.

Step 3. A 3 mm stab incision is placed in the right upper quadrant. An atraumatic grasper is then placed through this stab incision. A second stab incision is then placed in the left upper quadrant and an arthrotomy knife is placed through this incision. Under direct vision, the arthrotomy knife is opened to 3 mm depth. The pylorus is then stabilized with the left-hand atraumatic grasper.

Step 4. An incision is made along the length of the thickened pylorus, starting immediately proximal to the vein of Mayo and proceeding cranially to the distal antrum of the stomach. The arthrotomy knife is closed and removed from the abdomen. A pyloric spreader is advanced into the abdomen via the left upper quadrant incision. A pyloric spreader is a specific instrument that has small teeth on both inner and outer sides of each blade. The outer teeth help maintain the position of the blades between the layers of muscle. Both blades of the pyloric spreader are inserted into the incision on the pylorus. Using controlled tension, the outer fibers of the pylorus are gently spread, revealing bulging mucosa. The mucosa is exposed for the entire length of the pylorus, up to the level of the inner circular layer of the antral muscle. The two sides of the separated pyloric muscle should easily move independently of one another at the completion of the myotomy.

Step 5. The instruments are withdrawn from the abdominal cavity. All carbon dioxide is subsequently evacuated from the abdominal cavity, and the umbilical port is removed. The umbilical incision is closed at the fascia with a 3-0 absorbable suture. The umbilical skin is then approximated using a buried 5-0 absorbable suture. The bilateral upper quadrant stab incisions are closed with Dermabond®.

Postop

Infants generally do well with acetaminophen for postoperative pain control. Narcotic pain medications are contraindicated due to a high incidence of apnea and bradycardia in the immediate postoperative period. All infants younger than 50 weeks post conception age (gestational age in weeks at birth + number of weeks of life) require postoperative monitoring for apnea and bradycardia. Oral feeds are begun 6 hours after the completion of the operation. Postoperative vomiting is common for the first 24-48 hours after the pyloromyotomy is completed. The infants are discharged to home when they are tolerating 2 ounces of feeds every 2 hours.

Complications

Mucosal injury is a serious potential complication. Tachycardia is one of the earliest signs of distress in infants. Any unexplained tachycardia must be viewed seriously, and the infant will require an immediate upper gastrointestinal study to evaluate for a leak. In the event of a leak, immediate open exploration is necessary. Although leak is a very rare complication, it carries a high morbidity and mortality.

Follow-Up

The infant should be seen by the primary care physician within one week of discharge to ensure the child is maintaining discharge weight. The child will then return for one postoperative wound check to ensure proper healing of all incisions.

125

Chapter 126

Operation for Malrotation

Mary Beth Madonna

Embryology

Malrotation is a term used to define a group of congenital anomalies that result from abnormal intestinal fixation and rotation that occurs early in gestation. The incidence of symptomatic malrotation is estimated at 1:6000 live births, but at autopsy series as many as 0.5% of the population has an abnormal variant of rotation. During the fifth week of gestation, the midgut herniates into the yolk sac. While in the sac there is a 180° counterclockwise rotation of the bowel. As the bowel returns to the abdomen, the final 90° of counterclockwise rotation occurs which places the duodenum in the left upper quadrant and posterior to the superior mesenteric artery (SMA) and the cecum in the right lower quadrant and anterior to the SMA. These two areas then become fixed to the retroperitoneum. The remaining midgut is not fixed. This anatomic location allows the broadest base possible for the mesentery.

Associated Anomalies

Rotational anomalies are associated with many other congenital anomalies. Malrotation is almost universally found in patients with congenital diaphragmatic hernia, gastroschisis and omphalocele. In addition 30% to 60% of patients have anomalies of other organ systems such as cardiac, respiratory and neurologic. Additional GI anomalies are also seen.

Clinical Presentation

There are many varieties of anomalies in the rotation and fixation of the intestine but a few common clinical presentations. Approximately 90% of patients become symptomatic during the first year of life. Midgut volvulus is the concerning presentation. This is one of a few true surgical emergencies and occurs when there is torsion of the narrow mesenteric pedicle. Intestinal ischemia and necrosis rapidly occur. This is the presentation in up to 50% of patients with malrotation. Once the torsion occurs there is obstruction of the proximal duodenum, and bilious emesis occurs. An upper GI series is the gold standard for preoperative evaluation, but if the patient is in serious condition and there is a high index of suspicion then the patient should be taken to the operating room without a study. Occasionally, older children present with symptoms of intermittent emesis due to partial or intermittent volvulus. The second common presentation is duodenal obstruction due to adhesive bands traversing over the duodenum from the ascending colon, the so-called Ladd's bands.

Northwestern Handbook of Surgical Procedures, 2nd Edition, edited by Nathaniel J. Soper and Dixon B. Kaufman. ©2011 Landes Bioscience.

Preop

For infants presenting with midgut volvulus, early operative intervention is the key to prevent loss of bowel. Fluid resuscitation is also required with lactated Ringer's or saline boluses (20 ml/kg). Adequate intravenous access is often difficult to obtain, and a central line may be required. Nasograstric decompression of the stomach is preformed, and a Foley catheter is placed to monitor urine output. Prevention of hypothermia is important in any procedure performed on an infant. If the patient presents with chronic obstruction, the procedure is urgent but not emergent.

Procedure

The operative correction of malrotation is referred to as the Ladd procedure in honor of the father of pediatric surgery who first described the condition and repair.

Step 1. The entire abdomen is prepped and draped in sterile fashion. Often a protective plastic barrier is used on the abdomen to conserve heat.

Step 2. A right-sided supraumbilical incision is created approximately 2 cm above the umbilicus. Electrocautery is used to divide the rectus muscle and obliques. The peritoneal cavity is judiciously entered.

Step 3. The bowel is eviscerated from the abdominal cavity and examined for a midgut volvulus. To assist with this evaluation the root of the mesentery is identified just below and behind the greater curvature of the stomach. At the root of the mesentery, the volvulus is identified. The colon is noted to be wrapped around the root confirming the volvulus. The torsion of the bowel always occurs in a clockwise direction along the access of the SMA.

Step 4. Reduction of the volvulus is accomplished by careful rotation of the intestinal mass in a counterclockwise direction. This is performed in a stepwise fashion with maneuvers in 90-180° increments. The reduction is complete when the duodenum and ascending colon are parallel to each other with the duodenum on the right side of the colon.

Step 5. Intestinal viability is assessed. The bowel is observed and the intestinal loops placed in warm laparotomy pads and reassessed after a few minutes in questionable cases. If a large portion of the bowel is questionable after observation, a second look operation is planned in 24-48 hours and the skin is closed and the baby returned to the intensive care unit. If the entire midgut is frankly necrotic, a true abdominal catastrophe has occurred and closure without surgical correction is warranted. A frank discussion is undertaken with the family and comfort care measures are offered to the infant. Eventual bowel transplant can be considered as an option. The ethics committee should become involved in these difficult cases.

Step 6. The extrinsic compression of the duodenum by Ladd's bands is next addressed. These bands are identified coming from the ascending colon across the duodenum. Using the pylorus as a guide, the first portion of the duodenum is identified. The bands are divided adjacent to the wall of the bowel.

Step 7. The entire duodenum is then mobilized with a Kocher maneuver. This allows dividing any additional bands lateral to the duodenum and allows the duodenum to lie straight instead of tortuous as is usually seen initially.

126

Step 8. Additional bands between the left-sided cecum and right-sided duodenum are then divided taking care not to injure the mesenteric root. This will help convert the narrow mesenteric pedicle to a wider base. This dissection is carried from the root of the mesentery out toward the bowel wall. At the conclusion of this maneuver, the duodenum and cecum will be widely separated.

Step 9. If small areas of bowel are necrotic but the rest is viable, the necrotic areas are resected after ligating the mesentery with silk or Vicryl ties. The bowel is either anastomosed or an enterotomy is performed. Due to the small size of the infant, a hand sewn anastomosis is performed in either one or two layers depending on surgeon preference. If a stoma is to be created, this is matured in the main incision, usually at the lateral edge.

Step 10. Because 10% of patients with malrotation have an intrinsic duodenal obstruction, patency of the duodenum is next assessed. The simplest way is to pass a 10 F nasogastric tube through the pylorus and guide it through the entire duodenum.

Step 11. The appendix is removed with ligation of the vessels with a silk or Vicryl tie and double ligation of the appendix at its base.

Step 12. Prior to closure the bowel is placed carefully in the abdomen with the duodenum placed in the right gutter and the cecum placed on the left side of the abdomen.

Step 13. The fascia is closed in two layers with either PDS or Vicryl sutures (either 3-0 or 4-0 depending on the size of the patient).

Step 14. The skin is closed with a subcuticular suture using a 5-0 monofilament absorbable suture. Steri-strips and sterile dressing are applied over the site. If a stoma is created, a Vaseline gauze is placed on the stoma.

Postop

If the patients have significant bowel ischemia and/or loss they will have severe metabolic and hemodynamic derangements in the postoperative period and will require diligent management of fluid and electrolytes. Often they will require pressors to help maintain adequate blood pressure. In addition, most of these sicker babies will require ventilatory support. Those that have viable bowel will not have these severe problems. All patients who undergo malrotation surgery will require nasogastric decompression for many days as they will have a prolonged ileus from the extensive manipulation of the bowel. This may be removed when the aspirate clears and the patient has bowel movements. Feeds are slowly begun using an appropriate formula. If the patients have a significant loss of bowel, they will require an elemental type formula. Some infants will be discharged on parenteral nutrition, and the goal is to have a 12-hour cycle so the infants may be free during the day for appropriate neurologic development.

Complications

The majority of serious long-term complications result from delay in diagnosis and treatment and subsequent bowel loss. Early complications in this patient population include sepsis, shock and multiple organ system failure due to the inflammatory mediators released in response to the necrotic bowel. Short bowel syndrome is the most significant long-term complication in these patients. These patients are also at risk for sepsis, vascular access issues and liver failure.

Patients with malrotation remain at risk for recurrent intestinal obstruction from a variety of causes including recurrent midgut volvulus and adhesive bands. Patients may also have long-term GI motility issues.

Follow-Up

Once the infant is discharged from the hospital, they will require close follow-up to assess feeding tolerance and appropriate weight gain. Patients without bowel loss usually progress well. Those with bowel loss may benefit from a short bowel syndrome clinic.

Chapter 127

Intussusception Reduction: Laparoscopic and Open

David Rothstein

Indications

Operative intussusception reduction is indicated in patients with intussusceptions that either cannot be reduced by noninvasive means via air, saline or barium enema, or have contraindications to such attempts (i.e., intestinal perforation, peritonitis, sepsis). High-grade bowel obstruction with markedly distended small bowel loops and/or evidence of bowel perforation are relatively strong contraindications to attempting intussusception reduction laparoscopically, but they are not absolute.

Preop

Preoperative preparation is as for any abdominal operation where bowel resection is a possibility. The patient with intussusception is often dehydrated due to poor oral intake and bowel obstruction and must be aggressively rehydrated prior to institution of anesthesia. A nasogastric tube is placed, and if the patient is markedly dehydrated and/or septic, a urinary catheter is advisable as well. Broad-spectrum antibiotics are administered prior to incision. The patient is positioned supine. Reverse Trendelenburg may be helpful.

Procedure

Laparoscopic Reduction of Intussusception: Single-Site Laparoscopic Surgery

Step 1. A semi-circular incision is made in the inferior umbilical fold and a 5 mm trocar advanced into the peritoneal cavity. Insufflation pressures of 8-10 mm Hg usually suffice.

Step 2. Two additional 5 mm grasping instruments are advanced carefully into the peritoneal cavity through separate fascial incisions within the same umbilical skin entry. Low-profile trocars are often required to facilitate instrument placement. Three mm instruments should not be used as they make it difficult to grasp bowel safely while applying considerable tension.

Step 3. The site of intussusception is identified as an intraluminal mass, typically in the cecum or ascending colon. In patients who have had no attempts at reduction nonoperatively, the intussusceptum may actually terminate more distally, even in the sigmoid. In this case, the tip of the intussusceptum may need to be milked back toward the cecum by gently grasping and squeezing the colon in a "hand over hand" fashion. Another way to identify the area of intussusception is to run the small bowel distally until you are able to visualize its entry into the intussuscipiens.

127

Northwestern Handbook of Surgical Procedures, 2nd Edition, edited by Nathaniel J. Soper and Dixon B. Kaufman. ©2011 Landes Bioscience.

Step 4. Reduction is accomplished gradually by very carefully distracting the intussuscepted small intestine from the surrounding colon. An advantage of single-site placement of instruments is the ability to use the instruments in a crossed fashion to apply linear forces on the bowel in opposite directions. *Care must be taken to avoid tearing the bowel.*

Step 5. Successful reduction is confirmed by the complete withdrawal of the intussusceptum. The appendix is often part of the intussusception but does not need to be removed unless necrotic. The affected bowel is often markedly edematous and ischemic-appearing. Rarely does it require resection, but careful inspection is mandatory to look for an area of perforation.

Step 6. If at any point laparoscopic reduction results in tearing of the bowel or is otherwise unsuccessful, conversion to open procedure should be undertaken (see below).

Laparoscopic Reduction of Intussusception: Standard Laparoscopy

Step 1. A 5 mm trocar is advanced into the peritoneal cavity through an umbilical incision. Insufflation pressures of 8-10 mm Hg usually suffice.

Step 2. Two additional 5 mm grasping instruments are advanced carefully into the peritoneal cavity through separate fascial incisions in the left lower-quadrant and the suprapubic locations (similar to those for an appendectomy). Some surgeons prefer to place instruments in bilateral flank positions at the level of the umbilicus. Trocars facilitate instrument placement, but are not required. Three mm instruments should not be used as they make it difficult to grasp bowel safely while applying considerable tension.

Steps 3-6. Identical to those in *single-site laparoscopic surgery* (see above).

Open Reduction of Intussusception

Step 1. A 5 cm transverse incision is made over the right mid-abdomen. Extension over the midline is rarely necessary.

Step 2. The intussusception is delivered into the wound. Occasionally a distal intussusception must be milked back into the cecum to allow delivery.

Step 3. The intussusception is reduced primarily by squeezing the tip of the intussusceptum back through the ileocecal valve. The surgeon must be patient to allow the edema in the bowel wall to decrease gradually, facilitating the reduction and avoiding tears of the intestinal wall. This is very similar to the technique used in the reduction of an incarcerated inguinal hernia.

Step 4. Once delivered, the bowel is carefully inspected in a circumferential fashion to look for transmural tears. Although the delivered bowel can look markedly edematous and ischemic, it most often pinks up during the course of the operation.

Step 5. Bowel resection is required in less than 10% of all intussusceptions, in the case of free perforation or irreducibility. In most cases this requires an ileocecal resection, usually performed in an end-to-end fashion using interrupted or running long-term, resorbable sutures.

Step 6. The abdominal incision is closed by approximating the posterior and anterior rectus sheaths separately, followed by a two-layer skin closure.

Postop

In general, care is similar to that of a patient with appendicitis. If reduction is successful without bowel resection, a clear diet can be started shortly after the operation, and the patient can generally be discharged on the following day. Bowel resection may prompt a longer recovery period. Prolonged antibiotic administration is generally ill-advised except in the cases of free bowel perforation.

Complications

Recurrent intussusceptions may present in as many as 10% of patients after nonoperative reduction and less frequently after operative reduction. They usually occur within the first 24 hours after reduction. Postoperative fevers are the norm rather than exception, as idiopathic intussusceptions are thought to be sequelae of mesenteric and/or mural lymphatic hypertrophy in the setting of viral infection. However, particularly in the case of operative reduction and bowel resection, fevers should prompt evaluation for leak or infection. In the rare case of multiply recurrent intussusceptions, the patient should be investigated for a lead point.

Follow-Up

Patients requiring operative intussusception reduction should have a follow-up visit for wound checks and to assure return to normal bowel function. Long-term follow-up is not necessary.

More Handbooks in this Series...

$\mathcal{V}ademecum$

How to order:
- at our website www.landesbioscience.com
- by email: orders@landesbioscience.com
- by fax: 1.512.637.6079; by phone: 1.800.736.9948
- by mail: Landes Bioscience, 1806 Rio Grande, Austin, TX 78701